ELECTROMAGNETIC RADIATION IN SPACE

ASTROPHYSICS AND SPACE SCIENCE LIBRARY

A SERIES OF BOOKS ON THE RECENT DEVELOPMENTS
OF SPACE SCIENCE AND OF GENERAL GEOPHYSICS AND ASTROPHYSICS
PUBLISHED IN CONNECTION WITH THE JOURNAL
SPACE SCIENCE REVIEWS

VOLUME 9

ELECTROMAGNETIC RADIATION IN SPACE

PROCEEDINGS OF THE THIRD ESRO SUMMER SCHOOL IN SPACE PHYSICS, HELD IN ALPBACH, AUSTRIA, FROM 19 JULY TO 13 AUGUST, 1965

Edited by

J. G. EMMING
Space Research Laboratory Utrecht, The Netherlands

D. REIDEL PUBLISHING COMPANY

DORDRECHT-HOLLAND

ISBN-13: 978-94-010-3528-6 e-ISBN-13: 978-94-010-3526-2
DOI: 10.1007/978-94-010-3526-2

PREFACE

The subject of this volume in the Astrophysics and Space Science Library is Electromagnetic Radiation in Space. It is essentially based on the lectures given at the third ESRO Summer School which was held from 19 July to 13 August, 1965, in Alpbach, Austria. Fifty-eight selected students attended the courses representing the following countries: Austria (2), Belgium (1), Denmark (1), France (12), Germany (10), Italy (7), Netherlands (2), Spain (4), Sweden (6), Switzerland (3), United Kingdom (9), United States (1). Thirteen lectures courses and nine seminars were given by sixteen different scientists in total.

In this book the courses and seminars have been classified in three parts according to the kind of radiation which they mainly deal with: Ultraviolet Radiation, X-Radiation and Cosmic Radiation. These parts can be broken down further in theoretical and observational aspects, whereas in the first and second part solar as well as stellar ultraviolet- and X-radiation can be distinguished.*

Due to various reasons the publication of this volume had to be delayed; it was therefore judged appropriate to bring the text up to date.

The various lecturers have been asked to revise the manuscripts and to eventually add new information which has been acquired in this rapidly evolving field of space astrophysics. Most authors have responded positively to this request, some even have completely rewritten the manuscript.

In the first place this book is intended for students who want to obtain insight in the theory of the various production mechanisms of electromagnetic radiation from space and in its properties. Also post graduate students and space scientists and engineers will find much of their interest in this volume.

Since instrumentation plays such a large role in the space sciences most papers include a few sections on it, giving also recent results of experiments with balloons, rockets en satellites, while some are exclusively dealing with this aspect.

An extensive subject index has been incorporated and the reader who wants to study the original papers on special subjects will find numerous references at the end of each article.

* The lectures given Prof. S.R. Pottasch: On the theory of solar ultraviolet radiation; Prof. Dr. L. Biermann: Theoretical aspects of extraterrestrial X-rays of non-solar origin, and the seminars on Aspects of the LAS project by Prof. Dr. R. Lüst, Prof. Dr. C. de Jager and Dr. R. Wilson, as well as the seminar by Dr. H.E. Hinteregger: Some results and problems of observed solar extreme uv radiation are not published in this volume.

The editor would like to thank the students who have drawn up the first lecture notes, and the lecturers who have adapted these notes to the latest available information.

Utrecht, December 1967 J. G. EMMING

TABLE OF CONTENTS

INTRODUCTION TO ASTROPHYSICS

H. C. VAN DE HULST

University of Leyden, Leyden, The Netherlands

Abstract. A general review is given on astronomy, some of its historical developments and present problems. In the second section the stars, their classification and evolution are considered. The third section is devoted to interstellar matter. The last two sections deal essentially with radio-astronomical investigations and their results.

1. Introduction: Geography of the Universe

As science goes on, entire chapters of astronomy have to be rewritten. New methods and techniques emerge in different periods and make it necessary to reconsider concepts and definitions used before. The same is true for the techniques used in space research. Already now we find that our former concepts about the moon and interplanetary space and to some extent those about the planets, have radically changed, even though only very few space vehicles have been used.

For those who are not accustomed to think in terms of astronomical dimensions, it is interesting to review the way the size of the universe accessible to serious study has changed throughout the history of astronomy. Let us consider Table I.

Let us add a few comments.

1. The word "local" is always used in some context. In connection with the earth, the magnetosphere or the ionosphere are "local" but the moon is not. Similarly, when talking about stellar astronomy, local usually means the local spiral arm. There arises the philosophical question whether one should talk about *the* universe or the *local* universe.

TABLE I

Object	log size (cm)	Beginning of scientific investigations (year A.C.)
Lecture room	3.0	
Alpbach	5.5	
Tirol	7.3	
Austria	7.8	
Europe	8.6	0
Earth	9.1	1500
Solar system	13.2 = 1 AU *	1700
Local spiral arm	21.5 = 1000 parsecs *	1850–1900
The galaxy	23.0 = 30 kpc	
Local clusters of galaxies	24.5 = 3 M-lightyears	1930
The universe	28.0 = 10 G-lightyears	

* A star has a distance of 1 parsec if the radius of the earth's orbit as viewed from the star equals 1 second of arc. By measuring the apparent displacement of a star relative to the background we can obtain an estimate of the distance in parsec. – 1 AU (Astronomical unit) = 150 million km = distance Sun–Earth.

J. G. Emming (ed.), Electromagnetic Radiation in Space.

2. The changing concept of "world". In Table I the last column indicates the approximate time when that particular part of the universe was opened up for research and for a while seemed the largest physical system we live in. In Roman times Europe was considered the "World". During the Renaissance, after the discovery of America, the earth was considered the "World". After Galileo and Newton, when the laws governing the gravitational motion had been discovered, the system of the "world" coincided with the solar system. In those days the stars for all practical purposes were fixed points on a sphere.

In Copernicus' solar system the earth has an elliptical orbit around the sun, which would give a measurable parallax of sufficiently close stars. Tycho Brahe thought he could see the stars as disks with his naked eye. This would put them at a distance close enough to result in an appreciable parallax, provided their sizes were comparable to that of the sun. Since no parallax was observed, he concluded that Copernicus was wrong. But then Galileo observed with his *telescope* that the stars were point-like, which decided the issue in favour of Copernicus. The moral of the story is that even a simple, seemingly negative observation can supply far-reaching consequences.

Table I shows a difference of 7 orders of magnitude between the size of the solar system and the local spiral arm. The increasing difficulty to measure larger distances explains why it took so long before the local spiral arm was well studied. Even if we extend the basis of our parallax measurement by using the motion of the solar system, which is about four astronomical units in a year, we cannot measure directly larger distances than about a hundred parsecs. As we know now, the centre of the galaxy is about ten kpc away from us, and therefore beyond the reach of those methods. Around 1900 Kapteyn made his pioneering statistical studies to measure the size of our galaxy and found a diameter of about a thousand parsecs with the sun in the centre. The reason for this wrong result was that at that time he could not take into account the effect of interstellar absorption, and relied on the inverse-square law for the brightness of distant stars. Only at about 1930 interstellar absorption was studied extensively and could be taken into account. The wavelength-dependence of the absorption makes it possible to calculate its magnitude. The real break-through came when radio astronomy came into use after the second world war. Rapid progress was then made in studying the large-scale structure of the galaxy.

3. The word "cosmos" is equally telling. In classical times the word "cosmos" was associated with divine order. In present-day astronomy it has lost this meaning. For instance in the term "cosmic radiation", cosmic means extraterrestrial.

The classical concept was that cosmic bodies should be perfect. Galileo's dialogues use the word "incorruptible". This means perfect stationary spherical objects symbolizing self-containedness, constancy and an eternal fixedness. Beautiful "proofs" were found in the sun and the moon. New discoveries had to fit into this model or, if they did not, were discarded; this may have been one of the reasons why the sunspots were not discussed until the Renaissance.

Nowadays this concept of perfect order has retracted to the fundamental laws of nature. Instead, we now have the scientific concept of equilibrium, which means that a system with a certain mechanism to dissipate energy, if left alone for a sufficiently long time, reaches an equilibrium situation. No traces of the previous history can be found in this equilibrium situation. However, at an earlier stage, in a non-equilibrium situation, we can still see clear signs of what happened before. This idea may be applied to the processes taking place in the universe and in our solar system. The solar system has approached a state of equilibrium because we see nearly circular orbits in a plane. This evolution must have happened in its early history, because the dissipating forces at the moment are nearly zero. The galaxy is not so close to equilibrium. We can even now see the strong interaction between the young stars and the various components of the interstellar matter. Such interaction must have caused the orderly circular orbits and presumably the spiral arms. The interaction between a star, once it has been formed, and the interstellar matter is only weak. This explains the disorderly motion of older stars through the galaxy, and the fact that they are not confined to the galactic plane.

2. The Stars

The study of the stars is only one particular aspect of astronomy, since astronomy comprises the physical study of anything outside the earth. What makes stars especially interesting is, firstly, that they are sources of energy which make life possible on planets like the earth and, secondly, that they are discrete radiation sources which makes them particularly convenient to observe. It is wrong, however, to say that stars are the most important objects in the universe. Interstellar matter, for example, is at least as important, being the substance responsible for a good part of galactic dynamics.

Nearly all stars can be fitted into a certain spectral class according to features displayed in their spectra. If we plot the absolute magnitude of stars against their spectral class in the so-called Hertzsprung-Russell diagram we find a very strong correlation. This is an empirical diagram, since the absolute magnitude is found from the apparent magnitude and the distance, and the spectral type from spectral studies. In the diagram represented in Figure 1 we can distinguish between (1) the main sequence, (2) the red giant branch, and (3) the white dwarf stars.

If we take all stars within 100 parsec the great majority would be on the main sequence of the H–R diagram with very few red giants but many white dwarfs. The sun is an average main sequence star of spectral type G. However, if we use apparent magnitude criteria, we obtain many more red giants, since they are intrinsically brighter. Nowadays more refined spectral criteria have been found which make it possible to distinguish between dwarfs and giants from the spectrum.

Instead of absolute magnitude M_v against spectral type we can also plot the luminosity (L) against effective temperature (T_e), where L is defined as total energy output in the optical frequencies in erg/sec. T_e is then defined by:

$$L = 4\pi R^2 (\sigma T_e^4)$$

where R is the radius of the star, and σ the Stefan-Boltzmann constant.

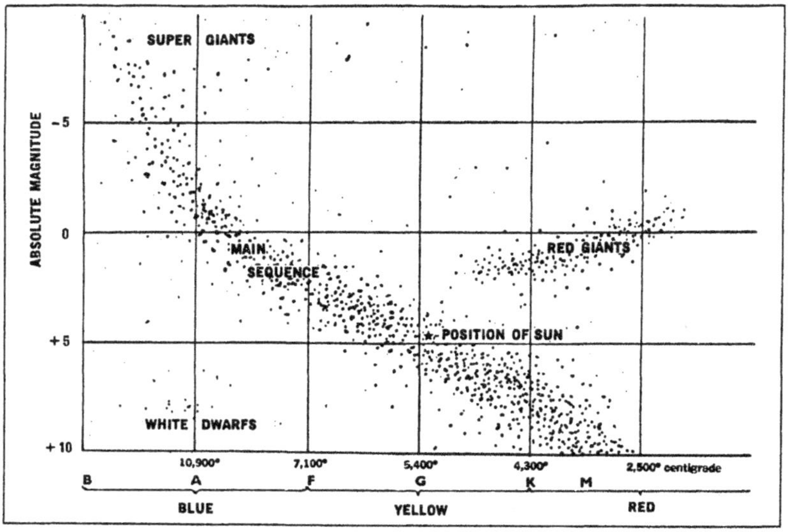

Fig. 1. The Hertzsprung-Russell diagram. (After D. ALTER, *Pictorial Astronomy*, p. 269.)

Since we can derive theoretically a relation between T_e and the spectral type and of course one between M_v and L, we can convert the H–R diagram into one using the more objective scales of L and T_e.

To explain some aspects of the H–R diagram we must first get an idea of how a star is built. A star normally consists of an inner sphere and a surrounding atmosphere. The radiation originating in the inner parts is absorbed and re-emitted many times before it reaches the surface. The atmosphere is defined as that part of the star that is partially transparent to light and therefore determines the observed spectrum. The atmosphere of the sun has a thickness of only 100 to 200 km whereas the total solar radius is about $7 \cdot 10^5$ km. Stars may have a corona on top of the atmosphere like the sun, and also a transition region between the atmosphere and the corona, alike the solar chromosphere. However, no direct evidence of the existence of these regions exists except in the case of the sun. Stars with more or less extended atmospheres have also been observed.

In the interior of a star there is virtually perfect thermodynamical equilibrium. However, in the atmosphere radiation escapes, and every spectral line implies deviation from thermodynamical equilibrium. The problem is how to calculate the escaping radiation for every frequency, i.e. to calculate the spectrum. The parameters involved are:

(1) πF, energy flux through one cm^2/sec ($\pi F = \sigma T_e^4$)
(2) the chemical composition (i.e. the number of different nuclei)
(3) the gravitational field.

The gravity g can be taken as constant throughout the atmosphere and does not vary much among the main sequence stars. If we keep the chemical composition constant, g and T_e give a two-dimensional classification of stars. If we also add the chemical composition as a variable it becomes three-dimensional. It turns out that the essential features in the spectrum are dependent upon T_e. Only the fine details of the spectrum depend on g and the chemical composition.

The O- and B-type stars are the brightest stars of the main sequence. They are interesting because

they are very bright, and can therefore be used in the study of the galactic structure;

they must be young because they spend their energy in a relatively short period;

they decay rather rapidly, so that their motions and spatial distribution may give information on stellar evolution.

A problem however is the fact, that they emit most of their energy and have most of their prominent lines in the UV-region. Since the earth's atmosphere is not transparent to the UV-light in which the maximum of emission of the O- and B-type stars occurs, it is necessary to use space vehicles to investigate them. Observations in the UV-part of the spectrum may also contribute largely to our understanding of dynamical processes in stars like convection.

Starting as a contracting cloud of interstellar matter, the star reaches rather gradually the main sequence, where it starts with nuclear energy production inside its nucleus.

The H–R diagram clearly invokes a discussion of the stellar evolution. After a star on the main sequence has burned its hydrogen into helium it leaves the main sequence and moves to the right. In this way the whole region to the right of the main sequence should be scattered with stars. However, a group of stars formed at the same time will result in a branch where the bright stars have already turned off the main sequence, and the red ones are still on it. The turning point occurs somewhere along the main sequence. From observations we know that many clusters of stars exhibit this phenomenon, each with its own turning point, dependent on the age of the cluster. In the nearer clusters this age can also be calculated using the proper-motion method: plotting the velocities in the cluster, all the stars seem to have come from one common point, just like the radiant of a meteorite shower. The age obtained from this starting point of the cluster agrees very well with those obtained from the turning point in the H–R diagram.

A temperature inversion occurs in the transition region between the atmosphere and the corona. The coolest layer is at about 4000 °K for the sun after which there is a sharp increase reaching 10000 °K in the chromosphere and increasing to about 2000000 °K in the corona. The existence of the corona had not been predicted by theory. It was observed and then interpreted. The high temperature of the corona must be due to a non-thermal process, possibly the dissipation of the energy of shock-waves, that travel upward through photosphere and chromosphere.

The solar corona is visible during an eclipse, and even then only up to several solar radii, since scattered light in the earth's atmosphere makes the sky too bright. Under favourable circumstances it is possible to observe much further extensions of the solar corona, called the zodiacal light, but only above 30° from the sun-earth line.

The intensity of the corona and the intensity of the zodiacal light plotted as functions of the angle from the sun can be smoothly connected. The initial gap between the two curves has by now been filled experimentally, mainly by observations from high-flying aircraft.

Polarization and spectral observations show that the zodiacal light arises mainly from scattering by interplanetary dust. The existence of an interplanetary gas was known earlier only from the behaviour of comet tails. This entire domain has now been opened to direct observations by space research. The interplanetary gas is commonly called "solar wind".

3. Interstellar Matter Locally

Interstellar space is not the same as interplanetary space, although these terms have often been confused, for instance, in talking about cosmic ray modulation. The scale of the solar system is about 100 AU, and the scale of the interstellar space locally is about 100 parsecs. Therefore one should be extremely careful in interpreting a directional effect in the cosmic rays in terms of the magnetic field in interstellar space.

Interstellar matter consists of:

gas

dust (or, better, smoke)

cosmic rays, including not only the cosmic rays we observe but also the low energy component, down to 1 MeV, which undoubtedly exists though it is not observable from the earth.

Interstellar matter is a typical non-equilibrium system. No model has been or can probably be made that is coherent with every aspect of interstellar space. Many phenomena observed must have a transient, randomly recurrent character.

3.1 Interstellar gas

Gas concentrations are called nebulae, whereas dust concentrations are referred to as dark clouds. Nebulae and the dark clouds have been known for a long time. The gas was discovered by its absorption lines, the dust because of interstellar absorption and reddening of light.

For studies of the gas dynamics we should like to obtain a six-dimensional map in phase-space. However, the position can only be measured perpendicularly to the line of sight, and the velocity only along the line of sight by means of the Doppler effect (radial velocity). In a few cases the radial velocity can be converted into distance, thus completing a three-dimensional map in ordinary space.

If in a uniform gas which consists mostly of hydrogen a hot O- or B-type star is formed, the star ionizes the hydrogen since it radiates most of its energy in wavelengths below the Lyman limit. Since the supply of ionizing quanta is limited and has to keep up with the recombination, we get a spherical region around the star which contains only protons and electrons, the H II region. The outside medium consists predominantly of atoms and perhaps molecules (H I region). The H II is heated by the ionizing process to about 10000 °K, whereas the H I region stays at about 50 °K. By the time thermal equilibrium has been reached there is no dynamical equilibrium, and the H II region will expand, pushing and compressing the H I region. Many examples of nebulae around stars have been found exhibiting these features or similar ones. Doppler measurements show that velocities within the nebula are of the order of 10 to 20 km/sec.

Direct study of the cool, non ionized H I regions is possible by means of the 21-cm line of atomic hydrogen, which is discussed in some detail in the next section. The question also arises whether molecular hydrogen, although not observed up to the present, occurs in interstellar space. Solid grains might act as very good catalyzers in forming H_2 from H. The strong lines of H_2 near 1100 Å would be the best indicator of the presence of the hydrogen molecule. The hope is that these lines will soon be observed by UV measurements from a spacecraft or sounding rocket.

3.2. Interstellar dust

After Spitzer the dust particles are now called interstellar grains. Their existence is inferred from

the interstellar extinction of light, i.e., the stars seem fainter than they should be

the interstellar reddening of light, i.e., the wavelength dependence of the extinction

the interstellar polarization of light, i.e., dependence of the extinction on the direction of polarization

scattering of light. There exist bright nebulae with stars cooler than B3 which must get their brightness from scattered light (reflexion nebulae).

If the ratio of the reddening to the extinction is known we can calculate the distance to the star from the apparent magnitude and its spectrum, which gives us the spectral class and therefore the real colour and the absolute magnitude. Using the Mie theory, i.e., scattering of light by homogeneous spheres with some refractive index, a theoretical curve can be found to fit the wavelength dependence of the extinction. Various model experiments in the microwave region have been made to determine also the polarization effect of non-spherical grains. Observations in the UV-region from stars will help a great deal in settling the many remaining problems. Any theory explaining the extinction and polarization has also to explain the near-uniformity of the law of reddening throughout the sky. The polarization is interpreted as indicating the existence of elongated grains whose axes of rotation are aligned to the magnetic field, connected with the spiral arms.

4. Galactic Structure

The study of galactic structure is intimately linked with radio astronomy, since the radio astronomy techniques have made it possible to extend the data and concepts based on the mostly local optical studies to the entire galactic system.

Radio emission from the galaxy (and beyond) occurs as a continuum and in spectral lines, which should be well separated in the discussion. The emission from a discrete source of continuum emission is usually expressed in terms of its flux density

$$S = B\Omega$$

where S is the energy flux received at the earth in erg/s/cm^2, B is the brightness and Ω is the solid angle under which the source is viewed.

$$\Omega = \pi \frac{a^2}{r^2}$$

where a is the radius of the source and r its distance from earth. We are justified in using the Rayleigh-Jeans approximation for the black body radiation:

$$B(T) = kT \frac{2\nu^2}{c^2}.$$

As most sources do not radiate as a black body, but do have a radio brightness, we use this relation to define their brightness temperature.

The fact that there must be various emission mechanisms shows most strikingly from spectral graphs. In such a graph the abscissa is proportional to the logarithm of the frequency. The ordinate is also on a logarithmic scale and we choose

S for discrete sources (Ω small and perhaps unknown),

B for extended sources.

The sun and the moon, for example, show radio spectra in good agreement with the thermal law over a large frequency range. However, for most radio sources S or B decreases when ν increases. The radiation from these sources is non-thermal; it is due to the synchrotron or magnetic bremsstrahlung process.

Observed spectral lines in the radio region are those of H at 21 cm and of OH, which were discovered recently. Deuterium has not been detected. We know from the efforts made in order to find it that it is as rare in interstellar space as it is on the earth.

The 21-cm line of atomic hydrogen was first observed in 1951, nearly simultaneously in the United States, in Holland and in Australia. In 1953 the first map of several spiral arms was drawn. The observations yield longitude, latitude and the Doppler shift, which gives the radial velocity. If the general velocity field in the galaxy is known, we can derive the distance from the radial velocity, thus yielding a three-dimensional map. The order of accuracy in this conversion is about 100 parsecs, so this method will only work in a large scale and not locally in the galaxy. In these studies it is normally assumed that a differential galactic rotation law holds with a velocity of about 200 km/sec at 10 kpc from the galactic centre. The angular velocity increases sharply for decreasing R (distance from galactic centre).

Close to the centre this assumption is definitely incorrect. An expanding spiral arm consisting of approaching gas has been observed in the direction of the galactic centre with a velocity of 55.3 km/sec. Other clouds have recently been observed at high latitude approaching us with velocities of 100 km/sec and more. No distance criteria for these clouds are available, but they are tentatively interpreted as extragalactic hydrogen clouds entering the galaxy.

5. Non-Thermal Phenomena

This concluding lecture of the introductory series should convey the message that astrophysics is by no means a closed subject. It is true that e.g. the thermal and mechanical equilibrium of a star are basically understood, so that the study of this subject has entered its second round of looking at refined details (by observation and theory). Many other subjects, however, are still basically puzzling. Among them are the non-thermal radiosources, cosmic rays, and a variety of strange objects beyond our own galaxy.

It is first necessary to examine our knowledge of cosmic *magnetic fields*. These could be ignored in older studies since

the planetary motions are not affected by them, and

the equilibrium of stationary stellar atmospheres is independent of magnetic fields.

However, nowadays many phenomena in astronomy are believed to be connected with the existence of magnetic fields. Whether spiral arms belong to this class of phenomena is not yet clear, since new theories using gravitational arguments only have recently been suggested. The observable effects of magnetic fields can be classified as follows.

Along the line of sight:

Optical polarization

Faraday rotation of radio waves, i.e., a rotation of the polarization plane caused by free electrons in a magnetic field.

In the source:

Zeeman effect in the 21-cm line, which, however, is still below the level of detection. This implies that the magnetic field must be weaker than 10^{-6} gauss

Synchrotron emission in both optical and radio frequencies.

Suggestive shapes:

The "plumes" in the corona which can be explained by uninhibited diffusion along the magnetic field lines

Peculiarly twisted planetary nebulae

Faint streaks in nebulae.

The magnitude of the magnetic field is still a matter of debate. Quoted values are field strengths of 10^{-5} to 10^{-6} gauss. The difference is very important, since in one case the magnetic energy density is higher, in the other case lower than the kinetic energy density of the particles present, so that the answer to the question "which drives which" depends on a decision between the two values.

Nowadays, the emission from nearly all radio sources is believed to be synchrotron radiation or in some cases radiation due to the inverse Compton effect. An electron moving in a magnetic field spirals around the field lines thereby emitting electromagnetic radiation. In the non-relativistic case ($v \ll c$), the radiation is emitted in all direction with a frequency strictly determined by the period of rotation. This frequency, is called the gyro-frequency. Electrons with a velocity near c, however, emit a very narrow beam of radiation, observed only when the electron moves towards the observer, and Fourier analysis shows that the observed spectrum has many higher harmonics of the fundamental frequency which merge into a continuous spectrum. The theory is discussed in detail in Prof. Lequeux's contribution. Here some illustrative numbers may suffice (cf. Table II).

TABLE II
Examples of Synchrotron Radiation

	Crab nebula		cosmic ray electrons
	optical	radio	
E (Energy in eV)	$2 \cdot 10^{11}$	10^8	$3 \cdot 10^9$
H (Magnetic field strength in gauss)	10^{-3}	10^{-3}	10^{-5}
$R = Ec/eHv$ (gyro-radius)	$\frac{1}{20}$ AU	3000 km	$\frac{1}{15}$ AU
$\nu_{\rm char} \sim E/H$ (characteristic frequency)	violet	160 Mc/s ($\lambda = 2$ m)	1500 Mc/s ($\lambda = 20$ cm)
$t \sim 1/H^2E$ (decay time)	50 years	10^5 years	$3 \cdot 10^7$ years

It is noted that the gyro-radii are always small compared to the region under consideration, and that the decay time is shorter for high energy particles than for those of low energy. The Crab nebula which is the remnant of a supernova explosion in 1054, still emits optical synchrotron radiation, although the decay time is only 50 years. Many theories have been proposed to explain the continuous injection of high energy electrons, but the problem is as yet unsolved. This production mechanism of radiation was seriously considered only after SHKLOVSKII's suggestion (1955) that synchrotron radiation might account for the optical emission of the Crab nebula was confirmed by detection of its polarization. Since then a successful search has been made for polarization of the general radio background emitted in the galaxy. The polarization found can be explained by assuming the presence of a field of 10^{-5} gauss and an electron energy spectrum similar to that of the cosmic rays. If the differential energy spectrum of the electrons satisfies the equation

$$N(E)\,\mathrm{d}E = \mathrm{const}\, E^{-\gamma}\,\mathrm{d}E$$

the spectrum of the emitted synchrotron radiation is given by

$$\varepsilon_\nu\,\mathrm{d}\nu = \mathrm{constant} \times H^{\alpha+1}\,\nu^{-\alpha}\,\mathrm{d}\nu$$

where

$$\alpha = (\gamma - 1)/2\,.$$

Experimentally it is found that $\gamma = 2.3$ and $\alpha = 0.65$, in reasonable agreement with the theory, but deviations from and variation in these laws are of the greatest significance.

In a bi-logarithmic diagram many non-thermal spectra, instead of being straight lines, show a drop at one or both ends. For low frequencies and bright sources this can be explained by self-absorption. A drop at high frequencies may occur because the high-energy electrons lose their energy faster than those of low energy. If this is the only cause, the turning point moves to lower frequencies as time goes on and may therefore give an estimate of the age of the source. However, many difficulties arise in an exact treatment, and observations are not always reliable enough to decide upon one particular explanation or another.

Some regions of the galaxy have now been found to display differential Faraday effect at two frequencies. Calculating back to zero rotation of the plane of polarization, the direction of the magnetic field thus obtained agrees rather well with that obtained from determination of optical polarization due to interstellar grains.

The non-thermal radiation of the general galactic background comes from an ill-defined region of space called the galactic *halo*. The optical halo of the galaxy, i.e., the region of the galaxy containing the older stars and globular clusters, is well known. The initial idea that coinciding with this halo there is a radio halo, responsible for the general galactic background radio emission is now questioned. Recent observations of the very similar Andromeda nebula show hardly any distinct radio halo.

In recent years studies of objects in the universe outside our galactic system have led to many surprises. The mildest of these is that many discrete double radio sources have been identified with optical galaxies of far stronger radio emission than our own. For a while the radio emission was thought to be due to the collision of two galaxies. The present idea is that the radio source responsible for the emission is caused by some violent explosions inside these radio galaxies. There are no optical criteria for whether a galaxy should be a strong radio source or not. Some do, however, show peculiar "jets".

Other surprising objects are the so-called quasi-stellar radio sources, quasars, of which about one hundred are known. They have been observed only fairly recently. They can be identified with faint starlike optical objects and have spectra that can only be explained by an enormous cosmological redshift. The size of their radio source is very small, one minute of arc or often less, and their radio spectra are often curved. Their brightness is unusually high and their total energy output under reasonable assumptions must have been about 10^{60} ergs, which is very hard to explain in terms of the presently known energy sources. A more numerous class of objects, the already known blue Zwicky-Humason objects, has now also come under suspicion. Many of them might have a similarly large redshift without being radio sources, thus also being extragalactic. There is also a possibility that the quasi-stellar sources are not at cosmological distances, but at more or less local distance (1–10 Mpc). No definite conclusion as to the nature of these objects can be drawn as yet.

This rather arbitrary selection of some topics of extragalactic research now at the frontier of man's knowledge will show that much remains to be gained by a combined effort of theory and observation with all available techniques, including space research. Any progress gained in understanding these far-away objects may provide new clues to the nature and evolution of the universe.

Bibliography

A thorough background in astrophysics is given in the following handbooks:

1. *Stars and Stellar Systems*, vols. I–IX (editor: G.P. Kuiper, associate editor: B.M. Middlehurst, contributors: many distinguished astronomers), University of Chicago Press, Chicago.
2. *Handbook of Physics*, vols. 50–54 (editor S. Flügge), Springer-Verlag, Berlin.
3. *The Solar System*, vols. I–IV (editors G.P. Kuiper and B.M. Middlehurst).

More recent developments are covered in the two annually appearing publications:

1. *Annual Review of Astronomy and Astrophysics*, vol. Iff. (editor L. Goldberg), Annual Reviews Inc., Palo Alto, Calif.

and

2. *Advances in Astronomy and Astrophysics*, vol. Iff. (editor Z. Kopal), Academic Press, New York,

while space science is covered in the monthly

Space Science Reviews (editor C. de Jager), Reidel, Dordrecht, The Netherlands.

A comprehensive survey is given every three years in the

Transactions of the International Union, various editors and publishers.

With respect to a few subjects specifically mentioned in this introduction there are on:

Stellar evolution:

Handbuch der Physik **LI** (editor S. Flügge), 1958, Springer-Verlag,

Gratton, L. (ed.): 1963, *Star evolution* (Proceedings of the International School of Physics 'Enrico Fermi': Course 28), Academic Press, New York.

Zodiacal Light and the Interplanetary medium:

Symposium proceedings of the symposium on Zodiacal Light and the Interplanetary medium, held in Honolulu, Jan. 30 – Feb. 2, 1967, (ed. by J.L. Weinberg), to be published.

Interstellar Matter:

Serkowski, K.: 1962, 'Polarization of Starlight', *Advances in Astronomy and Astrophysics* **1** (ed. by Z. Kopal), Academic Press, New York.

Greenberg, J.M.: 1963, 'Geometry and composition of interstellar grains', *Annual Review of Astronomy and Astrophysics* **1** (ed. by Leo Goldberg), Annual Reviews, Palo Alto.

Woltjer, L.: 1962, *Interstellar Matter in Galaxies*, Benjamin, New York.

Stars and Stellar Systems, vol. V, 1965, 'Galactic Structure' (ed. by A. Blaauw and M. Schmidt), University of Chicago Press, Chicago.

Kaplan, S.A. and Pikelner, S.B.: *Interstellar Medium*, Harvard, in press.

Stars and Stellar Systems, vol. VII, 1966, 'Nebulae and Interstellar Matter', University of Chicago Press, Chicago.

Kahn, F.D. and Dyson, J.E.: 1965, *Ann. Rev. Astr. and Astrophys.* **3**, 47.

Galactic Structure:

Stars and Stellar Systems, vol. V, 1965, 'Galactic Structure' (ed. by A. Blaauw and M. Schmidt), University of Chicago Press, Chicago.

Proceedings of the IAU Symposium No. 31, 'Radio Astronomy and the Galactic System', August 1966, Noordwijk, Netherlands (ed. by H. van Woerden).

Non-thermal phenomena:

GINZBURG, V. L. and SYROVATSKII, S. I.: 1964, *The Origin of Cosmic Rays*, Pergamon Press, Oxford.

SHKLOVSKY, I. S.: 1960, *Cosmic Radio Waves*, Oxford University Press, London.

MARAN, S. P. and CAMERON, A. G. W.: 1964, *Physics of Non-Thermal Radio Sources NASA SP–46*, Washington.

KAHN, F. D. and PALMER, H. P.: 1967, *Quasars*, University of Manchester Press, Manchester.

BURBIDGE, G. R. and BURBIDGE, E. M.: 1967, *Quasi-stellar Objects*, Freeman, New York.

Notes and bibliography compiled by J. Rosenberg and E. Soop.

PART I

ULTRAVIOLET RADIATION

THEORETICAL ASPECTS OF STELLAR ULTRAVIOLET RADIATION

ANNE B. UNDERHILL

University of Utrecht, Utrecht, The Netherlands

Abstract. The theory of stellar atmospheres is reviewed. The most important characteristic quantities appearing in the computation of a model atmosphere are discussed. The basic equations used for these calculations are briefly derived. The treatment is then applied to derive the UV spectrum of early type stars. The results of specific calculations are compared with observational data.

1. Introduction

In stellar astronomy there are two main types of problem, namely:

(a) Statistical problems in which the emphasis is on such topics as the distribution of stars in space, the number of stars in the visible universe of a certain type, etc.

(b) Problems concerning the individual characteristics of stars, i.e. the way in which stars function. Energy emissions and other parameters are studied in order to investigate the underlying physical processes.

A possible method of determining the physical processes occurring in stars is the observation of the energy distribution in a particular frequency range. Thus the technique is to observe stellar spectra, recording the intensity distribution F_λ/F_{λ_1} over a short wavelength region, F_λ being the flux at any wavelength λ and F_{λ_1} being the flux at wavelength λ_1 to which all other fluxes are normalized. Thus what is studied is relative energy and not absolute energy. To determine the latter one has to know the size of the star, its distance from the observer and the absorption in interstellar space and in the earth's atmosphere.

All radiation having a wavelength shorter than 3050 Å can be considered as making up the ultra-violet spectrum of a star. In the present discussion, the wavelength range between the Lyman limit at 911.6 Å and the ozone cut-off near 3050 Å will be studied. The parameters entering the theory will be those suitable for O and B type stars because it is known that stars of these spectral types are brighter in this spectral range than are stars of other types. The lower limit is imposed by the predicted strong discontinuity due to absorption at the Lyman limit of hydrogen. The theory can predict radiation at shorter wavelengths but the flux at these wavelengths emerging from normal O and B type atmospheres is expected to lie below the level of detection.

Stellar spectra consist of a continuous spectrum upon which are superimposed absorption lines and occasionally a few emission lines. The stars may be arranged in order of increasing temperature by considering which spectral lines are present. Types O and B correspond to the hottest stars known to have more or less normal spectra. The main elements contributing to the spectrum of an O star are H, He I and

J. G. Emming (ed.), Electromagnetic Radiation in Space.

He II while C III, O III, N III and N IV are less important. The temperature in the atmosphere of an O star will be of the order of 25000 °K to 40000 °K, and the mass of an O star is 20 to 30 times the solar mass. O stars are fairly rare in the universe (only about 2000 are known) the nearest being ζ Ophiuchi at a distance of 500 parsecs. Most of these stars are rapidly rotating and hence have diffuse spectral lines due to the Doppler effect. Their spectra are also complicated by interstellar clouds lying between the star and the observer.

The most important lines in the spectra of B stars are from H and He I. Also present but rather weaker are lines from C II, O II, N II, Si II, Si III, Si IV and Fe III; lines from Fe II, Ni II and Cr II are very weak. The temperature in a B type atmosphere is in the range 10000 °K to 25000 °K, this range being subdivided to give the nine groups B8 to B0. B stars are more abundant than O stars (about 6000 are observable) and the nearest B star is γ Cassiopeiae which is 29 parsecs distant. Most B stars lie beyond 100 parsecs, so that observations are complicated by interstellar absorption. The radii of such stars are 2 to 6 times the solar radius, and their masses are 4 to 20 times the solar mass. The electron pressure in a normal atmosphere of an O or B star is of the order of 10^2–10^4 dynes/cm^2.

It is found that stellar atmospheres are composed chiefly of hydrogen whose mass fraction, X, is close to 0.70, and of helium whose mass fraction, Y, is about 0.28. The remaining elements make up a mass fraction Z which is about 0.02. For studying the major properties of the atmospheres of O and B stars it is only necessary to consider the interactions between H and He and the radiation. At the temperatures and pressures which exist in these stars, the hydrogen is almost completely ionized, the fraction of neutral hydrogen present lying between 10^{-4} and 10^{-6}. The fraction of singly and doubly ionized helium varies in the O and B stars from nearly completely neutral helium at the coolest types, B9, to almost completely doubly ionized helium in the hottest stars, O5.

The theory of stellar atmospheres represents the stellar atmosphere as plane parallel layers of gas in mechanical equilibrium under the acceleration of gravity (the ratio of the atmospheric thickness to the radius of the star is about 10^{-4}). Thus one may write:

$$P_g = NkT\,, \qquad \text{(the perfect gas equation)}$$

where P_g is the gas pressure and T is the temperature and

$$\frac{\mathrm{d}P_g}{\mathrm{d}z} = -g\rho + \frac{\mathrm{d}P_r}{\mathrm{d}z}\,, \qquad \text{(the equation of mechanical equilibrium)}$$

where z is the geometric length chosen to increase outwards in the star, g is the gravitational acceleration, ρ is the gas density and P_r is the pressure of radiation.

Little account is taken of the velocities in the gas other than those due to thermal motion, though it is now becoming clear that much of what one observes in O and B type spectra is strongly influenced by the state of motion of the gases making up the stellar atmosphere.

Since normal stellar spectra consist of absorption lines, it is postulated that the temperature in a normal stellar atmosphere must increase inwards. A model for a stellar atmosphere is constructed by solving the above equations to obtain the pressure at each depth, z, in the gas. A temperature law $T(z)$ is adopted. In order to relate gas pressure P_g and electron pressure P_e, only X and Y are needed because hydrogen is nearly fully ionized in the stellar atmosphere and thus provides practically all the electrons. (To calculate the degree of ionization, Saha's law for an atmosphere in thermal equilibrium may be used. This thermal equilibrium assumption is reasonable for most stellar atmospheres at 20 000 °K.)

It is found that a more convenient way of measuring depth is by using the coordinate τ (optical depth) which is related to z by the equation:

$$d\tau = -(\kappa + \sigma)\,\rho\,dz$$

where κ is a characteristic mass absorption coefficient for a certain wavelength range, σ is the electron scattering coefficient and ρ is the density.

It is seen that τ increases from the outside (where it is equal to zero) inwards. Thus τ is a measure of how far one can see into the star at a certain wavelength. In past literature, the value of κ which would give a simple analytical solution to the equations, posed a problem. Since the advent of large computers, the equations can be solved numerically and there is little reason to select a particular theoretical value of κ.

The main sources of opacity in a stellar atmosphere are H, He, and electrons. The first two give line absorption and continuous absorption, while the electrons are efficient scatterers. The electrons are assumed to scatter coherently and isotropically. The mass absorption coefficient κ_ν at a frequency ν is the sum of all individual absorption coefficients i.e.

$$\kappa_\nu = \sum_{\text{all elements}} \kappa_\nu(\text{element}) = \kappa_\nu(\text{H}) + \kappa_\nu(\text{He I}) + \cdots.$$

If the temperature is greater than 25000 °K, then κ_ν (He II) has to be added, and if the temperature is less than 16000 °K, $\kappa_\nu(\text{H}^-)$ must be considered.

It is concluded that a model represents a stellar atmosphere when the radiation field in the atmosphere satisfies a selected constraint. The theory of stellar atmospheres tells how to compute this radiation field by using the rules of physics to describe the interactions which occur between radiation and matter. A star has a gravitational attraction (of the order of 10^4 cm/sec^2, depending on its mass and radius), a certain temperature variation with distance from its centre and a certain flux $F_\nu(\tau = 0)$ outside the star which is an observable quantity. However, the problem is complicated and approximations have to be made in the physics used and in the mathematical treatment.

One of the following three constraints is usually applied in order to find a model which will represent a star.

(i) The geometrical distance from the centre of the star being denoted by z, adopt that $T(z)$ which results in a simple analytical solution of the equations of radiative transfer, e.g. the approximate grey-body law:

$$T^4(\tau) = T_0^4 \left(1 + \tfrac{3}{2}\tau\right)$$

This is very rough and can only be justified when rather crude results are required.

(ii) Adopt that $T(z)$ which gives a predicted limb darkening which is like that observed for the sun. This constraint is useful only for the sun and for stars of about the same spectral type. The solar temperature law may be slightly scaled to fit somewhat hotter or cooler stars.

(iii) Adopt that $T(z)$ which predicts a radiation field which is in radiative equilibrium throughout the stellar atmosphere. (This assumes that there are no energy sources or sinks in the stellar atmosphere.) That is, the integrated flux is constant at all depths:

$$F(\tau) = \int_0^\infty F_\nu(\tau)\, \mathrm{d}\nu = \text{constant}. \tag{1}$$

The determination of $F_\nu(\tau)$ is difficult as it depends on κ_ν which varies with depth in the atmosphere and with frequency. The above constraint is used to find the temperature law in models of O and B type stars. The monochromatic fluxes entering Equation (1) are usually computed for the case where only continuous sources of opacity are considered. The opacity due to absorption lines is omitted, at least as a first approximation.

When a suitable model of a stellar atmosphere has been obtained, it is necessary to identify this model with a real star. This is done by predicting some details in the spectrum of the star and comparing predicted values with observed values. One can also predict the spectrum that the model will give in the spectral regions not accessible to observation from the earth.

2. Derivation of Basic Equations

The basic equations of the radiation field which define a model may be set up with the aid of Figure 1.

Consider a cone of radiation at an angle θ to the normal, the cone having a solid angle $\mathrm{d}\omega$. Then the energy of radiation between frequency ν and $\nu + \mathrm{d}\nu$ contained in solid angle $\mathrm{d}\omega$ crossing an element of area $\mathrm{d}\sigma$ at position x, y, z, in time $\mathrm{d}t$ is:

$$\mathrm{d}E_\nu(\theta, \phi)\, \mathrm{d}\nu\, \mathrm{d}t = I_\nu(\theta, \phi, x, y, z) \cos\theta\, \mathrm{d}\sigma\, \mathrm{d}\omega\, \mathrm{d}\nu\, \mathrm{d}t$$

where

$$I_\nu(\theta, \phi, x, y, z)$$

is defined as the specific intensity. Since

$$\mathrm{d}\omega = \sin\theta\, \mathrm{d}\theta\, \mathrm{d}\phi$$

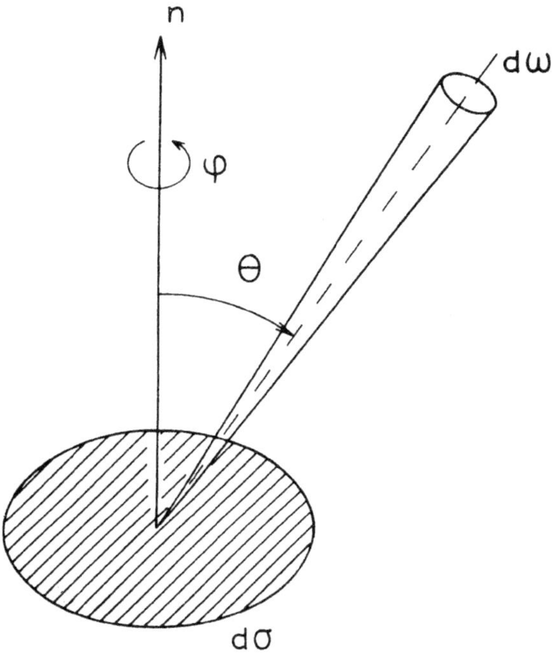

Fig. 1. The geometry of the problem of radiative transfer in plane parallel layers. The pencil of radiation in solid angle $d\omega$ is inclined at an angle θ to the outward pointing normal to the layers. The azimuthal angle is φ.

and the conditions are assumed to be independent of x and y,

$$dE(\theta, \phi, z) = I_\nu(\theta, \phi, z) \cos\theta \sin\theta \, d\theta \, d\phi \, d\sigma .$$

Thus the rate of energy flow across $d\sigma$ in unit time is:

$$\int_0^\pi \int_0^{2\pi} I_\nu(\theta, \phi, z) \cos\theta \sin\theta \, d\theta \, d\phi \, d\sigma$$

which is equal to

$$2\pi \int_0^\pi I_\nu(\theta, z) \cos\theta \sin\theta \, d\theta \, d\sigma$$

in the case where the specific intensity is independent of ϕ. The "flux", F_ν, at depth z is defined as:

$$F_\nu(z) = 2 \int_0^\pi I_\nu(\theta, z) \cos\theta \sin\theta \, d\theta .$$

It is seen that F_ν is equal to the real energy flow divided by π.

The flux to an observer outside the star is represented by:

$$F_\nu(0) = 2 \int_0^\pi I_\nu(\theta, 0) \cos\theta \sin\theta \, d\theta .$$

In order to obtain the equation of radiative transfer, one considers the monochromatic optical depth t_ν defined by the equation:

$$dt_\nu = -(\kappa_\nu + \sigma) \rho \, dz$$

so that

$$t_\nu = -\int_z^\infty (\kappa_\nu + \sigma) \rho \, dz$$

where $(\kappa_\nu + \sigma)$ is known as the monochromatic opacity.

The index of refraction in the atmosphere is assumed constant (this is true at optical wavelengths but not necessarily for radio wavelengths), and the change in I_ν in distance ds is then given by:

$$dI_\nu = -(\kappa_\nu + \sigma) \rho(z) I_\nu \, ds + j_\nu(z, \theta, \phi) \rho(z) \, ds .$$

The first term represents the loss of radiation in ds, the second term is the emission along the path element ds. The quantity $j_\nu(z, \theta, \phi)$ is termed the emissivity. After changing coordinates, one obtains the equation:

$$\cos\theta \frac{dI_\nu(\theta, \phi, t_\nu)}{dt_\nu} = I_\nu(\theta, \phi, t_\nu) - \frac{j_\nu}{\kappa_\nu + \sigma} .$$

The boundary conditions to be applied are:

(1) when $t_\nu = 0$ there must be no inward going radiation: $I_\nu(-\theta, \phi, 0) \equiv 0$.

(2) as $t_\nu \to \infty$, I_ν must not approach ∞ more rapidly than e^{t_ν}. In order to solve the above equations the term $j_\nu/(\kappa_\nu + \sigma)$ which is known as the source function, denoted by $S_\nu(t, \theta, \phi)$, must be known in terms of κ_ν, σ and the angles.

In the calculation of a stellar model, two cases can be considered:

(i) local cases, in which the source function depends only on conditions at a point. From the model one can find T, P_e and P_g at any point. The continuous spectra belong to this group.

(ii) non-local cases, where the source function at any one point depends on conditions in the deep stellar layers and therefore is difficult to obtain. The resonance lines belong to this group.

Up to the present, nearly all solutions have been for local cases. To obtain solutions two simple cases may be treated:

(a) the case of thermal equilibrium. Then we have

$$j_\nu = \kappa_\nu \rho \, B_\nu(t_\nu) ,$$

where $B_\nu(t_\nu)$ is the Planck function. Then

$$\cos\theta \frac{\mathrm{d}I_\nu}{\mathrm{d}t_\nu} = I_\nu - B_\nu(t_\nu),$$

$B_\nu(t_\nu)$ being a function of t_ν only as a result of its temperature dependence.

(b) the case of pure scattering (isotropic, coherent) when the contribution of radiation in the beam is supplied only by scattered light from other regions of the atmosphere. In this case

$$j_\nu = \sigma\rho \int I_\nu(\theta, \phi, t_\nu) \frac{d\omega}{4\pi}.$$

When I_ν is independent of ϕ,

$$j_\nu = \tfrac{1}{2}\sigma\rho \int_0^\pi I_\nu \sin\theta \, \mathrm{d}\theta = \sigma\rho \, J_\nu(t_\nu).$$

For early type stars, local thermal equilibrium (l.t.e.) is assumed to hold as well as isotropic, coherent scattering, so that the transfer equation becomes

$$\cos\theta \frac{\mathrm{d}I_\nu(\theta, t_\nu)}{\mathrm{d}t_\nu} = I_\nu(\theta, t_\nu) - S_\nu(t_\nu),$$

where $S_\nu(t_\nu)$ is given by

$$S_\nu(t_\nu) = \frac{\sigma}{\kappa_\nu + l_\nu + \sigma} J_\nu(t_\nu) + \frac{\kappa_\nu + l_\nu}{\kappa_\nu + l_\nu + \sigma} B_\nu(t_\nu). \tag{2}$$

Here l_ν is the line absorption coefficient.

One can obtain the following solutions by applying the boundary conditions:

$$I_\nu(\theta, t_\nu) = (\cos\theta)^{-1} \int_{t_\nu}^\infty \mathrm{e}^{-t_\nu/\cos\theta} S_\nu(t_\nu) \, \mathrm{d}t_\nu$$

and

$$J_\nu(t_\nu) = \tfrac{1}{2}\Big[\int_0^\infty S_\nu(t_\nu + t) \, \mathrm{E}_1(t) \, \mathrm{d}t + \int_0^t S_\nu(t_\nu - t) \, \mathrm{E}_1(t) \, \mathrm{d}t\Big]$$

where E_1 is the first exponential function and t is a running variable. The flux is then

$$F_\nu(t_\nu) = 2\Big[\int_0^\infty S_\nu(t_\nu + t) \, \mathrm{E}_2(t) \, \mathrm{d}t - \int_0^{t_\nu} S_\nu(t_\nu - t) \, \mathrm{E}_2(t) \, \mathrm{d}t\Big].$$

If we use the Λ-transformation notation, then

$$J_\nu(t_\nu) = \Lambda_{t_\nu}\{S_\nu\}. \tag{3}$$

We can substitute (3) in (2) obtaining an integral equation which can be used to obtain $S_\nu(t_\nu)$ by the method of successive approximations.

The total flux may be written:

$$F(\tau) = \int_0^\infty F_\nu(\tau)\,\mathrm{d}\nu\,.$$

When calculating models, one may sum up the partial fluxes in different intervals, and an approximation to the above integral is

$$F(\tau) = \sum_{i=1}^{n} a_i F_i$$

where a_i is a numerical quadrature weight and F_i is the flux at a frequency ν_i.

To find the total flux one must sum over at least 34 points to have any accuracy, but if one really wants to consider absorption lines, then one also has to sum over the shapes of the chosen lines. These calculations depend on κ_ν and l_ν. The line absorption coefficient l_ν may be written:

$$l_\nu = \frac{\pi e^2}{mc} f N^* G_\nu \tag{4}$$

where e is the charge of the electron, f is the f-value (oscillator strength) and N^* the number of atoms in lower level of line.

For normal line shapes one has:

$$G_\nu = \frac{\gamma}{4\pi^2(\nu - \nu_0)^2 + \left(\frac{\gamma}{2}\right)^2}$$

where ν_0 is the frequency in the middle of the line and γ is the damping constant.

Accounting also for thermal motions and turbulence one obtains

$$G_\nu = \int_{-\infty}^{+\infty} \frac{\gamma W(\xi)\,\mathrm{d}\xi}{4\pi^2\left(\nu - \nu_0 - \frac{\xi}{c}\nu_0\right)^2 + \left(\frac{\gamma}{2}\right)^2}$$

where each atom or ion has the velocity ξ in line of sight and where $W(\xi)$ is the distribution of velocities (normalized). The Doppler effect gives a displacement

$$\Delta\nu_D = \frac{\xi}{c}\nu_0\,.$$

Assuming a Maxwell distribution of velocities and substituting back in (4) we find that:

$$l_\nu = \frac{\sqrt{\pi}\,e^2 f}{mc\Delta\nu_D} N^* H(a, v)$$

where $\Delta\nu_D$ = Doppler shift in frequency

$$a = \frac{\gamma}{4\pi\Delta\nu_D}, \qquad v = \frac{\nu - \nu_0}{\Delta\nu_D}$$

and $H(a, v)$ is the Voigt function.

If $W(\xi)$ depends on θ, as it might in moving atmospheres, then I_ν will also depend on θ, and the solutions will, in practice, be difficult to obtain.

3. The Theoretical UV Spectrum of O and B Stars

When studying the lines in the UV region, the spectrum may be divided into two parts:

Region I: ranging from 911.6 Å to 1900 Å. Here one finds resonance lines and lines from low levels (<4 eV from the ground level). The *f*-value is large for these lines. Important lines are from H, He II, C II, C III, C IV, Si IV.

The only significant elements not having resonance lines in this region are He I and He II, O II and O III, Ne I and Ne II, and Mg II.

It is impossible to predict lines accurately in this region by the theory sketched above because the assumptions concerning l.t.e. and the form of the source function are not to be justified.

Region II: ranging from 1900 Å to 3000 Å. Here one finds subordinate lines from levels rather near the ionization limit. The level populations are close to thermal equilibrium and the *f*-values are low. The resonance lines of Mg II occur in this region as well as important lines from C II, C III and Si II, Si III, and He I and lines from the second and third spectra of the metals. Reliable prediction of lines is possible in this region.

The present models are mostly calculated without regard to the blanketing effect; it is only in Region I of the UV spectrum, that the spectra predicted from unblanketed and blanketed models differ from each other by a significant amount.

Table I gives some information about the existing models, as of 1965.

The diagram in Figure 2 shows some calculated continuous spectra for types O8–B4 (Underhill, 1963). It is evident that the cooler stars up to B2 are very weak in the UV region. No lines have been considered in these calculations.

TABLE I

Comparison between Unblanketed and Blanketed Models

unblanketed models		blanketed models
	Opacity given by H, He I, He II	
H^- and σ_{el}		H^-, σ_{el} and lines for $\lambda < 1500$ Å chiefly
	Models calculated by	
Underhill O9–B9		Strom (H lines only)
Mihalas O8–A2		Morton and Mihalas (lines in detail)
Strom B6–A2		Guillaume (schematic lines)

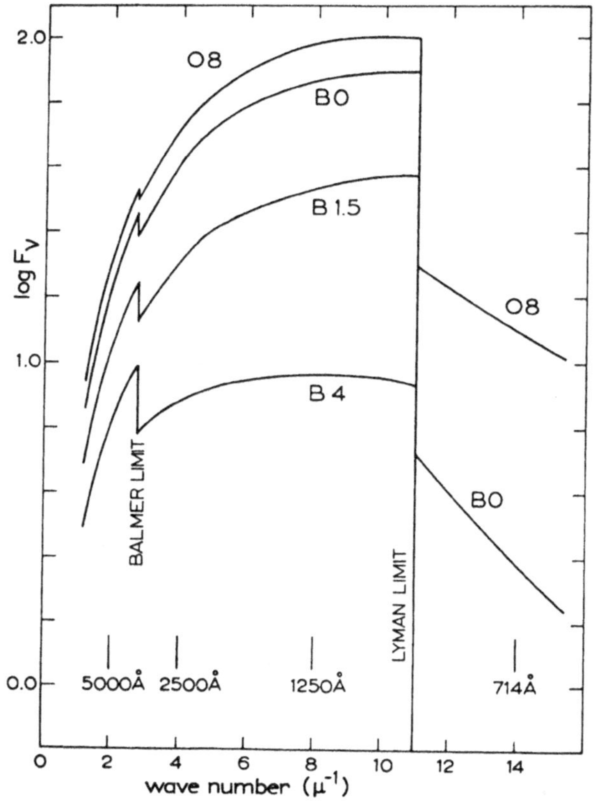

Fig. 2. The predicted emergent spectrum from model atmospheres of types B4 to O8. Line blanketing has not been allowed for in these calculations. The emergent flux F_ν is measured in 10^{-4} ergs cm^{-2} sec^{-1} per unit frequency interval.

Figure 3 shows a plot of the ratio $F_\lambda / F_{\lambda_1}$ versus wavelength, F_λ being the computed flux at any wavelength and F_{λ_1} being the flux at $\lambda_1 = 3650$ Å.

One finds that the observed UV spectrum from some stars (Stecher and Milligan) differs from the predicted one. The great difference between observation and theory at wavelengths shorter than 2400 Å is not to be understood by normal theory. These observations should be repeated to confirm or disprove the presence of an unexpected experimental error.

Profiles of some of the strong resonance and other lines between 911.6 Å and 1500 Å have been predicted by MORTON (1965), using the theory sketched above. A typical part of the results near 1200 Å is shown in Figure 4. These absorption lines are very deep and wide and remove much light from the spectrum. Lines in the spectral region between 1900 Å and 3000 Å have been predicted by GUILLAUME, VAN RENSBERGEN and UNDERHILL (1965), by GUILLAUME (1966) and by ELST (1967).

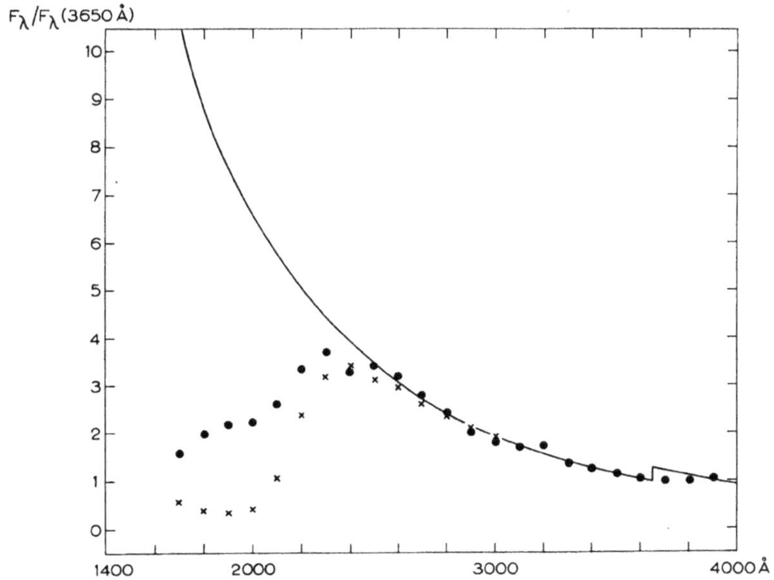

Fig. 3. The relative energy distribution per unit wavelength in a model atmosphere of type B1 compared with the observations of Stecher and Milligan for β Canis Majoris and γ Orionis (filled circles) and for ε Canis Majoris (crosses).

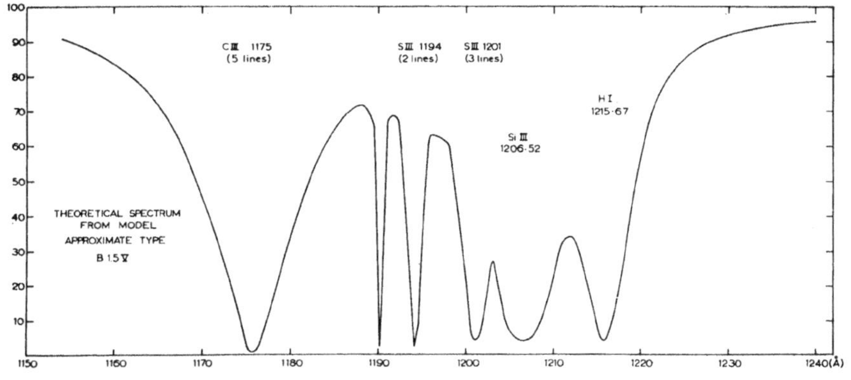

Fig. 4. The profiles of some strong resonance lines near 1200 Å in a model of type B1.

Some typical examples for a B1.5 star are shown in Figures 5 and 6. The Ne II line has a relatively high excitation potential and it is formed deep in the atmosphere. The other lines have rather low excitation potentials and are formed in the outer parts of the atmosphere. These profiles have been computed with a simulated microturbulence such that the Doppler halfwidth of the line absorption coefficient is four times the thermal Doppler halfwidth. Experience has shown that the profiles observed in the spectrum of an average "sharp-lined" star are fitted best in this way. There are many ways in which average velocities of the order of 10 to 15 km/sec could arise in normal stellar atmospheres. Very few B stars have truly sharp lines.

In general the predicted lines in the spectral region observed with ground-based instruments agree well with the observed ones after introduction of thermal and

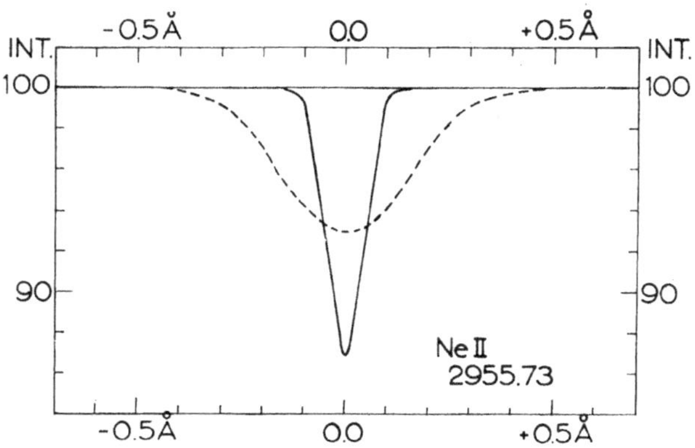

Fig. 5. The profile of a line of Ne II in the spectrum of a model of type B1.5. The broken line indicates how the profile is changed when there is microturbulence.

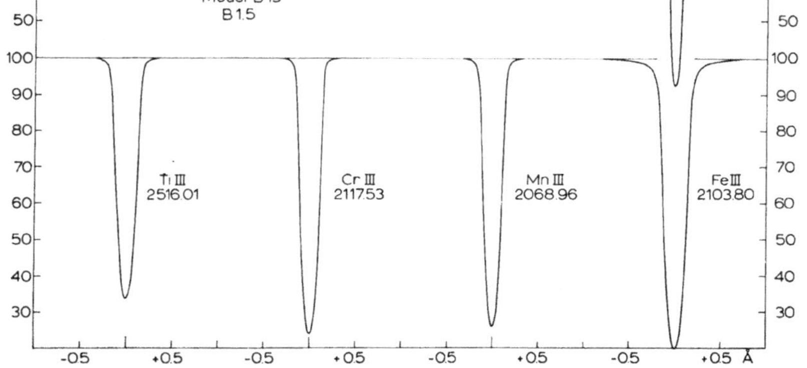

Fig. 6. Profiles of lines of the second and third spectra of the metals in a model of type B1.5.

turbulent motions. To identify a model with a star, which is a great problem, studying the lines is the most convenient way of proceeding. It is difficult to say anything about the lines below 1500 Å, but in the interval between 2000 Å and 3000 Å one may expect to find three types of lines:

(1) type Ne II – broad and shallow lines, from levels of high excitation potential; difficult to observe

(2) type Fe II – deep narrow lines from low excitation levels; difficult to observe

(3) type Cr III – very deep and rather broad lines. Probably observable, particularly if there is microturbulence in the atmosphere.

The spectra from B stars in the interval 2000 Å–3000 Å will probably be similar to F type spectra in the visible region consisting of many weak lines and a moderate number of deep, strong lines.

References

UNBLANKETED MODELS

Underhill, A. B.: 1962, *Publ. Dominion Astrophys. Obs.* **11**, Nos. **23** and **24**.
Mihalas, D. M.: 1965, *Astrophys. J., Suppl.* **9**, 321.
Strom, S. E. and Avrett, E. H.: 1965, *Astrophys. J., Suppl.* **12**, 1.

BLANKETED MODELS

Strom, S. E. and Avrett, E. H.: 1964, *Astrophys. J.* **140**, 1381.
Mihalas, D. M. and Morton, D. C.: 1965, *Astrophys. J.* **142**, 253.
Guillaume, C.: 1966, *Bull. Astron. Inst. Neth.* **18**, 175.
Mihalas, D. M.: 1966, *J. Quant. Spectr. Radiative Trans.* **6**, 581.

ULTRAVIOLET SPECTRA

Underhill, A. B.: 1963, *Space Sci. Rev.* **1**, 749.
Morton, D. C.: 1965, *Astrophys. J.* **141**, 73.
Guillaume, C., Rensbergen, W. v., and Underhill, A. B.: 1965, *Bull. Astron. Inst. Neth.* **18**, 106.
Guillaume, C.: 1966, *Bull. Astron. Inst. Neth.* **18**, 175.
Elst, E. W.: 1967, *Bull. Astron. Inst. Neth.* **19**, 90.
Underhill, A. B.: 1967, *J. Roy. Astron. Soc. Can.* **61**, 9.

ULTRAVIOLET ASTRONOMY: OBSERVATIONAL ASPECTS

R. WILSON
Culham Laboratory, Abingdon, Berkshire, England

1. Introduction

The several disciplines which are incorporated under the heading 'space research' can be placed into the two broad categories of geophysics and astronomy. No hard and fast demarcation can be drawn and a considerable region of overlap exists, including, for example, the moon and planets. There is a basic difference between the two subjects in the experimental sense since most geophysical experiments use the spacecraft as a probe to make measurements pertinent to its immediate environment whereas astronomical experiments use the spacecraft as an observational platform to make measurements of the radiation emitted by very remote objects. In the latter case it is essential to know the exact attitude of the space vehicle and, for any advanced measurements, it is also essential to control the attitude. It is mainly because of this requirement for a stabilized platform that advances in astrophysics have tended to be slower than for geophysics in the space research field.

The primary reason for making astronomical observations from space vehicles is to avoid the spectral attenuation of the earth's atmosphere. Whereas in the case of the sun and planets virtually the whole range of the electromagnetic spectrum becomes accessible, in the case of stars and galaxies limitations are still imposed by the interstellar gas. Thus strong absorption will occur from the Lyman limit of atomic hydrogen at 912Å down to the soft X-rays short of about 100Å. In what follows, the word astronomy will be used in the limited sense to refer to studies of objects beyond the solar system. A clear division therefore exists between UV (900–3000Å) and X-ray (1–100Å) astronomy.

This large extension of the accessible spectrum is in itself sufficient justification for making observations from space vehicles. The whole history of science shows that major advances usually occur when a new dimension is introduced into the character of the observations. A close parallel is the development of radio astronomy which in a relatively short time interval has revolutionized astronomy. There is little doubt that UV and X-ray astronomy will similarly lead to major break-throughs.

UV and X-ray astronomy are now in their infancy, only pioneering observations of a rudimentary character having been made so far. It is apparent that they have a very wide future and that some of the major scientific implications are probably unknown at this time. The present situation in X-ray astronomy is closely analogous to that in radio astronomy in its early days; thus X-ray sources have been discovered, but identifications with optical objects are limited to one or possibly two sources and the emission mechanism is not understood. In UV astronomy however, the position is much firmer in that a wide range of important investigations can be predicted now

J. G. Emming (ed.), Electromagnetic Radiation in Space.

from existing information. One of the main reasons why an extension of observations into the ultraviolet will be important is the fact that most of the resonance lines of the elements in their neutral or ionized states lie shortward of the visible wavelengths. Since these lines are the most sensitive pointers to the physics and chemistry of the sources their observation is of paramount importance. An example of this is the interstellar gas for which UV spectroscopic observations will result in a determination of chemical abundance, including the very important one of the molecular hydrogen content, and physical parameters such as electron density.

2. Observational Parameters

Any particular observation is defined by the values of the following parameters:

(i) Wavelength range $\lambda\lambda$.

(ii) Spectral resolution $\delta\lambda$.

(iii) Photometric accuracy ε.

(iv) Polarization p.

The desired values of these parameters will be set by the astrophysical aims of the observation, but in practice it will usually be necessary to adopt some compromise with the technical difficulties posed. For example, a high spectral resolution may result firstly, in a reduced wavelength range because of a dimensional restriction in the space vehicle, or secondly, in a reduction in photometric accuracy imposed by the size of telescope and the brightness of the object being studied.

An upper limit to the wavelength range $\lambda\lambda$ is already determined by the transmission properties of the earth's atmosphere and the interstellar gas. The upper limit is at 3000 Å where the atmospheric transmission of about 1% makes ground observations possible. The lower limit is set by the interstellar atomic hydrogen, the absorption edge of the hydrogen continuum being at 912 Å. The cross section for the hydrogen absorption continuum is given by $\sigma \approx 8 \times 10^{-18} (\lambda'/\lambda_0)^3$ cm^2 where λ_0 is 912 Å. From a knowledge of the density of interstellar atomic hydrogen this gives the interstellar absorption coefficient at the Lyman edge to be about 25 parsec^{-1}, and hence no stellar radiation can be expected at or immediately below this wavelength. At shorter wavelengths a calculation of the absorption coefficient must include the contribution of other components in the interstellar gas, in particular helium, but apart from the characteristic discontinuities introduced by these elements, in general the cross section decreases with decreasing wavelength until at about 10 Å the mean free path becomes of the order of the diameter of the galaxy. An exceptional case which allows studies in the inaccessible zone below 912 Å is afforded by studies of distant galaxies where the large recessional velocities will shift wavelengths into the accessible region above 912 Å.

The spectral resolution $\delta\lambda$ is defined as the half intensity width of the apparatus function. The possible range of values of $\delta\lambda$ for various astronomical observations is immense and the selection of this parameter is probably the most important in affecting the design of the subsequent instrumentation.

The photometric accuracy ε can be divided into two parts. Firstly, the statistical accuracy ε_S due to the fundamental statistics of the observed count rate and, secondly, the absolute photometric accuracy ε_A, i.e. the error in the calibration factor which transfers the observed signal count to an absolute monochromatic flux in erg cm^{-2} sec^{-1} Å^{-1} incident on the space vehicle.

The statistical accuracy ε_S depends on N_λ the total number of signal counts accumulated over an integration time τ for the spectral element $\delta\lambda$, and n_λ the corresponding value for the background count. Thus

$$\varepsilon_S = \frac{\sqrt{N_\lambda + 2n_\lambda}}{N_\lambda}.$$

The subsequent error in the measurement of the intensity W of an absorption line, expressed in equivalent Ångstroms, is given by

$$\varepsilon(W) = \varepsilon_S \frac{\delta\lambda}{W}.$$

For a particular instrument the observational value ε_S will often be a compromise depending on the brightness of the observed object, but a desirable value is usually 1%.

The absolute accuracy ε_A depends ultimately on the accuracy of the photometric standard from which the calibration is derived. Although considerable advances are being made in this area a great deal of work still remains to be done in order to achieve a desired figure of 10% or better.

The linear polarization p is defined by

$$p = \frac{I_\pi - I_\sigma}{I_\pi + I_\sigma}$$

where I_π is the intensity in the plane of polarization where the electric vector amplitude is a maximum – at some angle θ to a reference plane – and I_σ is the intensity in a perpendicular plane. For a full specification of the polarization properties, the phase difference δ between the π and σ components must also be known and can result in circular or elliptical polarization.

Polarization measurements are important for certain specialized areas of study such as interstellar extinction, synchrotron radiation and resonance lines excited by electron streams. The astronomical possibilities for UV studies have been outlined by ÖHMAN [1] and the instrumental possibilities by WILSON [2].

3. Survey of Astrophysical Possibilities

Any particular investigation will define optimum values for the observational parameters. Since the number of astrophysical possibilities is very wide a broad survey will be given here, particularly to indicate the optimum values of wavelength range and spectral resolution. To assist in this the possible investigations of ultraviolet astronomical spectra have been placed into five broad categories, each of which to some extent represents different observational problems and therefore reflects on the general design and operation of the scientific equipment.

A. STELLAR PHOTOSPHERES (FRAUNHOFER TYPE SPECTRUM)

This category includes stars with a Fraunhofer type spectrum, i.e. a continuous spectrum with spectral lines in absorption. This general form of spectrum is expected from the photospheres of normal stars and a reasonable theoretical basis now exists for an interpretation in terms of the physics and chemistry of the atmosphere.

Measurements of the energy distribution in the ultraviolet continuous spectrum and of the strengths of spectral lines will considerably extend the work carried out in the visible to study the structure and composition of stellar atmospheres. Broad band photometric observations of bright stars in the ultraviolet have already been achieved from sounding rockets [3–11] and have allowed checks on the continuous fluxes predicted by model atmosphere calculations [12,13].

Spectroscopic measurements with fine resolution will allow detailed studies of stellar atmospheres. In general, the stronger lines will give information on the physics of the atmospheres; thus departures from the fundamental hypotheses of the classical model atmosphere will be indicated by a tendency to emission in resonance lines which are much more populous in the ultraviolet. Faint lines, on the other hand, will generally give information on chemical abundances and the presence of resonance lines will enable the observation of the rarer elements to be made.

The statistical consideration of such results will have an important bearing on questions of stellar evolution. The determination of spectral types, and hence temperatures and luminosities, can be made with moderate resolution thus enabling fainter stars to be observed.

Broad band observations will extend the range over which the spectral energy distribution in stars can be measured and will allow even fainter stars to be observable with a consequent increase of the statistics and sampling of the Galaxy. Such observations should include studies of stars in clusters.

A very large amount of theoretical work has been carried out to interpret the Fraunhofer spectra of stars obtained by ground based observations. Model atmospheres are based on the assumptions of radiative and hydrostatic equilibrium and the equation of radiative transfer is solved assuming local thermodynamic equilibrium. An extension of these calculations into the ultraviolet to consider the effects of spectral lines has been made by Gaustad and Spitzer [14] and Stromgren [15] and detailed predictions have recently been made for early B main sequence stars by Morton [16, 17], Mihalas and Morton [18] and Guillaume, Van Rensbergen, and Underhill [19]. These calculations are at present very restricted in spectral type, luminosity and character of the star. A considerable extension is needed in preparation of the analysis of the data to be expected in this field.

There is a wide range of stars which are expected to emit strongly in the UV and should be studied under the present heading. Predominant among these are population I stars of spectral classes O, B and A, and of luminosity classes I–V inclusive. Other suitable objects are the sub-main-sequence stars such as the largely degenerate white dwarfs, the hot sub-dwarfs and the intensely hot nuclei of planetary nebulae. A survey of these sub-dwarfs has been published by GREENSTEIN [20].

B. EXTENDED STELLAR ATMOSPHERES (EMISSION LINES)

This category includes stars expected to show an emission line spectrum in the ultraviolet irrespective of whether a Fraunhofer type spectrum is also present. They are separated from the previous category because of the observational differences inherent in the study of an emission line spectrum compared to an absorption line spectrum. However, an astrophysical difference also exists in that the presence of emission lines indicates some gross non-equilibrium process and the model atmosphere calculations discussed above are inapplicable to their study.

1. *Early Type Stars*

A significant proportion of the early type stars show emission lines in the visible, indicating that they are surrounded by extensive envelopes of gas at low pressure in addition to the usual photospheric layers. Since the region 900–3000 Å embraces several resonance lines of the more abundant elements, observations in this range will greatly extend our studies of such phenomena. The early type stars with extended atmospheres fall into a number of groups of which the main ones are as follows:

Be and Shell Stars. Be stars have the spectrum of a rapidly rotating B star with superposed double hydrogen lines in emission. The emission lines are interpreted as coming from an extended atmosphere rotating with the star. Shell stars also have a Be spectrum but with additional sharp absorption lines, characteristic of a later spectral type, due to a greater amount of material in the envelope than in the case of Be stars. Many Be and shell spectra are slowly variable on a time scale of the order of 5–10 years.

P Cygni Stars. These are generally supergiants of type B showing some sharp emission lines accompanied by violet-displaced absorption cores. This pattern can be explained by the presence of an expanding envelope of moderate density.

The Be, shell and P Cygni stars have been studied extensively by ground based observations and for a survey of the work the reader is referred to the review article by UNDERHILL [21]. They are all near spectral type B and are similar in many respects to the 'normal' B stars listed under category (I). Since B stars are expected to emit very strongly in the ultraviolet, the emission lines from their extended atmospheres may not be observable against the strong background of continuous flux over most of the range 900–3000 Å, in the same way as the chromospheric and coronal emission lines are not observable in total solar radiation in the visible where the sun's continuous spectrum is at a maximum. However, model atmosphere calculations [16] for an early B star indicate very strong line blanketing at short wavelengths below about

1250 Å. Hence the range 900–1250 Å, which includes several resonance lines, becomes important for the study of emission B stars whereas it is not so important for normal B stars.

Wolf-Rayet Stars. The visible spectra of these objects are dominated by very broad emission lines, the total energy in which is usually of the order of that in the continuum. Two classes of spectra are evident: the WC stars in which carbon emission lines dominate, and the WN stars in which nitrogen lines dominate. The classic early work in this field was carried out by BEALS [22, 23] who postulated a rapidly expanding atmosphere (of the order of 1000 km/sec corresponding to tens of keV in energy) as the source of the emission lines and a real anomalous abundance to explain the carbon and nitrogen sequences. Obviously the latter possibility is of fundamental significance in terms of energy generation processes and the evolution of stars but arguments by UNDERHILL [24] have shown that a physical rather than a chemical interpretation may be possible. Since the excitation mechanism for the line emission is unknown the solution to this problem will probably be revealed by observations of the resonance lines in the extreme ultraviolet.

A link between the Wolf-Rayet and O stars is probably afforded by the Of stars which, in addition to a normal O-type absorption spectrum, show emission lines due to He II, N III and C III. These lines have been shown by WILSON [25, 26] and UNDERHILL [27] to have very extensive wings indicating an embryonic Wolf-Rayet atmosphere. A review of Wolf-Rayet and Of stars has been published by UNDERHILL [21].

Planetary Nebulae. These are very extensive, usually expanding envelopes around intensely hot central stars some of which are of the Wolf-Rayet type. Although distant objects, the nebulae can appear as rings in direct photographs, but their angular diameters are generally less than one arc min. Different lines tend to show a different spatial distribution, usually in the sense that the highly ionized species are from less extensive layers than the lowly ionized species. The central stars (nuclei) can show absorption and emission lines, and the objects are expected to emit strongly in the extreme ultraviolet.

The accepted model of a planetary nebula is of an intensely hot nucleus whose ultraviolet emission photoexcites the surrounding nebula. The interpretation of the nebula emission spectrum in the visible, based on the theory of ZANSTRA [28] shows that the central stars are among the hottest known, and their expected flux in the ultraviolet has been investigated, for example, by HUMMER and SEATON [29]. The form of the nebula spectrum in the ultraviolet has been considered by CODE [30], ALLER [31] and OSTERBROCK [32] and it is expected that the strongest lines will be the permitted lines of the abundant elements in early stages of ionization, in contrast to the observed visible spectrum where forbidden lines dominate.

2. *Chromospheric and Coronal Features in Later Type Stars* (G, K *and* M)

In addition to the early type (or hot) objects discussed above, emission lines are expected to be present in the ultraviolet spectra of later type stars due to the presence of chromospheric and coronal features. This is illustrated by the case of the sun in which the chromospheric and coronal emission line spectrum dominates the extreme ultraviolet at wavelengths below about 1800 Å. Evidence of chromospheric effects in G, K and M stars is afforded by the presence of central emission components lying in the deep absorption profiles of the H and K lines of Ca II which are accessible to ground based observation. The width of the emission component has been shown by WILSON and BAPPU [33] to be correlated with luminosity and independent of spectral type. The width increases with luminosity and reaches values of the order of 100 km/sec for the supergiants. The intensity of the emission component could not be correlated with any observable parameters until very recently when WILSON and SKUMANICH [34] showed that a correlation existed between intensity and age in main sequence stars. The intensity decreases with age, showing that the younger stars have the most active chromospheres. The observations of Wilson and Skumanich were insufficient to establish any relation with chemical composition as would be expected from the correlation with age. The intensity may also be linked to magnetic field strength as is shown by the observations of plages on the sun. These are bright in the Ca II K-line and are regions of enhanced magnetic field strength [35, 36] as well as increased X-ray emission [37, 38, 39].

As a rough rule, the region 900–2000 Å will reveal mainly chromospheric features in G, K and M stars (emission lines between 2000–3000 Å will probably be swamped by photospheric emission with the important exception of the Mg II resonance lines) while X-ray observation would reveal coronal features. These stars will probably emit Lyman α strongly but absorption by interstellar hydrogen is likely to prevent its detection with the possible exception of the closest stars.

It is apparent from the above that studies of stellar chromospheres will have an important bearing on questions of stellar evolution and may lead to an understanding of the processes which cause their production and heating. Such studies will be greatly extended by observations in the ultraviolet.

No detailed predictions exist of the chromospheric spectra of stars in the range 900–3000 Å and the best guide is the observed spectrum of the sun [40]. Several emission lines exist in this spectral range and of particular importance are the Mg II resonance lines near 2800 Å. These may be expected to behave in a similar fashion to the Ca II H and K lines but are very much stronger. Thus their observation in stars will be possible over a much greater range than for Ca II H and K and allow a considerable extension in probing the relation of width and intensity with luminosity, age and chemical abundance.

C. VARIABLE STARS

The observational difference between this category and the others is that the character of these observations also depends on the time scale of the stellar variations. This is particularly so in the case of the short period variables and of stars which show sudden outbursts.

A number of classes of variable stars exist; the following are some types suitable for study in the extreme ultraviolet.

1. *Cepheid Variables*

A review of these variables is given by KRAFT [41]. The Cepheids are a broad class of stars whose spectral and light variations show them to be pulsating. Periods range from an hour and a half to more than two years, the changes observed being in brightness, radial velocity and spectral class. The periods of these stars are correlated with their luminosities. Many Cepheid types show hydrogen lines in emission.

The Cepheids can be broken down into two broad categories, the RR Lyrae stars and the classical Cepheids. The RR Lyrae stars have periods of less than 1 day and an average spectral type of A6; they are generally of population II. The classical Cepheids have periods greater than 1 day and spectral types of the range F6–K2; those in the galactic disk and Magellanic clouds are of population I and those in the halo of population II.

Investigations of Cepheids are important for both evolutionary and galactic studies. Since they are not of early spectral type, they will not emit strongly over the whole of the extreme ultraviolet range but such observations may be important in the light of possible chromospheric activity. In this connection ground based observations of the Ca II emission in the spectra of classical Cepheids have been made by HERBIG [42], JACOBSEN [43] and KRAFT [44]. The results show that the emission is transitory, appearing at a phase just after minimum light. Further, measurements of the Ca II width by KRAFT [41] give luminosities from the Wilson-Bappu relation which are two magnitudes different from those obtained from the well established period-luminosity law. It is apparent that studies of the chromospheric effects in these stars are important as was the case for the non-variable later type stars, and that observations in the ultraviolet, particularly of the Mg II lines will considerably extend our understanding.

2. *β Cephei Stars*

These are a small, homogeneous group of pulsating stars clustered around B1–B2 in spectral type and II–IV in luminosity class. Periods are in the range 4–5 hours with radial velocity variations between 10 and 50 km/sec. The light is variable with the same period but $\pi/2$ out of phase. The spectral line profiles also vary and for β Cephei a tendency to emission, which varies over long periods, has been observed in Hα by WILSON and SEDDON [45].

3. *Magnetic Stars*

The discovery and much of the observation of stellar magnetic fields has been carried out by BABCOCK [46]. All stellar magnetic fields adequately observed are found to be variable and the majority vary irregularly. Fields ranging up to a maximum of 34 kG occur. Most of the known magnetic stars are sharp-line A stars (within the limits B8–F0) and many carry the designation 'peculiar' denoting that the Ca II K line is weak and lines of certain other elements, such as Mn, Sr, Cr, or the rare earths, are unusually strong.

The magnetic A stars have been grouped into three classes according to the characteristics of their magnetic variation. The α variables show periodic variations with reversals in polarity, accompanied by variations in the spectrum of the same period. The β variables show magnetic fields with irregular fluctuations including reversals of polarity, but no pronounced variations in the spectrum. The γ variables also show irregular magnetic variations but maintain the same polarity; again there are no pronounced variations in the spectrum.

The magnetic stars are hot enough to allow a significant extension of their Fraunhofer spectrum by observations in the ultraviolet. A study of their spectral variability is important, particularly in the α variables and the presence of resonance lines will allow an extension of the study of abundance anomalies as is also the case for the peculiar A stars.

4. *Novae*

A survey of the studies of novae by ground based observations has been published by McLaughlin [47]. These phenomena are caused by stellar explosions which produce sudden and very large increases in brightness of the order of 12 to 13 magnitudes. This is followed by a slow decline in brightness over periods of several years. The spectrum is generally very complex and shows large changes during the life history of the nova. Initially an absorption line spectrum can be observed with line profiles often showing several components. As this declines the emission lines increase in brightness and are mainly due to permitted transmissions. Eventually forbidden emission lines become the dominant feature of the spectrum as the system approaches the post nova stage. At this point the spectrum is similar to a gaseous nebula except for the large doppler broadening of the lines. The ultraviolet region should contain emission lines of the common elements in their first and second stages of ionization.

5. *Eclipsing Binaries*

There is a group of specialized objects which are important for ultraviolet studies. These are eclipsing binary systems, one member of which is a late type giant and the other an early type dwarf. The latter will dominate the emission in the ultraviolet and as it passes behind successive layers of the atmosphere of the giant star its absorption spectrum will allow these layers to be probed at various heights. This rather unique situation allows a study of the physics and chemistry of the giant atmosphere with spatial resolution using the technique of absorption spectroscopy. The interpretation of this is much less ambiguous than emission spectroscopy. Such objects are not very numerous, but three reasonably bright systems (ζ Aurigae, 31 Cygni and 32 Cygni) comprising a K super-giant and B companion have been studied extensively in the visible. A bibliography can be obtained from the review article by Wilson [48]. Studies in the ultraviolet where several resonance lines occur will be of great value in this type of study.

The above selection of variable stars is not meant to be exclusive but does give a reasonable coverage of the hottest of these objects. The dwarf variable stars are worthy of mention as possible objects for observations, but have not been included as a specific sub-group because of their very inhomogeneous character and wide variation in spectral type. They include the rapid irregular variables such as T Tauri, the flare stars and the SS Cygni variables.

D. INTERSTELLAR MEDIUM

In addition to stellar features the observed ultraviolet spectra of stars will also include the absorption effects of the interstellar medium. This has two main components.

1. *Interstellar Dust*

The interstellar medium includes small solid particles which cause extinction of the light from stars which is wavelength dependent and produces a reddening. Ground based observations of the extinction curve have been carried out over a wide spectral range by STEBBINS and WHITFORD [49] using six colour photometry, and the classical analysis of this curve was carried out by VAN DE HULST [50, 51] who concluded that the particles were dielectric with a size of the order of 1 micron. However, other interpretations based on smaller metallic [52] and graphite [53, 54] particles can also explain the observed extinction curve. Our knowledge of the chemistry of these particles is therefore extremely vague. In this connection broad absorption bands extending over about 100Å have also been observed in the visible spectra of reddened stars, but have yet to be identified. A review of these features has been published by WILSON [55]. A further effect of the interstellar grains is the polarization of star-light, discovered by HALL [56] and HILTNER [57]. This indicated that the dust particles are asymmetric in shape and aligned in interstellar magnetic fields by a mechanism proposed by DAVIS and GREENSTEIN [58]. The particles therefore act as tracers to indicate the magnetic field of the galaxy.

Observations in the ultraviolet should contribute to the understanding of the interstellar particles by an extension of the extinction curve and a search for other diffuse bands. It is also important to understand the extinction curve in the ultraviolet in order to estimate the actual flux from stars. Some theoretical considerations of the ultraviolet extinction by possible cosmic particles have been made by PECKER [59] and by HOYLE and WICKRAMASINGHE [60]. Much of this work can be carried out with broad band photometry and from observations that have already been made from unstabilized rockets [3–11], and an ultraviolet extinction curve has been derived by BOGGESS and BORGMAN [61]. The scientific aspects of ultraviolet studies of the interstellar dust have been outlined by BORGMAN [62].

2. *The Interstellar Gas*

Existing knowledge of the space and velocity distribution of the interstellar gas and of its physical state and chemical composition is largely based on observations of the interstellar absorption lines of neutral sodium and ionized calcium in the visible spectra of hot stars and radio observations of the 21 cm line emitted by interstellar atomic hydrogen. The resulting information is extremely limited and insufficient is known about the physics and chemistry of this important medium, whose study is linked to the whole question of the evolution of the galaxy. Observations of the several important resonance lines of interstellar molecules, atoms and ions which lie in the region 900–3000 Å will greatly extend our information and allow us for the first time to study the medium in great detail. Thus it should be possible to determine physical parameters like the electron density and the chemical abundance of the common elements including the particularly important one of the molecular hydrogen content. The possibilities of investigating the interstellar gas by ultraviolet spectroscopic measurements have been reviewed by SPITZER and ZABRISKIE [63] and is the primary aim of the Princeton experiment scheduled for the third Orbiting Astronomical Observatory (O.A.O.). The types of observations can be broken down as follows:

Atomic Hydrogen. The Lyman series of atomic hydrogen will be very strong even for the nearest O, B stars (which will have a total number of atoms in the line of sight of the order of 10^{20} cm^{-2}) and will show strong radiation damping. Checks of the Lyman cut-off will be possible together with estimates of the atomic hydrogen density, important as a standard for normalising the abundances of other elements.

Molecular Hydrogen. Resonance lines of the band $^1\Sigma_g^+ - ^1\Sigma_u^+$ occur between 1040 Å and 1120 Å and should be observable in the window between Lyman α and Lyman β in the nearest O, B stars. Such observations will give an estimate of the abundance of this particularly important molecule of which there is at present a complete lack of information. Further, since the isotope shift in the line of the HD molecule is of the order of 3 Å, the possibility of determining the abundance of deuterium, or at least an upper limit, exists.

Abundant Elements C, N, O, Mg *etc.* Many of the resonance lines of the more abundant cosmic elements occur in the far ultraviolet and observations of their interstellar absorption are therefore possible. This will enable their abundances to be determined and measurements of the ionization balance e.g. MgI/MgII will also enable the electron density (another important parameter, essentially unknown) to be determined.

E. EXTENDED SOURCES

All extended sources are included in this category even though they cover objects of widely different character, e.g. diffuse nebulae and external galaxies. They are grouped together because of their common observational problems. They are different from objects in the preceding categories in that they do not require the ultra high pointing control necessary for detailed stellar studies, accuracies of about one arc min being sufficient as compared to a fraction of an arc sec. On the other hand, their study at high efficiency requires a more complex optical design in order to achieve the same spectral resolution.

A selection of extended sources, whose study in the ultraviolet is considered important, is listed below. Although physically similar to some of these objects, the planetary nebulae have been discussed above since they are usually less than one arc min extent.

1. *Diffuse Nebulae*

In regions of the Galaxy near hot stars bright gaseous nebulae such as the Orion Nebula are observed which emit an emission line spectrum by the ZANSTRA [28] process due to the presence of ionizing radiation. The ultraviolet implications have been discussed by POTTASCH [64]. The line spectrum will extend to the ultraviolet where observations of resonance lines will advance their physical and chemical interpretation in the same way as with the planetary nebulae where the processes are similar but of higher excitation. A detailed knowledge of the ultraviolet spectra of both diffuse and planetary nebulae will also assist the study of external galaxies which are expected to have somewhat similar spectra but for which identifications may be difficult because of wavelength displacements due to the cosmological red shift.

2. *Supernovae Remnants*

Remnants of supernovae are usually radio sources and emit a continuous spectrum due to synchrotron radiation produced by relativistic electrons moving in magnetic fields [65]. This radiation is observed at optical as well as radio frequencies in the case of the Crab nebula which is also the only definitely identified X-ray source [66] outside the solar system*. The synchrotron flux in the ultraviolet has been estimated by WOLTJER [67] and observations in this region are needed in the further study of these objects which can also be expected to emit a nebula type line spectrum in the ultraviolet.

3. *External Galaxies*

The importance of studying the ultraviolet spectra of external galaxies has been considered by BURBIDGE [68].

In normal galaxies (elliptical, spiral and irregular) the radiation is dominated by stars no hotter than the sun and the ultraviolet spectrum is expected to be similar to that of the diffuse nebulae which exist in our own Galaxy and have been dealt with above. The emission lines of common elements in the first and second stages of ionization can be expected in the range 900–3000 Å but displaced with the appropriate red shift.

In abnormal galaxies, a violent release of energy is apparent. Objects under this heading include the very strong radio emitters such as Cygnus A, the quasi-stellar objects and the Seyfert Galaxies (a class of spiral galaxies with a bright nucleus). The optical spectra of abnormal galaxies show emission lines over a wide range of ionization and in the Seyfert galaxies, for example, forbidden lines of O I and Fe VII

* Since delivering these lectures, the X-ray source Sco X-1 has been identified with a very blue 13th magnitude star by SANDAGE [72].

have been observed. Hence the ultraviolet spectra should contain emission lines of the abundant elements from low to high stages of ionization.

One of the very important emission lines to consider in connection with the external galaxies is the Lyman α line of neutral hydrogen. Although this will probably be unobservable for stars within our own Galaxy due to the absorption of interstellar hydrogen, the red shifts of most external galaxies will displace the light sufficiently to avoid the galactic absorption. This problem has been discussed by OSTERBROCK [69], MUNCH [70] and COOK [71]. Munch concluded that Lyman α would be observable for galaxies with red shift velocities > 1000 km/sec, whereas Cook with more detailed calculations concluded that velocities in excess of 375 km/sec would be sufficient in certain directions. Hence a wide range of external galaxies will be observable in Lyman α should this line be emitted in sufficient strength.

In the normal galaxies the main source of emission in Lyman α is expected to be chromospheres of stars somewhat similar to the sun. However, if any interstellar dust is present the solid particles will act as strong absorption centres and will convert Lyman α energy to heat if there is any appreciable optical depth in the interstellar neutral hydrogen. Hence for any galaxy with an appreciable dust content, such as the normal spiral galaxy and some irregular galaxies like M82, the observation of Lyman α is doubtful. The calculations by BURBIDGE [68] are based on the assumption that component stars emit like the sun and give an estimated flux in Lyman α of 2×10^{-12} erg cm^{-2} sec^{-1} for a normal dust-free galaxy at 10 Mpc distance.

The spectra of the abnormal galaxies show strong emission lines in the visible. In the quasi-stellar sources and particularly the Seyfert galaxies the lines are very broad corresponding to velocities up to 1000 km/sec. From estimates of the Balmer α emission in these objects, Burbidge derives estimates of the expected Lyman α flux. These calculations, like those for the normal galaxies, are necessarily crude but give values of fluxes for the brighter abnormal galaxies of about 10^{-10} erg cm^{-2} sec^{-1} at the earth.

It is apparent that observations of the ultraviolet spectra of external galaxies will be of considerable importance, first, to gain new information on their physical condition, and secondly, to extend the red shift measurements. Hence such observations should contribute to cosmological studies.

The above breakdown of UV astrophysics into five main categories involves a considerable degree of overlap but is intended primarily to represent different observational problems which reflect on the general design of any scientific payload. From the considerations discussed above Table I has been compiled, and gives, for the main categories and their sub-division, the important spectral range for the required observations and the expected spectral widths or shifts of the features involved. These are necessarily somewhat rough and are sometimes based on rather inadequate evidence.

The above brief survey indicates the very wide range of astronomical investigations that would benefit from ultraviolet spectroscopic observations from the appropriate spacecraft. The range is wide both in the astrophysical sense and in the sense of the quantity of observations. Thus the astrophysics has ranged from detailed studies of the cold interstellar gas to studies of hot stars, from early type emission line objects such as the Wolf-Rayet stars to chromospheric activity in late type stars, from studies of variable stars such as the magnetic variables and eclipsing binaries to studies of external galaxies. The amount of observational work involved is immense; for example, the number of O and B stars alone that are brighter than visual magnitude 9 is of the order of 3×10^4.

From Table I it is apparent that the whole wavelength range 900–3000 Å should be covered to extract the full information available in UV astronomical spectra. This is rather a wide range to be covered by a single spectrometer and will react with the aims of spectral resolution and efficiency. As a guide to any possible optimization within this wavelength range, Table I shows that observations in the region below 1250 Å may be somewhat limited due to the absorption of interstellar hydrogen lines and the possible low flux from all but the hottest stars. Most investigations listed in Table I will probably receive their major contribution from observations above 1250 Å with one notable exception: the detection of molecular hydrogen is only possible between 1040–1120 Å.

Estimates of the intrinsic widths of spectral features are given in Table I for the various astronomical sources discussed above. These form a basis for determining the desired spectral resolution for each particular investigation, since a resolution which is much less than the intrinsic width is needed for a profile measurement, and a resolution equal to the intrinsic width is required for a line intensity measurement. The range of observations that could be carried out at a particular resolution can therefore be determined from the table, and an indication of their astronomical significance can be derived from the text. As an example, the possibilities of broad band ($\delta\lambda \approx 100$ Å), moderate resolution ($\delta\lambda \approx 1$ Å) and high resolution ($\delta\lambda \approx 0.1$ Å) are now considered.

Broad band measurements will not give detailed information of UV spectra and the range of investigation will be restricted to a consideration of the UV flux emitted by the hotter stars. The consequent studies are an extension of the work in the visible on the continuous spectra of stars and the interstellar extinction curve. It is obvious that the study of any particular object will be much more advanced by using a spectral resolution capable of measuring spectral lines. However, broad-band measurements have great practical advantages deriving from the greater energy per spectral element which allows observations to be made in much shorter times. Thus observations can be made from simple unstabilised rockets. For stabilised vehicles a larger number of stars, or fainter stars, can be measured within a given observing time than for higher resolution studies. They are therefore important for studies of stellar evolution which require considerable statistics, and for galactic studies which require an extensive distance. For studies of the former, observations of clusters should be included with the attendant technological difficulty of discrimination of objects in a dense star field.

A moderate resolution of 1 Å takes the measurements into the domain of resolved spectroscopy. Spectral lines can be detected and measurements made for the profiles

TABLE I

Observing categories

An indication of the important spectral range, and spectral widths to be expected in the UV spectra for various astrophysical groupings

			Important Spectral Range	Spectral Widths	Remarks
Stellar Photospheres (absorption lines)	*Early Type Stars* O B A Giants		1220–3300 Å[a]	Strong line 1–50 Å	Radiation damping; several resonance lines covered
				Faint line ~ 0.4 Å	Turbulence broadening
	Main Sequence	Broad line	1220–3300 Å[a]	Strong lines 2–50 Å	Radiation damping; several resonance lines covered
				Faint lines ~ 2 Å	Rotation broadening
		Narrow line	1220–3300 Å[a]	Strong lines 1–50 Å	Radiation damping; several resonance lines covered
				Faint lines > 0.05 Å	Lower limit from thermal broadening; no rotation or pole-on viewing
	Sub-main Sequence		1220–3300 Å[a]	Strong lines 1–50 Å	Radiation damping; also Stark broadening in H, He I, He II
				Faint lines > 0.05 Å[b]	Lower limit from thermal broadening
Extended Atmospheres (emission lines)	*Early Type Stars* O, B, A P Cygni Stars (supergiants; expanding envelope)		912–3300 Å	Absorption line shift ~ 1 Å	
	Be Stars Shell Stars	(Rotating envelope)	912–3300 Å	Photospheric absorption line ~ 4 Å Shell emission line ~ 2 Å Shell absorption line > 0.05 Å	Emission generally variable on a time scale of some years
	Planetary Nebulae		912–3300 Å	Emission lines 0.1–0.3[c]	UV flux may be very high
	Wolf-Rayet Stars		912–3300 Å	Emission lines ~ 20 Å	Energy in emission lines roughly equal to that in continuum
	Later Type Stars G, K, M Chromospheric, Coronal Features		912–2000 Å[d]	Mg II lines 0.3–3.0 Å[e] Coronal lines ~ 0.5 Å[f]	Lyman α, if observable, ~ 1 Å

(Table I, cont.)

Variable Stars	Cepheid Variables	1500–3300 Å[g]	Strong lines 1–50 Å Faint lines > 0.1 Å	Radiation damping; several resonance lines covered Lower limit mainly from turbulence
	β Cephei Stars	1200–3300 Å	Strong lines 1–50 Å Faint lines 0.2–0.8 Å	Radiation damping; several resonance lines covered Rotation broadening
	Magnetic Stars	1200–3300 Å	Strong lines 1–50 Å Faint lines > 0.1 Å	Radiation damping; several resonance lines covered Lower limit from thermal plus Zeeman broadening
	Novae	912–3300 Å	Various up to 20 Å	Great variations in spectrum; absorption and emission lines; complex structure
Interstellar Medium	*Interstellar Dust* Early type O, B stars	1220–3300 Å[h]	Broad band ~ 100 Å	
	Interstellar Gas Atomic Hydrogen	912–1220 Å	Lyman $\alpha \sim 5$ Å Lyman $\beta \sim 1.5$ Å	Values for nearest stars, $N = 10^{20}$ atoms cm^{-2}
	Nearest O, B Stars — Molecular Hydrogen	1040–1120 Å	$1\Sigma_g^+ - 1\Sigma_u^+ \sim 0.5$ Å; HD shift ~ 3 Å	Abundance unknown, stated value for 10%. HD observation also possible
	Relatively near O, B Stars — abundant elements C, N, O, Mg etc.	1100–3300 Å	Resonance lines ~ 0.1–0.4 Å[i]	Values for $N = 5 \times 10^{20}$ atoms cm^{-2}
External Sources (External Galaxies)	Normal Galaxies (Elliptical, Spectral, Irregular)	1250–3300 Å	Emission lines: Lyman $\alpha \sim 1$ Å Others ~ 0.1 Å	Lyman α flux ~ 10^{-12} erg cm^{-2} sec^{-1}[j]
	Abnormal Galaxies (Strong Radio Emitters, quasi-stellar objects, Seyfert Galaxies)	1250–3300 Å	Emission lines up to ~ 20 Å	Lyman α flux ~ 10^{-10} erg cm^{-2} sec^{-1}[j]

Notes see page 46.

of the strongest lines and the intensities of the moderate lines. Investigations are then possible for all astrophysical groups in Table I except those particularly requiring the measurement of narrow lines. The particular observations which would be severely degraded are (a) studies of the narrow line early type stars, important for detailed model atmosphere investigations; (b) studies of the shell stars, planetary nebulae and chromospheric activity in later type stars as afforded by profile measurement of the Mg II lines; (c) studies of the Cepheid variables and magnetic stars; (d) studies of the interstellar gas. It is apparent however, that a wide range of observations will be possible at this particular resolution.

An improvement of the resolution beyond 1 Å gradually allows an incorporation of all the possible observations listed in Table I and discussed in the text. A resolution in the range 0.1–0.2 Å would allow the inclusion of important observations such as stellar chromospheres, magnetic and peculiar A stars, and the interstellar gas.

4. Basic Scientific Instruments

The main scientific instruments can be divided into four groups, i.e. telescopes, spectrometers, detectors and polarisers. Each of these groups will be treated separately.

A. TELESCOPES [73–74]

For space applications, refracting telescopes can be ruled out because of the limited range of transmission in the UV of the materials available, e.g. fused silica transmits

---

[a] Heavy blanketing from model atmosphere calculations [17] for B2 stars. This blanketing should extend to A type but may be much lessened in O type stars.

[b] Lower limit ignores possible contribution of pressure broadening which may be significant for some lines.

[c] These values are obtained from expansion velocities which often cause a splitting of the line into two components. In some cases, much higher velocities are observed for lines of low stages of ionization.

[d] Chromospheric emission lines will probably be swamped by photospheric radiation at wavelengths above 2000 Å with the exception of Mg II resonance lines at 2796 Å and 2803 Å.

[e] Obtained from Wilson-Bappu relation [33] between width and luminosity for Ca II lines.

[f] Obtained from values of the sun.

[g] Since the spectral types involved are mainly late A, K and early F, photospheric emission should be appreciable only above 1500 Å. Observations of Mg II lines will be important.

[h] Observations below 1200 Å may be difficult for distant stars due to absorption of interstellar atomic hydrogen. Further, one would expect to observe a maximum in the extinction curve within this range except for particularly small particles ($< 0.1\ \mu$).

[i] The intrinsic line widths of very faint lines are very narrow, but the components due to individual clouds are shifted by the order of 0.1 Å due to the random motions of the clouds.

[j] Estimated Lyman α fluxes obtained from BURBIDGE [68]. The Galaxy will be transparent to this line for red shifts $v/c > 3 \times 10^{-3}$.

Added Note: The several groups listed in the table are not meant to be exhaustive and do not include all the possible types of investigation discussed in the text, some of which were excluded because of their unusual nature. Thus the eclipsing binaries are not listed in the table because of the composite and varying nature of the spectrum. The important spectrum component to isolate is the absorption spectrum of the K giant atmosphere produced by the B companion and in this case the narrowest width expected is about 0.1 Å, the lower limit being set by turbulence.

down to about 1800 Å. The synthetic optical crystals (e.g. CaF_2 or LiF) which transmit further into the UV are at present only available in small sizes. There are other technical reasons in favour of this choice which are the same as have led to a preference for reflecting telescopes in ground based observations. Hence it can be expected that reflecting instruments will also be used for astronomical research within the range accessible to silica and only such telescopes will be considered here.

The two main classes of instrument that can be distinguished are those which focus on-axis objects only and those which focus objects over a wide field.

1. *On-axis systems*

The basis of this class is the focussing property of conic sections of revolution. The various types of configuration are shown in Figure 1. If the mirror can be shaped and operated exactly to the required conic section the resolution is only limited by diffraction. In this case the angular resolution is equal to the value given by the Rayleigh formula – $1.22\ \lambda/D$ – where D is the diameter of the largest mirror and λ is the wavelength. As the object moves off-axis the resolution gets rapidly worse owing to the optical aberration known as coma. The value of the coma is given approximately by $0.2\ (D/F)^2\ \theta$, where F is the focal length of the system and θ is the off-axis angle.

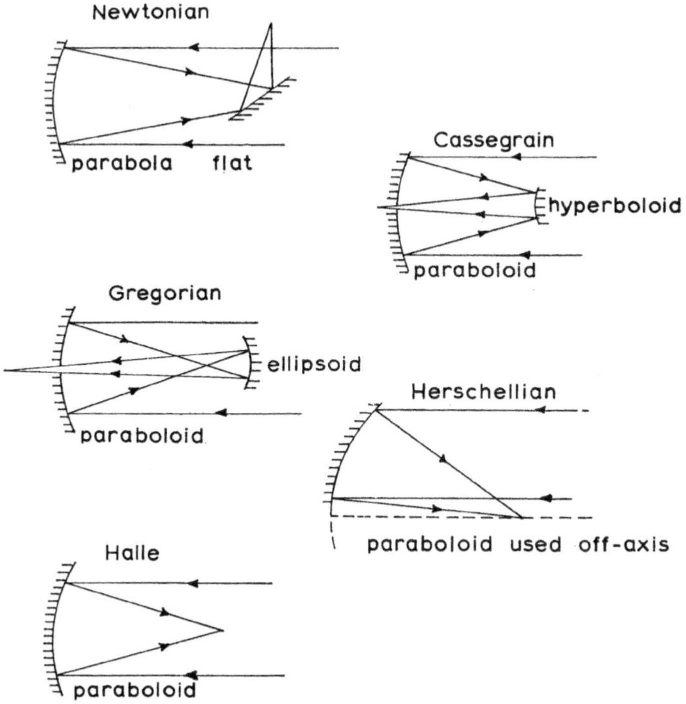

Fig. 1. On-axis telescopes.

By allowing the coma to equal the diffraction limit the field of view is defined by:

$$\Delta\theta = 6\lambda/D\ (F/D)^2 .$$

For example, when $F=500$ cm and $D=50$ cm then $\Delta\theta=1$ arc min in the visible. The instruments discussed above are therefore only adequate for small fields of view and if a wider field is required other types of telescopes must be used.

2. *Wide-field systems*

In these systems, the coma is minimised by changing the shape of the components or by introducing extra components. The choice of shape is restricted by the need to avoid the introduction of spherical aberration. The three principal types are shown in Figure 2. In the Schmidt system, a spherical mirror is used as the primary, with the aperture stop at its centre of curvature. This means that all parallel beams of light entering the system are on-axis, i.e. there can be no coma. The shape of the corrector plate is made in such a way as to eliminate the spherical aberration introduced by the primary. This instrument is limited in wavelength by the transmission of the corrector plate.

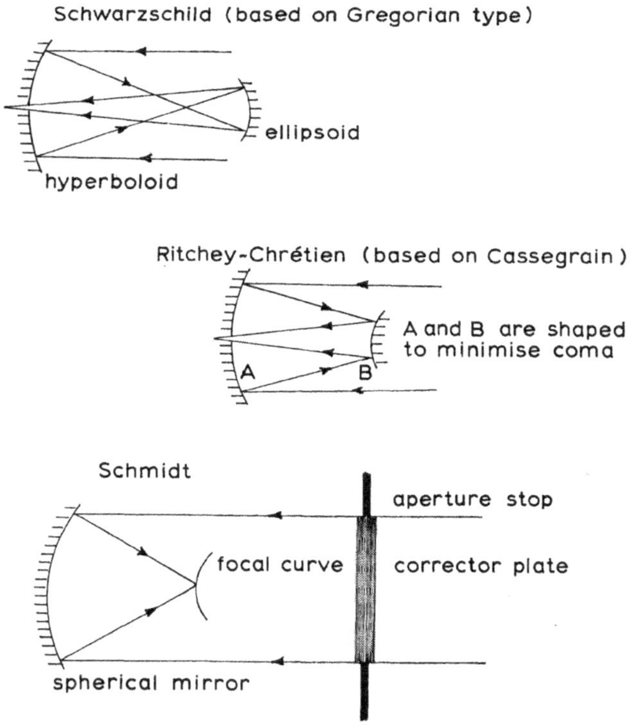

Fig. 2. Wide field telescopes.

Reflecting coatings. Considerable advances have been made during the past few years in the development of reflecting coatings for use in the extreme ultra-violet. A review of these developments and the relevant literature is given by HUNTER [75]. Of the several kinds of coating investigated, four types are particularly valuable for use in the wavelength range embraced by UV astronomy: Al+MgF_2, Al+LiF, Pt and ZnS. Overcoating of Al with MgF_2 has achieved a reflectivity of about 80% down to about 1200 Å, where it drops steeply. Overcoating of Al with LiF has achieved a reflectivity of about 60–70% down to about 1000 Å. Both these overcoatings are subject to damage by radiation. In addition, LiF is mechanically fragile. Platinum has a reflectivity of only 20–30% but this extends below 1000 Å down to about 600 Å. ZnS also has a reflectivity of about 25% which extends down to about 900 Å; it can also be used to discriminate against light of longer wavelengths, having a low reflectivity in the visible, particularly when overcoated on Ge.

B. SPECTROMETERS [76–78]

Prism spectrometers are not considered, mainly on transmission considerations, and attention is confined to plane and concave grating instruments.

Concave gratings. The positions of P and P′ (see Figure 3) are chosen to produce a

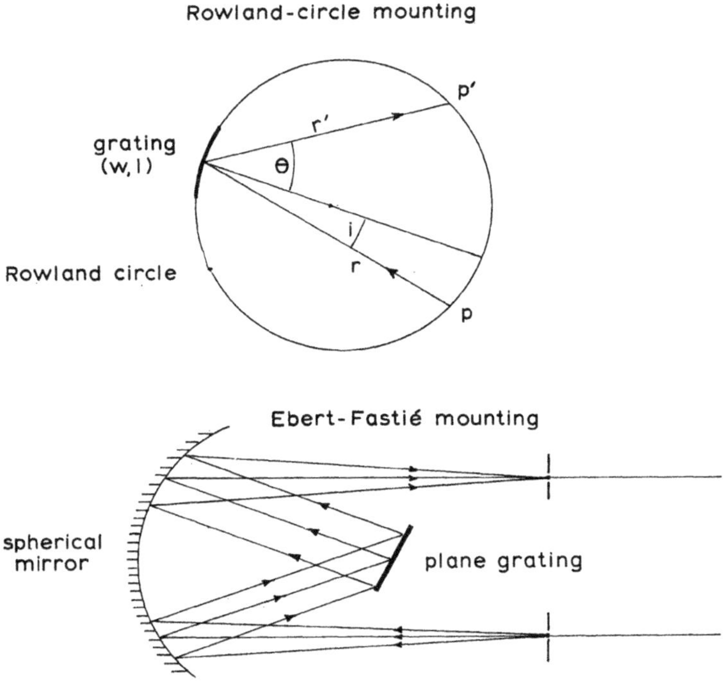

Fig. 3. Grating spectrometers.

focussing condition. A particular solution is:

$$r = R\cos i \quad \text{and} \quad r' = R\cos\theta$$

defining a circle with diameter R, equal to the radius of the grating; this circle is the Rowland circle and is the basis of most mountings. The discussion below is restricted to this type.

The relation between the position of an entrance slit and the consequent wavelength distribution along the Rowland circle is given by the diffraction formula:

$$\sigma(\sin i + \sin\theta) = m\lambda$$

where σ is the spacing between the grating rulings and m is the order of the diffracted spectrum.

The above expressions allow a determination of the geometry of the system only. The determination of its spectral resolution requires a consideration of the limits imposed by diffraction and primary aberrations:

(i) The diffraction limit is determined by $\lambda/\delta\lambda = mN$ where N is the total number of ruled lines in the grating.

(ii) Astigmatism: the length of the astigmatic line, l_a is given by:

$$l_a = 2l\left(\sin^2 i\left(\frac{\cos\theta}{\cos i}\right) + \sin^2\theta\right)$$

where l is the half-length of the rulings. For normal incidence, $i = 0°$ and this reduces to

$$l_a = 2l\sin^2\theta\,.$$

The value of the astigmatism increases with θ for a fixed value of i. For $i = 0°$, $w = 4$ cm, $l = 4$ cm, $R = 100$ cm and $N = 8000$ one finds:

$l_a/2l$	10^{-2}	4×10^{-2}	10^{-1}
λ	1000 Å	2000 Å	3000 Å

(iii) The spherical aberration given by:

$$\Delta l_s = \frac{w(w^2 + l^2)}{2R^2}\left\{\frac{\sin^2 i}{\cos i} + \frac{\sin^2\theta}{\cos\theta}\right\}.$$

In the case $i = 0$ this reduces to:

$$\Delta l_s = \frac{w(w^2 + l^2)}{2R^2}\sin\theta\tan\theta\,.$$

It is seen that spherical aberration increases rapidly with the aperture and this eliminates the use of low f-number gratings for good resolution work.

(iv) The coma is given by:

$$\Delta l_c = \frac{l^2}{2R}(\sin\theta\,\tan^2\theta + \sin i\,\sin^2\theta\,\tan^2\theta - \sin 2i\,\sin\theta\,\tan\theta - \sin^3 i).$$

For normal incidence, this reduces to:

$$\Delta l_c = \frac{l^2}{2R}\sin\theta\,\tan^2\theta\,.$$

A comparison of the various factors affecting resolution can now be made for the particular cases cited above for astigmatism. This is a realistic system based on an $f/12$ concave grating used at normal incidence and giving a linear reciprocal dispersion of 10 Å/mm. The defining parameters are $i=0°$, $w=l=4$ cm, $R=100$ cm and $\sigma=10^{-4}$ cm. The various factors are given in terms of a spectral width expressed in Å:

diffraction: $\Delta\lambda_D = 1.2\times 10^{-5}\lambda$

spherical aberration: $\Delta\lambda_s = 6.4\times 10^{-1}\sin\theta\,\tan\theta$

coma: $\Delta\lambda_c = 8\sin\theta\,\tan^2\theta$

The values of these quantities are given below for three different wavelengths which embrace the range of interest to UV astronomy. A full assessment of resolution must, of course, also include the geometric effects of finite entrance slit width.

λ	1000 Å	2000 Å	3000 Å
$\Delta\lambda_D$	0.01 Å	0.02 Å	0.04 Å
$\Delta\lambda_s$	0.01 Å	0.02 Å	0.06 Å
$\Delta\lambda_c$	0.01 Å	0.06 Å	0.24 Å

Despite the large value of $\Delta\lambda_c$, this is not necessarily the limiting factor in resolution since coma and astigmatism combine to form a curved line image and this may be received with a curved slit. Ray-tracing techniques are needed to fully elucidate this point.

C. TYPES OF MOUNTINGS

1. *Rowland circle*

If a photographic plate or some other image storage device is used, then the spectrometer components can be set up on the Rowland circle in fixed positions without any moving parts. The angle of incidence can be selected between the range 0°–90° but since a grazing incidence situation is not required for reflectivity reasons in the range 900–3000 Å, most instruments will be used near normal incidence ($i\sim 0°$) where aberrations are least troublesome.

For photoelectric recording employing an exit slit, mechanical methods for scanning the spectrum relative to the detector (or bank of detectors) must be devised. Several such mountings have been designed, usually as monochromators, and one or more of the various components are constrained to move in a manner which fulfills, either approximately or exactly, the Rowland circle condition. Some of the Rowland circle mountings are described briefly below.

Rowland: i varies, θ fixed $\approx 0°$; the entrance slit P is fixed and the grating G and entrance slit P' move (see Figure 3).

Abney: i varies, θ fixed $\approx 0°$; P moves, G and P' fixed. This mounting meets the Rowland circle condition exactly.

Paschen-Runge: i fixed $\approx 0°$, θ varies; P and G fixed, P' varies. This is the inverse of the Abney mounting and like that, meets the Rowland circle condition exactly.

Eagle: i and θ vary but $i \approx \theta$; P fixed, G and P' move.

Seya-Namioka: i and θ vary such that $i + \theta \approx 70°$; P, P' fixed, G rotates.

2. *Wadsworth*

In addition to the Rowland circle discussed above, another solution to the focus condition for a concave grating is given by:

$$r = \infty; \quad r' = \frac{R \cos^2 \theta}{\cos i + \cos \theta}.$$

Spectrometers based on this configuration are known as Wadsworth mountings. A spectrometer using an entrance slit requires a collimating mirror in order to illuminate the grating with parallel light and hence the number of optical components is increased by a factor 2 relative to Rowland circle systems. Since the focal plane is not a circle any scanning system for photoelectric recording will necessarily be complex. For stellar UV observations the collimating mirror may be dispensed with by using a slitless arrangement and using a telescope to direct a collimated beam onto the grating, or removing the telescope completely and allowing the star light to fall directly onto the grating. In the latter case a particularly simple arrangement results consisting of only one optical component. However, since the grating now acts as a telescope as well as a dispersing element, the energy collection area is limited to the size for which concave gratings can be ruled, usually less than 10 cm. This restriction in aperture will be somewhat compensated by the improved reflectivity of the whole system.

3. *Ebert-Fastie*

A number of spectrometer designs can be based on plane gratings. These do not suffer from the troublesome aberrations inherent in concave gratings, and also have the advantage that they can be used with much larger apertures, since they can be ruled in larger sizes than concave gratings. However, they do require auxiliary optics in order to collimate the beam from the entrance slit and then focus the beam emergent from the grating.

One particular plane grating spectrometer is the Ebert-Fastie mounting which is depicted in Figure 3. The single large spherical mirror both collimates and focusses, and the main aberration, coma, is cancelled out because of the symmetry of the configuration. The system forms an excellent monochromator in which the spectrum is scanned by a simple rotation of the plane grating.

4. *High resolution spectrometers*

The volume restriction imposed by operation in space vehicles necessitates the design of compact instruments. Hence for work with high spectral resolution, i.e. $\delta\lambda < 0.1$ Å, spectrometers with a high intrinsic angular dispersion are required, and this necessitates the use of instruments working in very high orders. Two such systems will be described briefly. The first is based on echelle gratings which are limited to plane systems because of practical ruling problems. Since they operate in a high order (20–100) the profile of the grooves must be of a very high quality and as a result the rulings are usually well spaced, i.e. about 200–300 lines/mm. Like the normal plane grating they require both collimating and focussing elements and additionally a coarser dispersing element operating in a direction perpendicular to that of the echelle in order to separate orders. The minimum number of optical components is three which can be achieved by using a spherical collimator mirror and a spherical diffraction grating which provides both the focussing and crossed dispersion. The echelle then operates in an Ebert-Fastie type mounting and the grating in a Wadsworth configuration. In addition to its high spectral resolution it has the added advantage of a wide spectral range. This is limited by aberrations but could cover 1000 Å for an instrument suitable for rocket launching.

The second type of high resolution instrument is the Fabry-Pérot interferometer which provides a very high spectral resolution over a very restricted spectral range. This also requires collimating and focussing optics plus some means for order isolation which usually takes the form of a narrow band filter. It is a very compact and efficient instrument, but since it uses transmission optics the wavelength range over which it can be operated is restricted by the transmitting properties of given optical materials. The lowest wavelength at which such a system has been operated in the laboratory is ≈ 1800 Å where BRADLEY *et al.* [79] used high quality 'Spectrosil' coated with $Al + MgF_2$ as an optical material for the interferometer plates, together with an interference filter consisting of layers of Al and MgF_2 deposited on fused quartz substrates.

Discussion. The above description of spectrometers has been necessarily brief and limited to an indication of the various types which may be of value in UV astronomy. The actual design adopted for any particular space experiment depends on a wide range of factors, starting with the astronomical aims and balancing these against the payload constraints and technological difficulties. The high resolution instruments will normally only be used when the scientific objective demands very high resolution ($\delta\lambda < 0.1$ Å) and the choice between an echelle or a Fabry-Pérot will be determined by the relative advantages. Thus any work below 1800 Å will demand the former. However, some significant advantages can accrue by using a high resolution system for problems requiring less than their intrinsic resolution, since the tolerances imposed on the overall design of an astronomical space observatory can be relieved since these scale with the dispersing power of the spectrometer. Thus, the use of an echelle spectrometer for operation in the range of spectral resolution 0.1 Å–1.0 Å can reduce the stringency on stabilisation by a factor greater than an order of magnitude and similarly relieve the mechanical and thermal designs.

Of the Rowland circle mountings, any choice depends largely on a balance between the optical performance (resolution and efficiency) and the mechanical complexity involved in any scanning system. The Abney mounting can be removed from the discussion since it involves a movement of the entrance slit with unacceptable consequences for operation in a space vehicle. The Seya-Namioka provides the simplest scanning, involving a rotation of the grating only. On the other hand, it also gives the poorest optical performance due to aberrations induced by operation at angles away from normal incidence. Of the remaining mountings, the Paschen-Runge provides the simplest scanning action plus the advantage of being easily adaptable to include several channels. Of the Rowland circle mountings, only the Seya-Namioka and Paschen-Runge will be compared with other types of mounting as representative of their class even though other types of mounting (including possible modifications of those discussed above) may be used in different circumstances.

The last type of concave grating mounting is the Wadsworth. This has an advantage of low astigmatism but other aberrations impose a limit of about $f/15$ on the system for reasonable resolution. Its scan mechanism is complex and will probably not be used in UV astronomy except in the special circumstances, mentioned above, when collimated light from the telescope or directly from a star is used to illuminate the grating.

The plane grating system described above is the Ebert-Fastie mounting. This has considerable advantages of low aberrations and a simple scan mechanism obtained by rotating the grating. Its one disadvantage is its three components which results in a degradation of total transmission which is most severe below 1200 Å where very high reflecting coatings do not exist.

The three systems selected – Paschen-Runge, Seya-Namioka and Ebert-Fastie – are reasonably illustrative of the possibilities for UV astronomical spectrometers for operation at spectral resolutions in the range 0.1 Å–1.0 Å. A comparison of the relative efficiencies of these has been made by Namioka [76] for optical components coated with $Al + MgF_2$ and is given in the following table.

	T	A	f/N	η
Paschen-Runge	0.8	1.0	$f/8$	0.1
Seya-Namioka	0.8	0.4	$f/12$	0.027
Ebert-Fastie	0.5	1.0	$f/5$	0.1

T is the overall reflectivity of the instrument derived by assuming 80% for the coating; this is only valid above 1200 Å. The effect of grating blaze is not included since this is common to all instruments. A is an astigmatic factor derived by assuming an exit slit of 1.5 cm in a 1 metre instrument. f/N is the limiting f-number of the system which can still maintain a resolution of 1 Å. The relative efficiency η of each mounting is then defined by:

$$\eta = TA/N$$

It is seen that the Paschen-Runge and Ebert-Fastie are equivalent on this basis with the Seya-Namioka substantially inferior. However, the values refer to wavelengths above 1200 Å and below this the relative efficiency of the Ebert-Fastie will be significantly reduced so that the Paschen-Runge is the best mounting for coverage of the full wavelength range of interest (900 Å–3000 Å). On the other hand, the Ebert-Fastie has advantages in its simpler scanning system, involving the grating rather than the exit slit detector combination.

D. DETECTORS

Photographic plates

The great advantage of photographic emulsion is its image-storing capacity and its extreme simplicity. For use with UV radiation special emulsions are required because of the absorption of the gelatine; these are Ilford QI, QII, QIII, Kodak SWR and Kodak Pathé SC5 and SC7 (in the manufacturing of Kodak Pathé SC5 and SC7, the photographic grains are brought to the surface of the gelatine by a centrifugal method). As an example of the sensitivity of this type of emulsion measurements [80] show that for the fastest – SC7 – 0.05 erg/cm^2 are required at Lyman-α to produce unit density.

The disadvantages of the photographic plate are:

(i) The signal is not of an electrical character. Therefore, for all practical purposes, its use in satellites is ruled out.

(ii) The quantum efficiency is only about 10^{-3}, compared with 10^{-1} for some photo-electric devices.

(iii) The response is non-linear.

(iv) The dynamic range is limited.

(v) The photometric accuracy is poor compared to photoelectric measurements.

(vi) The wide wavelength range of response leads to problems of background light, i.e. scattered visible light will fog the UV part of the spectrum.

Photomultiplier tubes

Photomultiplier tubes have now reached an advanced state of development and a wide selection is available commercially which are robust enough for space use and with a wide coverage of spectral response. The photomultiplier is an excellent detector giving high gain, low noise and reproducibility. Its only disadvantage for application to UV astronomy lies in its single channel character and hence the number of channels which can be built into any system is limited by the physical size of the tube. An extensive literature exists and the reader is referred to the review article by DUNKELMAN [81].

(a) *Sealed tubes.* A large number of these are available and have a high reliability and quantum efficiency [82]. For example, a CsI photocathode with LiF window responds in the range 1000–1600 Å. A CsSb photocathode with CaF_2 window responds uniformly in the range 1200–3000 Å. Photocathodes can be made insensitive to long wavelengths and hence reduce scattered light problems, e.g. RbTe responds only to wavelengths less than 3000 Å.

The short wavelength limit of the response of sealed tubes is set by the window, while the longer wavelength limit is set by the photocathode. Various combinations of these allow a variation in the response curve and tubes with selective characteristics can be selected. Such tubes are important for (a) broad band observations which can then be made without a dispersing system, (b) discrimination against background light scattered in a spectrometer, and (c) rejection of overlapping spectra of unwanted orders.

The major inherent cause of dark count in a well-designed photomultiplier is thermionic emission from the cathode and dynodes, that from the cathode being predominant because of the difference in gain throughout the tube. Cooling markedly reduces the dark count but no significant improvement seems to occur below about $-20\,°C$, when, for visible-sensitive cathodes, a residual nonthermal component, typically of about 1 count cm^{-2} sec^{-1} is observed [83]. Higher work function cathodes – the "solar blind" types, such as Cs-I and Cs-Te – exhibit very little dark count even at room temperature.

(b) *Open-structure types.* These must be used for wavelengths shorter than about 1100 Å where transmitting windows are not available. Some are of the same form as the sealed tubes, with dynodes, but recently two new types have been developed, i.e. the Bendix magnetic strip multiplier and the channel photomultiplier.

The Bendix magnetic strip multiplier is shown in Figure 4. The continuous dynode strip is made of a high resistance material and is matched by a similar field strip.

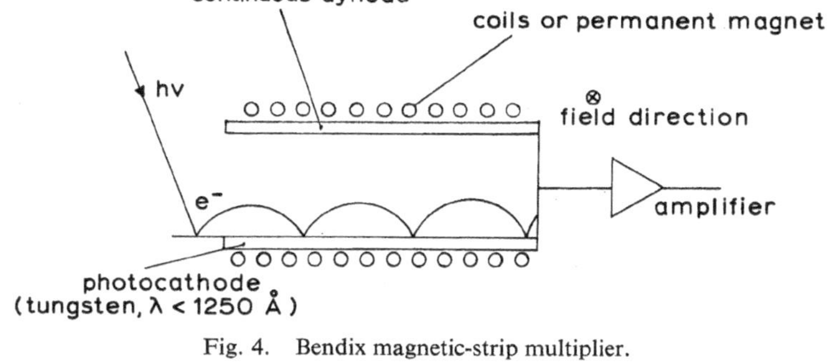

Fig. 4. Bendix magnetic-strip multiplier.

External voltages are applied to the ends of the strip which establish an electric field between the strips and a voltage gradient along them. A magnetic field is maintained perpendicular to the plane of the figure. Photoelectrons ejected from the photocathode follow curved trajectories in the electric and magnetic field configurations in such a manner as to undergo many collisions along the dynode strip with a subsequent multiplication due to secondary electrons. Although the voltage drop per stage is rather low, the number of effective stages is very high. A current gain of 10^8 can be obtained with an overall voltage drop of $\sim$2000 V and a low dark count of $\sim$0.1 sec^{-1} can be realised [84].

There is a limit to the width of the device since the electrons must not strike the upper surface and this restricts the maximum number which can be incorporated in a typical spectrometer to about 10.

The channel photomultiplier [85] is comprised of a glass tube of about 1 cm length with an aperture of about 1 mm, coated on the inside with a highly-resistive continuous dynode. An electric field is applied along the length of the tube, and the inner surface of the open input end constitutes the photocathode. Photoelectrons cascade down the length of the tube with several collisions with the inner surface causing secondary emission and producing gains in excess of 10^5. The small size of these tubes makes them potentially important for UV astronomy since they give the possibility of using a large number of channels in a restricted space. The number that can be accommodated in a spectrometer is limited, not by their size, but by the fact that each must have its own electronic circuitry.

Image Storage Tubes. The image storage tube combines the major advantages of the photographic plate and photomultiplier tube. Its imaging and storage properties lead to a large information capacity whereas the electrical character of the signal allows the telemetry of the information and hence its use in satellites. As a comparison with conventional photomultiplier tubes, let us consider an ultraviolet astronomical observation involving a spectrometer. A spectrum scan with photomultipliers will probably be limited to the order of 10 channels, although this number could be increased substantially with the full development of channel multipliers. The number of effective channels in an image storage device is, on the other hand, limited only by its own resolution which determines the number of picture elements. Hence, for a correctly designed system covering the wavelength range 1000 Å–3000 Å, the number of image points will be 2×10^3 for a spectral resolution of 1 Å and 2×10^4 for 0.1 Å. The gain to UV astronomy which will accrue from a full development of these devices to space requirements is therefore substantial.

The functional layout of an image storage tube is shown in Figure 5. The observed radiation is focussed on the photocathode and the liberated electrons are then focussed onto the storage plate. This stage may also include intensification, for example, by the acceleration of the photoelectrons which then produce an intensified charge image on the storage plate by the liberation of secondary electrons. The image is now contained on the storage plate in an electrical form and the next functional stage of

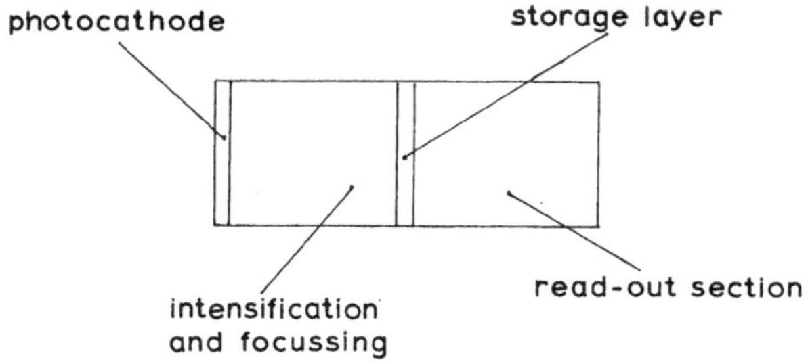

Fig. 5. General layout of an image tube.

the device is concerned with the read-out process, which may be by means of a raster scan by an electron beam of appropriate diameter so as to avoid degradation of resolution. The information can then be processed to the desired form in preparation for telemetry.

A review of the possible application of image tubes to ultraviolet astronomy has been made by BOKSENBERG [86]. Their development for use in this field is dependent on the fulfilment of rather demanding requirements, as follows:

(i) A spectral response covering most and preferably all of the wavelength range 900–3000 Å.
(ii) A long storage time – from minutes up to several hours depending on the particular use.
(iii) A low background to permit long integration times.
(iv) A high spatial resolution.
(v) Good reproducibility.
(vi) Compatibility with spacecraft.
(vii) Low weight.
(viii) Low power consumption.
(ix) Rugged design.

Requirements (i)–(v) are the main design criteria for laboratory development, whereas the other specifications are imposed as a result of operation in space vehicles. Requirements (vi)–(viii) indicate a preference for an electrostatic system rather than a magnetic system which would be heavier, more demanding on power and may introduce adverse affects on the vehicle due to magnetic field coupling.

An ultraviolet sensitive tube, known as a Uvicon [87], is being developed by Westinghouse Research Laboratories for the Celescope experiment [88] scheduled for flying in one of the American Orbiting Astronomical Observatories (OAO). The imaging and read-out sections are both entirely electrostatic. The tube is sealed, the entrance window being of LiF and carrying the photocathode on its rear surface. In addition to this particular device, which is aimed specifically at a space experiment, a wide range of laboratory developments are being carried out for application to many fields of work [86].

E. POLARIZATION DEVICES

Polarization measurements, both for laboratory physics and astrophysics, have been largely confined to the visible, near infrared and near ultraviolet regions of the spectrum, and a number of elegant and sophisticated devices are available which allow the full polarization characteristic of radiation in these ranges to be determined. In the wavelength range embraced by UV astronomy the discussion of polarization devices falls naturally into two parts, those for wavelengths above and below about 2000 Å. For the longer wavelengths well developed optical polarizers are now available whereas at shorter wavelengths the situation is in a state of research and development.

1. ≈2000 Å *to* 3000 Å

Some doubly refracting crystals still transmit over part of this range (e.g. Calcite with a cut-off wavelength of 2100 Å) and quartz, with a cut-off wavelength of 1800 Å, transmits over the whole range. Thus the use of quartz devices allows polarization measurements to be made in this wavelength range with the same degree of sophistication as in the visible; further, quartz is a good robust optical material which should stand up to the environment of a space vehicle.

Two types of linear polarizing device can be used. Firstly, the Rochon prism which transmits an undeviated ordinary ray and a deviated extraordinary ray; this device is valuable if modulation is required, since it can be simply spun about its axis. Secondly, a Wollaston prism deviates both the ordinary and extraordinary ray by equal amounts; this is a symmetrical device and is valuable for a non-modulated measurement when both components of polarization can be monitored continuously. Both these polarizers are composite devices composed of quartz crystalline prisms which are usually cemented together, but for work down to 2000 Å, cement may have to be dispensed with and the prisms mounted mechanically. Retardation plates can be obtained with sheets of quartz cut parallel to the optical axis, thus giving a fixed retardation (say $\lambda/4$) for a fixed wavelength, but a composite prism such as a Babinet-Soleil compensator is generally more useful. This device has an adjustable retardation, and, operated as a quarter-wave plate, can be tuned to each wavelength.

These devices, used with reflecting or fluorite optics should allow the design of an optical system required to make any desired measurements in the wavelength range down to 2000 Å, or if cemented optics are avoided, a little lower. They are described in standard text books on optics (e.g. JENKINS and WHITE [89]). The shortest wavelength to which they have been applied is in the work of JAHODA *et al.* [90] who set up instrumentation to measure the Zeeman effect in the C v line at 2271 Å emitted by a hot plasma.

The only restriction imposed by quartz devices in this wavelength range is one of size, and crystals bigger than about 5 cm are difficult to produce. Such sizes should be adequate for almost all purposes, but where larger sizes are needed, plastic films of the type produced by Polacoat Inc. may have to be used. Such films have been investigated by McDERMOTT and NOVICK [91] who showed that they had polarizing properties down to 2150 Å. Further, retardation plates can be obtained by using stretched polyvinyl alcohol (PVA) which have given quarter-wave retardation down to 2280 Å. Such films are obviously optically inferior to quartz and should only be considered if large apertures are essential. Some development work would be required, particularly for use in a space vehicle, since the films are very fragile and would need to be backed on to a quartz plate with some protective overcoat.

2. 1200 Å *to* ≈2000 Å

Few materials are transparent within this spectral range, and only one, LiF is transparent over the whole range. Polarization work in this region has been carried out by BURTON [92] who constructed a polarization analyser consisting of a LiF plate which reflected the radiation being investigated at the Brewster angle. This was used to measure the polarization of a vacuum ultraviolet monochromator from the visible down to 1216 Å. This simple device is limited in its efficiency, but can be improved considerably by using a stack of plates to convert it into a transmission as well as reflection polarizer. The polarization in the beam transmitted by n parallel plates made of material of refractive index μ and set at the Brewster angle is given by:

$$\frac{I\pi - I\sigma}{I\pi + I\sigma} = \frac{n}{n + \left(\dfrac{2\mu}{1 - \mu^2}\right)^2}.$$

This expression is only applicable if there is no interference between the various reflecting beams. Such interference, in piles of thin plates, could improve the transmission of the system.

Such transmission polarizers allow a rotation of the plane of polarization without moving the detector and therefore offer a considerable advantage on the reflection type. The development of an experimental system based on a stack of LiF plates has been carried out by WALKER [93]. For a system of 6 plates Walker measured polarization efficiencies of 82% at 1200 Å and 65% at 1600 Å. However, the transmission was rather low, the respective values being 4.3% and 21.5%.

Studies of the polarization properties of a number of materials in the extreme ultraviolet is being carried out by HUNTER [94] as a basis for developing both reflecting and transmitting polarizers. He indicates that reflecting polarizers should be possible with efficiencies up to about 30% whereas transmission systems using multilayer unbacked metal films will have a lesser degree of polarization. The possibilities of using thin films has also been considered by ABELES [95].

F. PHOTOMETRIC CALIBRATION

Although considerable advances have been made in the past few years in absolute calibration techniques, a large amount of work remains to be done before reliable, fully checked, accurate systems will be generally available for the range 900–3000 Å. Absolute calibration of a complete observing instrument requires the determination as a function of wavelength of the factor which converts the recorded signal from the source into erg cm^{-2} sec^{-1} $Å^{-1}$ at the space vehicle. It is assumed that the detector characteristic is either linear or can be determined and although a knowledge of this characteristic will be sufficient for some cases, e.g. spectral line intensities which are normalised to the adjoining continuum, the absolute determination of UV flux will in general be necessary for full advances in UV astronomy.

A review of calibration techniques in regard to solar UV studies has been published by HINTEREGGER [96]. Here only the basic approaches to the problem will be indicated. Every calibration must ultimately be based on a laboratory standard which can either be in the form of a source or a detector.

1. *Standard Sources*

The setting up of a standard source with sufficiently high flux over the whole wavelength range 900–3000 Å is the ideal solution to the problem. However, no such source exists in the sense of a fully tested primary radiation standard. An extension of the standards used in the visible, based on black-body radiation, is not feasible because of the prohibitively high temperatures required to produce sufficient flux in the ultraviolet. However, other possibilities exist, still requiring development.

Synchrotron Radiation. This is a potential primary radiation standard since the flux distribution can be calculated precisely from the values of magnetic field and electron energy spectrum. The highly collimated geometry of the emission presents some problems to its full application.

Čerenkov Radiation. This is another potential standard, the emitted flux depending on the velocity of the electrons and the dielectric constant of the material through which they are passing. Severe problems may be encountered due to competing radiation processes such as fluorescence.

Branching Ratio [97]. This is a secondary standard which depends on a transfer of a visible standard into the ultraviolet. The method depends on simultaneous observations of two spectral lines emitted by a common upper level of various atoms and ions from plasmas. The ratio of such lines is in the general case a complex function of the various plasma parameters but in the extreme case of optically thin lines the ratio reduces simply to the ratio of the Einstein transition probabilities. Hence having set up an optically thin situation and using pairs of transitions which give one line in the visible and the other in the ultraviolet, it is possible to transfer an absolute calibration made with the visible line to that in the ultraviolet.

2. *Standard Detectors*

The establishment of a standard detector allows the measurement of the absolute flux incident upon it. Its use, therefore, additionally requires the measurement of the monochromatic transmission of the instrumental system feeding the detector. This is normally carried out by using an appropriate source and monochromator in tandem with the instrument being calibrated. The beam emergent from the source monochromator forms the incident beam for the instrument. This flux is measured and compared with a measurement of the flux emergent from the instrument, their ratio giving the monochromatic transmission at that particular wavelength. Because of possible polarization effects it may be necessary to rotate either the source monochromator or the instrument being calibrated about the optical axis. This transmission measurement can be carried out without any photometric standard. The addition of a standard detector now gives a fully calibrated instrument, but since in many cases the standard detector will not be used in flight its calibration must be transferred to the operational detector.

Two types of standard detector have been used in the wavelength range under consideration.

Thermopiles. These are well developed detectors for use at longer wavelengths. It is necessary that all incident radiation (or a substantial fraction of it which can be accurately measured) be absorbed by the detector and transformed to heat. Problems may arise in the UV due to competing processes such as a production of photoelectrons. Its sensitivity is somewhat limited and will usually require a bright source for its application.

Ionization Chambers. These depend on the photoionization of a gas contained in a tube with a central wire electrode which maintains a potential relative to the walls. It may be used as an ionization chamber or a photon counter. The gas has sufficient optical depth above its ionization limit to ensure the absorption of each photon, which then produces an ion-electron pair which can be measured by suitable electronics. For full calibration the monochromator transmission of the window must also be measured, and the spectral response of the system is limited on the long wavelength side by the ionization potential of the gas, and on the short wavelength side by the transmission of the window. Various gases can be used up to wavelengths of about 1600 Å and the lower limit in wavelength for a fully sealed, robust tube is set at 1050 Å by the transmission cut-off of LiF. Such tubes have been flown in a number of space experiments. For operation down to 900 Å either very thin films of metals or plastics can be used or the window dispensed with completely and a continuous flow system set up with the appropriate vacuum precautions. Efficiencies of about 90% can be achieved with these devices.

Accurate absolute photometry is always a difficult task and demands considerable care even for visible measurements in the laboratory. In UV astronomy we have the combined difficulties imposed by the requirements to extend absolute calibration into the UV and to maintain this in a space vehicle. The above discussion is largely based on an approach to the laboratory problem of deriving an absolute standard and having obtained this it is very desirable to transfer the calibration to a secondary standard source which is both reproducible and portable. This eases considerably the problems of calibrating instruments in the laboratory.

For space work the use of such groundbased calibration demands that it should not vary from the period between laboratory and flight operation. Since in some cases this can be an extended period, during which the environment of the instrument varies considerably the possibility of in-flight calibration should be seriously considered. This requires flying a primary or secondary standard source which must then meet the attendant demands imposed by operation in a spacecraft. One possibility is a Čerenkov source produced by LiF irradiated by a β emitter coated on its back. Ultimately the best solution to this whole problem is the setting up of a comprehensive system of standard reference stars and specific experiments directed to this one goal would be of great importance to the development of UV astronomy.

5. Present Observations

At the time of giving these lectures all observations in UV astronomy had been made from unstabilized sounding rockets in which the sky was scanned by the motion of the vehicle. The limited observing time per star therefore restricted the measurements to broad band flux determinations from the brightest hot objects [3–11]. A review of these results and their analysis has been made by SEATON [12]. In general the UV flux radiated by hot stars is found to be less than that predicted by theoretical model atmosphere calculations, the deficiency becoming more pronounced at the shorter wavelengths. An analysis has also been made to extend the interstellar extinction curve down to 2200 Å by BOGGESS and BORGMAN [98] using the measurements of BOGGESS [10].

The first UV astronomical observations were secured in a flight during 1955 by BYRAM *et al.* [3] which was followed by further observations by the NRL group [6, 7, 11]. The early observations of these workers were made by using broad band selective photomultipliers with mechanical collimating devices. These were later replaced by prime focus telescopes of 10–15 cm diameter with a field of view of 2°–3°. Ionization chambers were introduced as detectors using, for example, nitric oxide with calcium fluoride windows, and operating in the range 1200–1500 Å. Several photometers are flown in one payload with a viewing direction perpendicular to the rocket axis, a sky scan being obtained by the roll motion of the vehicle.

BOGGESS [4, 10] has made observations in the 2000–3000 Å region using photoelectric photometers fed by Cassegrain telescopes. ALEXANDER *et al.* [9] made observations in the southern hemisphere using a mechanical collimator with a field of view of $\approx 1\frac{1}{2}°$ to feed a photomultiplier with a gold photocathode and fused silica window giving a band width of $\approx$400 Å centred at 1950 Å. GULLEDGE and PACKER [8] used photomultipliers with interference filters to give observed bands centred at 2985 Å and 2100 Å with respective band widths of 52 Å and 160 Å.

The first dispersing system was used by STECHER and MILLIGAN [5] who flew a spectrometer of the configuration shown in Figure 6. The mechanical collimator directs the radiation onto the large plane grating, the diffracted light then being focussed onto the entrance slit by means of two mirrors and measured by a photomultiplier.

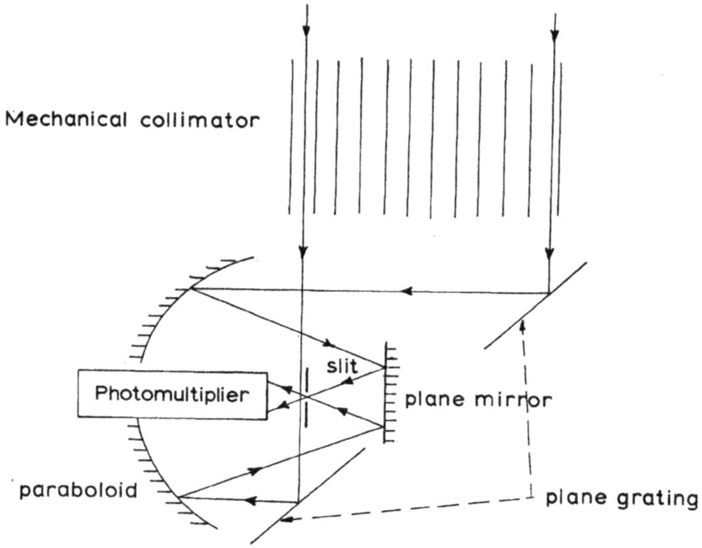

Fig. 6. General layout of spectrometer flown by STECHER and MILLIGAN [5].

The spin axis of the rocket is normal to the plane of dispersion and the spectrum of a star in the field of view is scanned along the slit by the motion of the rocket. The field of view is 3° and coma restricts the ultimate resolution to about 40 Å at the extremes. The operating spectral resolution was set at about 100 Å and the observations were limited to stars brighter than about third magnitude. A typical flux for a bright early type star at 2000 Å is 10^{-8} erg cm^{-2} sec^{-1} Å^{-1}.

None of the observations reported above had sufficient resolution to detect spectral lines. However, since these lectures were given MORTON and SPITZER [99] and MORTON [100] have obtained ultraviolet spectra of some bright early type stars with spectral resolutions of about 1–3 Å. The payloads were flown in Aerobee rockets equipped with Space General attitude control systems which operated with a bias error of a few degrees and a limit cycle of $\pm\frac{1}{4}°$. The experimental payloads consisted of objective grating spectrographs each of which used a plane grating to disperse the radiation into an $f/2$ Schmidt camera covering a 10° field. Corrector plates of different materials were employed to give various cut-off wavelengths. The spectrographs were attached to a large gyroscope rotor which maintained a finer stabilization in the direction parallel to the dispersion, the limit cycle being reduced to ±16 arcsec. The first flight [99] obtained spectra of the early B stars δ and π Scorpii over the respective ranges 1260 Å to 1720 Å and 1260 Å to 2180 Å. The spectra are not of sufficiently good quality to allow any intensity determinations, but are good enough to detect and identify several spectral lines, all in absorption. Identifications include C II 1335 Å, Si IV 1394 Å, 1403 Å, C IV 1549 Å and He II 1640 Å. Some interstellar features due to O I 1302 Å, Si II 1526 Å and Al III 1671 Å were also suspected. In a later flight [100] the UV spectra of δ, ε, ζ, η, ι and κ Orion were obtained to wavelengths down to 1200 Å using LiF correctors in the Schmidt camera. These spectra are of much better quality and permitted intensity estimates to be made. In addition to a number of absorption lines, unshifted emission lines of Si IV 1403 Å and C IV 1550 Å were detected in ε and ζ Orion, together with absorption components shifted to short wavelengths by about 1900 km/sec. These observations are interpreted as being due to a rapidly expanding shell in which these particular lines are formed. The velocities of expansion are two or three times the escape velocity from the photosphere and indicate a substantial loss of mass from the stars. In addition to these stellar features an absorption line in Lyman α is also observed which is attributed to an interstellar origin.

6. Future Observations

At the present time UV astronomy is in its very early stages and major advances can be expected over the next decade with a number of major projects coming to fruition. In America the orbiting astronomical observatories (OAO) have been planned with a series of payloads in which the spectral resolution progresses from broad band observations of UV flux to high resolution studies of stellar and interstellar spectra [101]. In Europe ESRO is planning its major project to be a large astronomical satellite for detailed studies in this field. In addition to these projects a wide range of experiments are being planned in several countries using smaller satellites and sounding rockets. The essential requirement for major advances in UV astronomy is for stabilized vehicles which can lock the observing instruments onto the object being studied. In addition to this such payloads will make a heavy demand on the size, weight and power provision of the spacecraft.

A. STABILIZATION

The pointing accuracy required by an astronomical payload can be derived by expressing spectral resolution, $\delta\lambda$, of the system in terms of its projected angle in the telescope field.

$$\delta\omega = \frac{G}{T} D\delta\lambda$$

where $\delta\omega$ is the angle of the telescope field subtended by a spectral element $\delta\lambda$; G and T are the diameters of the dispersing element and telescope respectively, and D is the angular dispersion. $\delta\omega$ can now be interpreted as the peak to peak amplitude of the limit cycle of the stabilization system and the expression represents a criterion for maintaining a spectral resolution $\delta\lambda$ in a slitless system. For an arrangement involving a slit spectrometer, the criterion becomes one for efficiency only. A payload consisting of a telescope and grating spectrometer can be represented by the following typical values – $G/T = 10^{-1}$, $D = 25$ arcsec Å^{-1}, giving

$$\delta\omega\,(\text{arcsec}) \leqslant 2.5\delta\lambda\,(\text{Å}).$$

Hence, $\delta\omega = 2.5$ arcsec for $\delta\lambda = 1$ Å and $\delta\omega = 0.25$ arcsec for $\delta\lambda = 0.1$ Å.

B. ROCKETS AND SATELLITES

It is of value to consider the relative advantages of sounding rockets and satellites for carrying UV astronomical experiments. The advantages of a sounding rocket are:

(i) greater versatility
(ii) shorter development time
(iii) possibility to use photographic plates
(iv) possibility to test satellite equipment.

The advantages of a satellite are:

(i) longer observing time
(ii) results cheaper per bit of information
(iii) fainter stars can be observed.

It is apparent that any balanced programme in UV astronomy will involve both stabilized rockets and satellites, and although the vehicles will often perform different functions, it is informative to compare them for the same type of observation – say spectroscopy of single stars in the 900–3000 Å region, using photographic recording in the rocket and scanning photomultipliers in the satellite. Systems operating at *f*/12 with grating rulings of 1200 line/mm are capable of a spectral resolution of 0.2 Å and for the volume restrictions involved the telescope and grating sizes shown in the following table will result for the two types of vehicle. The required stabilization for full efficiency is the same in each case – 0.4 arcsec.

	Rocket (Skylark)	Satellite (LAS)
Diameter of telescope (T)	40 cm	80 cm
Diameter of grating (G)	4 cm	8 cm
Collecting Area (A)	1.2×10^3 cm^2	4.8×10^3 cm^2
Quantum eff. (Q)	10^{-3} (photographic)	5×10^{-2} (photoelectric)
No. of detectors (N)	10^4	10
Observing time (τ)	3×10^2 sec (above 100 km)	2×10^3 sec/orbit (dark period)

A rough figure of merit is represented by the product $A \times Q \times N \times \tau$ which is 4×10^6 per launch in the case of a rocket and 5×10^6 per orbit in the case of a satellite, i.e. one rocket flight is equivalent to one satellite orbit.

This comparison is based on the use of the most appropriate detectors which are currently fully developed to space standards. The development of electronic image storage devices in the future will drastically change the figures of merit, resulting in a gain of 50 for the rocket and of 10^3 for the satellite. These gains may not be concurrent in time, however, since it is conceivable that these devices may be developed to rocket use before satellite use where the requirements are much more demanding.

C. CALCULATION OF OBSERVING TIMES

The expressions for observing times will be derived for the UV spectroscopic payloads discussed above, i.e. a Cassegrain telescope feeding a Paschen-Runge slit spectrometer employing scanning photomultipliers and photographic recording.

Let

F_λ = stellar flux at earth at wavelength λ (photon cm^{-2} sec^{-1} Å^{-1})
A = Collecting area of primary mirror (cm^2).
M_λ = Reflectivity of primary mirror at λ.
m_λ = Reflectivity of secondary mirror at λ.
S = Fractional transmission through entrance slit.
G_λ = Efficiency of grating at λ.

and for photoelectric instrument,

Q_λ = Quantum efficiency of detector at λ.
P = Number of detectors.
$\Delta\lambda$ = Full spectral range (Å)
N_λ = Total number of signal counts at λ.

and for photographic instrument,

d = Linear dispersion of spectrometer (Å cm^{-1})
w = Width of spectrum (cm)
E_λ = Energy density to produce unit photographic density (erg cm^{-2}). Then for photoelectric recording, the integration time τ_λ to record N_λ counts at wavelength λ with exit slit width $\delta\lambda$ (Å) is given by:

$$\tau_\lambda = \frac{N_\lambda}{F_\lambda A M_\lambda m_\lambda S G_\lambda Q_\lambda \delta\lambda} \text{ sec}.$$

The total observing time T required to scan the full spectral range $\Delta\lambda$ is given by:

$$T\,(\text{photoelectric}) = \frac{\Delta\lambda\tau_\lambda}{\delta\lambda P} \text{ sec}.$$

For photographic recording, the observing time is set by the time to produce an optimum photographic density (say unity) and is given by:

$$T\,(\text{photographic}) = \frac{E_\lambda w}{F_\lambda A M_\lambda m_\lambda S G_\lambda d} \text{ sec}.$$

We will now apply this formula to the photographic rocket payload with the parameters given above. The stellar fluxes F_λ, including the effect of interstellar extinction, are obtained from the calculations of HOUZIAUX [102, 103]. These are obtained from model atmosphere calculations together with an extrapolated interstellar extinction curve and probably cannot be regarded as being better than a factor of 2. The coatings of the mirror components are assumed to be $Al + MgF_2$ and the values of reflectivity M_λ and m_λ are obtained from the measurements of HASS and TOUSEY [104]. The grating efficiency G_λ is obtained from the measurements of BURTON [105] for a Bausch and Lomb grating overcoated with $Al + MgF_2$. w is taken as 0.05 cm and S assumed to be 0.5. The radius of curvature of the grating is 40 cm giving d=200 Å/cm. A spectrum width $w = 0.05$ cm is adopted and the value $E_\lambda = 0.05$ erg cm^{-2} to produce a photographic density = 1.0 for SC-7 Kodak Pathe emulsion is taken from BURTON and RIDGELY [80].

The calculations are presented in the form of the number of stars and limiting magnitude for different spectral types that can be observed photographically with a spectral resolution of 0.2 Å over a wavelength range 1200–3000 Å during a stabilized rocket flight of 300 sec above 100 km.

Star Type	No. of Stars	Limiting m_v
W–R, O	9	4.4
B0–B4, Be	51	4.4
B5–B9	12	3.3
A	18	2.4

Hence, a total of 90 stars can be observed over a wide wavelength range using stabilized sounding rockets. The limiting magnitude is set by the flux at the short wavelength limit of 1200 Å and causes the fall off in numbers of stars as the spectral type advances. Many more of these later type stars could be observed by accepting wavelength limits above 1200 Å. Observations can also be made between 900 Å and 1200 Å but full optimisation in this range will require reflecting coatings other than $Al + MgF_2$ e.g. Pt.

Because of the equivalence discussed above the table also gives the range of observations that a large astronomical satellite using scanning photoelectric recording could accomplish for one orbit to each object. This equivalence depends on equating the number of integrated counts N_λ to the number of developed grains per spectral element in the SC-7 emulsion. In the case of the satellite, however, the longer observing times can be used to reach much fainter stars. The rocket-satellite comparison was also carried out for a spectral resolution $\delta\lambda = 0.2$ Å and since the observing time scales as $(\delta\lambda)^{-2}$, a degradation of resolution can extend the magnitude limit further. Ultimately, the faintest object observable will be determined by the prohibitively long observing time or the dark count due to photomultiplier noise and radiation background. In practice, the limit may be set for some time by the technical problem of acquiring faint objects.

References

[1] ÖHMAN, Y.: 1964, *E.S.R.O. Scientific Report* No. 1, p. 182.
[2] WILSON, R.: 1964, *E.S.R.O. Scientific Report* No. 1, p. 190.
[3] BYRAM, E. T. *et al.*: 1957, *Astron. J.* **62**, 9.
[4] BOGGESS, A.: 1961, *Mem. Soc. Roy. Sci. Liege* **4**, 459.
[5] STECHER, T. P. and MILLIGAN, J. E.: 1962, *Astrophys. J.* **136**, 1.
[6] FRIEDMAN, H.: 1962, *J. Quant. Spectrosc. Rad. Trans.* **2**, 547.
[7] CHUBB, T. A. and BYRAM, E. T.: 1963, *Astrophys. J.* **138**, 617.
[8] GULLEDGE, I. S. and PACKER, D. M.: 1963, *Astron. J.* **68**, 537.

[9] ALEXANDER, J. D. H., BOWEN, P. J., and HEDDLE, D. W. O.: 1964, *Proc. Roy. Soc.* **A279**, 510.
[10] BOGGESS, A.: 1964, *IAU Symp. No.* 23, *Liege*, p. 173.
[11] BYRAM, E. T. *et al.*: 1964, *IAU Symp. No.* 23, *Liege*, p. 177.
[12] SEATON, M. J.: 1964, *IAU Symp. No.* 23, *Liege*, p. 133.
[13] AVRETT, E. H. and STROM, S. E.: 1964, *IAU Symp. No.* 23, *Liege*, p. 147.
[14] GAUSTAD, J. E. and SPITZER, L.: 1961, *Astrophys. J.* **134**, 771.
[15] STROMGREN, B.: 1964, *Rev. Mod. Phys.* **36**, 532.
[16] MORTON, D. C.: 1964, *IAU Symp. No.* 23, *Liege*, p. 163.
[17] MORTON, D. C.: 1965, *Astrophys. J.* **141**, 73.
[18] MIHALAS, D. M. and MORTON, D. C.: 1965, *Astrophys. J.* **142**, 253.
[19] GUILLAUME, C., VAN RENSBERGEN, W., and UNDERHILL, A. B.: 1965, *Bull. Astron. Inst. Neth.* **18**, 106.
[20] GREENSTEIN, J. L.: 1961, *Stellar Atmospheres*, University Chicago Press, ch. 19, p. 676.
[21] UNDERHILL, A. B.: 1961, *Stellar Atmospheres*, University Chicago Press, ch. 10, p. 411.
[22] BEALS, C. S.: 1929, *Mon. Not. Roy. Astron. Soc.* **89**, 202.
[23] BEALS, C. S.: 1930, *Publ. Dominion Astrophys. Obs.* (*Victoria*) **4**, 271.
[24] UNDERHILL, A. B.: 1957, *Mem. Soc. Roy. Sci. Liege* **20**, 17.
[25] WILSON, R.: 1955, *Observatory*, **75**, 222.
[26] WILSON, R.: 1957, *Mem. Soc. Roy. Sci. Liege* **20**, 85.
[27] UNDERHILL, A. B.: 1957, *Mem. Soc. Roy. Sci. Liege* **20**, 91.
[28] ZANSTRA, H.: 1931, *Publ. Dom. Astrophys. Observatory* (*Victoria*) **4**, 209.
[29] HUMMER, D. G. and SEATON, M. J.: 1961, *Mem. Soc. Roy. Sci. Liege* **4**, 539.
[30] CODE, A. D.: 1960, *Astron. J.* **65**, 278.
[31] ALLER, L. H.: 1961, *Mem. Soc. Roy. Sci. Liege* **4**, 535.
[32] OSTERBROCK, D. E.: 1963, *Planet Space Sci.* **11**, 621.
[33] WILSON, O. C. and BAPPU, M. K. V.: 1957, *Astrophys. J.* **125**, 661.
[34] WILSON, O. C. and SKUMANICH, A.: 1964, *Astrophys. J.* **140**, 1401.
[35] BABCOCK, H. W. and BABCOCK, H. D.: 1955, *Astrophys. J.* **121**, 349.
[36] BABCOCK, H. W. and BABCOCK, H. D.: 1958, Proc. Int. Astron. Symposium No. 6, Stockholm, August 1956, Cambridge University Press.
[37] BLAKE, R. L. *et al.*: 1963, *Astrophys. J.* **137**, 3.
[38] RUSSELL, P. C.: 1965, *Nature* **205**, 684.
[39] BLACK, W. S. *et al.*: 1965, *Nature* **206**, 654.
[40] TOUSEY, R.: 1964, *Quart. J. Roy. Astron. Soc.* **5**, 123.
[41] KRAFT, R. P.: 1961, *Stellar Atmospheres*, University Chicago Press, ch. 9, p. 370.
[42] HERBIG, G. H.: 1952, *Astrophys. J.* **116**, 369.
[43] JACOBSEN, T. S.: 1956, *Publ. Dominion Astrophys. Observatory* (*Victoria*) **10**, 145.
[44] KRAFT, R. P.: 1957, *Astrophys. J.* **125**, 336.
[45] WILSON, R. and SEDDON, H.: 1956, *Observatory* **76**, 145.
[46] BABCOCK, H. W.: 1961, *Stellar Atmospheres*, University Chicago Press, ch. 7, p. 282.
[47] MCLAUGHLIN, D. B.: 1961, *Stellar Atmospheres*, University Chicago Press, ch. 17, p. 585.
[48] WILSON, O. C.: 1961, *Stellar Atmospheres*, University Chicago Press, ch. 11, p. 436.
[49] STEBBINS, J. and WHITFORD, A. E.: 1945, *Astrophys. J.* **102**, 318.
[50] VAN DE HULST, H. C.: 1946, *Rech. Astron. de l'Observatoire d'Utrecht* **11**.
[51] VAN DE HULST, H. C.: 1949, *Rech. Astron. de l'Observatoire d'Utrecht* **11**.
[52] GUTTLER, A.: 1952, *Z. Astrophys.* **31**, 1.
[53] CAYREL, R. and SCHATZMAN, E.: 1954, *Ann. Astrophys.* **17**, 555.
[54] HOYLE, F. and WICKRAMASINGHE, N. C.: 1962, *Mon. Not. Roy. Astron. Soc.* **124**, 417.
[55] WILSON, R.: 1964, *Publ. Roy. Obs. Edinburgh* **4**, 67.
[56] HALL, J. S.: 1949, *Science* **109**, 166.
[57] HILTNER, W. A.: 1949, *Nature* **163**, 283.
[58] DAVIS, L. and GREENSTEIN, J. L.: 1951, *Astrophys. J.* **114**, 206.
[59] PECKER, J. C.: 1963, *Space Sci. Rev.* **1**, 729.
[60] HOYLE, F. and WICKRAMASINGHE, N. C.: 1963, *Mon. Not. Roy. Astron. Soc.* **126**, 401.
[61] BOGGESS, A. and BORGMAN, J.: 1964, *Astrophys. J.* **140**, 1636.
[62] BORGMAN, J.: 1964, *ESRO Scientific Report* No. 1.
[63] SPITZER, L. and ZABRISKIE, F. R.: 1959, *Publ. Astron. Soc. Pacific* **71**, 412.

[64] POTTASCH, S. R.: 1961, *Mem. Soc. Roy. Sci. Liege* **4**, 563.
[65] SHKLOVSKY, I. S.: 1960, *Cosmic Radio Waves*, Harvard University Press.
[66] BOWYER, S. *et al.*: 1964, *IAU Symp. No.* 23, *Liege*, p. 227.
[67] WOLTJER, L.: 1964, *Astrophys. J.* **140**, 1309.
[68] BURBIDGE, E. M.: 1964, *IAU Symp. No.* 23, *Liege*, p. 181.
[69] OSTERBROCK, D. E.: 1962, *Astrophys. J.* **135**, 195.
[70] MUNCH, G.: 1962, *Space Age Astronomy*, Academic Press, New York, p. 219.
[71] COOK, A. H.: 1963, *Quart. J. Roy. Astron. Soc.* **4**, 203.
[72] SANDAGE, A. R.: 1966, *Astrophys. J.* **146**, 316.
[73] BORN, M., and WOLF, E.: 1959, *Principles of Optics*, Pergamon Press, London.
[74] BOUNTRY, G. A.: 1944, *Optique Instrumentale*, Masson, Paris.
[75] HUNTER, W. R.: 1962, *Optica Acta* **9**, 255.
[76] NAMIOKA, T.: 1961, *Space Astrophysics*, McGraw-Hill, New York.
[77] WELFORD, W. T.: 1964, *Progress in Optics* **4**, North-Holland, Amsterdam.
[78] BEUTLER: 1945, *J. Opt. Soc. Amer.* **35**, 311.
[79] BRADLEY, D. J., BATES, B., JUULMAN, C. O. L. and MAJAMDAR, S.: 1964, *Nature* **202**, 579.
[80] BURTON, W. M. and RIDGELEY, A.: 1966, *Internal Culham Report*, to be published.
[81] DUNKELMAN, L., 1962, *J. Quant. Spectr. Rad. Transf.* **2**, 533.
[82] ROME, M., 1964, *IEEE Trans. Nucl. Sc.* **NS-11**, 93.
[83] SHARP, J.: 1964, *EMI Report*, No. CP5475.
[84] HEROUX, L. and HINTEREGGER, H. E.: 1960, *Rev. Sci. Inst.* **31**, 280.
[85] GOODRICH, G. W. and WILEY, W. C.: 1961, *Proc. Image Intensifier Symp.*, Fort Belvoir, Va.
[86] BOKSENBERG, A.: 1965, *ESRO Memorandum* SM-5.
[87] SKORINKO, G., DOUGHTY, D. D. and FEIBELMAN, W. A.: 1961, *Westinghouse Res. Lab. Sci. Paper* 912-*J*902-*P*1.
[88] DAVIS, R. J.: 1962, *Smithsonian Astrophys. Obs. Special Report No.* 110.
[89] JENKINS, F. A. and WHITE, H. E.: 1950, *Fundamentals of Physical Optics*, McGraw-Hill.
[90] JAHODA, F. C., RIBE, F. L. and SAWYER, G. A.: 1963, *Phys. Rev.* **131**, 24.
[91] MCDERMOTT, M. N. and NOVICK, R.: 1961, *J. Opt. Soc. Amer.* **51**, 1008.
[92] BURTON, W. M.: 1964, *Internal Culham Memorandum.*
[93] WALKER, W. C.: 1964, *Appl. Optics* **3**, 1457.
[94] HUNTER, W. R.: 1964, *Proc. Conf. Photo. Spec. Optics* (Tokyo and Kyoto), p. 520.
[95] ABELES, F.: 1964, *Proc. Conf. Photo. Spec. Optics* (Tokyo and Kyoto), p. 517.
[96] HINTEREGGER, H. E.: 1965, *Space Science Rev.* **4**, 461.
[97] GRIFFIN, W. G. and MCWHIRTER, R. W. P.: 1962, *Proc. Conf. Optical Inst. Techniques*, Imperial College, London, p. 14.
[98] BOGGESS, A. and BORGMAN, J.: 1964, *Astrophys. J.* **140**, 1636.
[99] MORTON, D. C. and SPITZER, L.: 1966, *Astrophys. J.* **144**, 1.
[100] MORTON, D. C.: 1966, *Princeton Univ. Report.*
[101] ROGERSON, J. B.: 1963, *Space Science Rev.* **2**, 621.
[102] HOUZIAUX, L.: 1964, *ESRO Scientific Report* No. 1, Chapter 3.
[103] HOUZIAUX, L.: 1964, *ESRO Memorandum* SM-3.
[104] HASS, G. and TOUSEY, R.: 1959, *J.O.S.A.* **49**, 593.
[105] BURTON, W. M.: 1962, *Internal Culham Report.*

OBSERVATIONS OF SOLAR ULTRAVIOLET RADIATION*

H. E. HINTEREGGER

Air Force Cambridge Research Laboratories, Bedford, Mass., U.S.A.

Abstract. In view of the fact that the subject of his lectures is well covered by a number of recent review articles including one by himself, the author restricts himself to a brief discussion of the main scientific objectives of present-day solar UV observations. The most important instrumental problems are reviewed. The opinion is stated that several comparatively simple satellites seem to be better suited for investigating the solar UV spectrum than a very complex spacecraft carrying many sophisticated experiments.

1. Introduction

Early knowledge of the electromagnetic radiation of the sun in the XUV region of the spectrum (XUV denoting, cumulatively, soft X-rays and the far vacuum ultraviolet, often called extreme ultraviolet, or EUV) was limited to inferences based on observations of its ionising effects on the earth's upper atmosphere and extrapolations from spectroscopic observations at longer wavelengths, i.e. for 3000 Å. The existence of appreciable fluxes of solar radiation in the XUV part of the spectrum appeared quite certain on these grounds and, consequently, various experiments were conducted from high mountain tops and stratospheric balloons in an effort to extend observations of the solar spectrum toward shorter wavelengths. However, these early attempts were essentially unsuccessful simply because the altitudes reached were much too low.

About twenty years ago, a new era was initiated by the two historically most important events, namely (a) the discovery of rather high temperatures of the solar corona by Edlen, allowing at least some rough estimates of the expected line and continuum emissions and (b) the introduction of rockets allowing observations at sufficiently high altitudes such that the remaining absorption in the atmosphere above the point of measurement became acceptably low. A third historical marker, perhaps less spectacular but most important experimentally, was the successful development of a solar pointing control allowing the entrance axis of the observing instrument on a rocket to be held in solar alignment within several minutes of arc (Rense, 1953; Neupert *et al.*, 1964). Since then, i.e. over the past 12 years, further improvements in rocketry as well as the development of new types of rocket spectrographs and telemetering XUV monochromators have led to a wealth of new information on both qualitative and quantitative aspects of the solar XUV emission spectrum. The year 1962 marks another great technological step accomplished by the successful use of an orbiting vehicle allowing for the first time observations of a part of the solar XUV spectrum to be made continuously over the period of several solar rotations, whereas hitherto rocket experiments, each of a useful duration of only a few minutes, had

* Seminar given at the ESRO Summer School, in Alpbach (Austria), on 6 August 1965.

J. G. Emming (ed.), Electromagnetic Radiation in Space.

been all that could be done. The full impact of this new capability of doing solar XUV spectroscopy from a long-lived orbiting vehicle has perhaps made itself felt not yet as drastically as one might have expected. So far not much was done notably better from the OSO–A and OSO–B satellites than was achieved with corresponding rocket experiments. It would be entirely wrong, however, to interpret this largely accidental fact as if it indicated any serious limitation of the value of more extensive satellite experiments. Rather, we must conclude that we have hardly begun to take advantage of the new experimental capability and that its true scientific impact, though reserved for the future, unquestionably will be great.

It is not intended here to give any systematic review of the observational aspects of solar UV research, since this topic has been covered in adequate detail in various recent articles, e.g. by TOUSEY (1961, 1963), TOUSEY *et al.* (1965), FRIEDMAN (1961 and 1963), BOYD (1965) and the present lecturer (1961 and 1965). The present discussion is therefore limited to a few comments on selected topics of broader interest.

2. Review of Experimental Tasks

Some years ago, the principal goals of observations of solar XUV radiation were to explore the general character of the solar emission spectrum for progressively shorter wavelengths, to improve spectral resolution and wavelength calibration leading to reliable identification of at least the stronger lines and continua, to achieve a modest degree of absolute photometric accuracy, and to conduct some exploratory measurements allowing spatial resolution, i.e. monochromatic images of the sun, at least in some of the stronger and sufficiently well isolated emission lines. Truly high spectral resolution allowing a detailed determination of line shapes has so far only been accomplished adequately (i.e. with $\Delta\lambda \lesssim 0.03$ Å) for the hydrogen Lyman-alpha line which, with an intensity of about 5 erg cm^{-2}sec^{-1} (incident on top of the earth's atmosphere), is by far the strongest solar emission line in the entire XUV range of the spectrum (the relative magnitude of this H Ly-α emission is best illustrated by the fact that the integrated intensity of all wavelengths below 1200 Å is not greater than that of H Ly-α alone).

Today, most of these goals are still valid with the addition of certain modifications such as extending identifications to include weaker lines, upgrading the absolute accuracy of photometric calibrations, accomplishing better image formation at a greater number of wavelengths and, last but not least, realising spectrophotometric monitoring as a permanent and spectrally complete programme. The latter is needed primarily for purposes of the physics of the upper atmosphere (aeronomy), but solar physicists are, of course, also interested in establishing an appropriate programme of solar XUV flux monitoring.

At present, our knowledge of temporal intensity variations in the solar XUV emission spectrum is still extremely meagre (about 2 months of data on the 160–400 Å region from OSO-A in 1962 and a relatively small number of rocket data on the 60–1300 Å region scattered over the years 1959–1965). Both the OSO-A data (NEUPERT *et al.*, 1964) and the rocket data (Hall *et al.*, 1965) agree that relatively strong temporal variations appear to occur only in lines emitted from elements in relatively high stages of ionisation (e.g. the Fe XV line at 284 Å). Monochromatic images of the sun in some of these lines accomplished with spectroheliographs from rockets have shown clearly that their brightness distribution over the solar disk is particularly strongly localised in active regions, whereas the emission of lines such as H Ly-α and He II Ly-α can be observed with notable intensity from the whole solar disk.

Ideally, one would wish to conduct observations of the sun in the XUV range of the spectrum with

(a) complete coverage of the spectrum, say, from 2000 Å to 1 Å,

(b) sufficiently high resolution to identify line blends and to separate true continuum contributions from otherwise unresolved densely grouped weak lines,

(c) sufficient suppression of any instrumental and natural sources of background which would otherwise jeopardise the practicability of task (b) above,

(d) truly "high" resolution, i.e. that required to dètermine the actual shapes of many more solar emission lines besides that of the strong H Ly-α line,

(e) good absolute photometric accuracy including occasional recalibration after launch (at least for long-lived experiments aboard satellites),

(f) image formation of the solar disk in many different XUV wavelengths and with a high spatial resolution,

(g) high photometric stability to detect relatively small variations over long periods of time,

(h) high-speed photometry to establish amplitudes and time constants of fast fluctuations, if any,

(j) a fairly high duty cycle of actual measurements and for many of those listed under (a) to (h) above, one would prefer continuous observations.

The accomplishment of every one of these tasks is clearly desirable scientifically and essentially practicable experimentally. On the other hand, experimenters tend to agree that the construction of a single large observing platform, attempting the instrumentally combined pursuit of all of these tasks, should not be recommended, at least not at this time. This negative suggestion appears to be motivated by several arguments. The first is the fear of creating some excessively complicated technical monstrosity, at a rather high cost and with too little assurance that it would really work as planned. The second argument against such a plan is that it would lead to a deplorable loss of individuality among the participating scientists or groups of scientists. The third reason is an understandable reluctance regarding major problems of possible interference and expected conflicts of priorities or authorities. Finally, it is the wide-spread experience of experimental physicists and engineers that, in general, any special type of measurement is best performed with some appropriately specialised apparatus and not with some multipurpose instrument designed to accomplish many different types of measurements, falling short, however, of optimisation for each individual task, which would otherwise be possible. We should admit, however, that all of the foregoing arguments are of a fairly subjective nature. Therefore, we should keep an open mind for revising our present strong preference for less highly organised observing facilities, perhaps on the grounds of new practical experiences – perhaps even on the basis of new analyses of the handling of complicated experiment facilities in an advanced manner that may remove much of our present fears regarding the crowding of all experimental tasks on one common platform. Being concerned with the problem of preparing long-lived satellite observatories in an appropriate form, we must not forget the continued importance of individual rocket experiments. Right now, a rather vigorous pursuit of the latter seems to be particularly important to safeguard against major disappointments in the course of preparing satellite observatories for the near future.

3. State of the Experimental Art

No attempt is made here to review the development of experimental techniques of solar XUV observations following the advent of rockets as instrument carriers some 20 years ago, since this development has been covered in the recent articles cited in the introduction. It suffices to summarise briefly a few general aspects of this development:

(a) The use of diffraction gratings has been extended successfully to shorter wavelengths of the solar spectrum, almost overlapping with the long wavelength limit of solar X-ray observations with crystal spectrometers.

(b) In spite of considerable progress in photoelectric recording techniques, the photographic plate, thus far, remained superior for any precise wavelength measurements, accomplished resolution, and for well-defined and simple image formation of the sun in various prominent solar emission lines in the XUV range.

(c) For absolute photometry, however, photographic techniques have been largely abandoned in favour of photoelectric techniques of measuring the dispersed radiation.

(d) The transition from short-lived rocket experiments to the execution of similar experiments from orbiting vehicles has not uncovered any basically new technological complications and it is now justified to expect that the quality of carefully prepared satellite observations should clearly surpass the standards set by corresponding experiments from rockets (regardless of any occasional disappointments en route).

(e) The accomplished quality of solar pointing of the instrumental entrance axis is now considered adequate, for many types of solar observations; accuracies of biaxial control within minutes of arc have been achieved regularly for some time, and triaxial control within much narrower tolerances is now available.

(f) The quality of instrumental calibration in terms of absolute photon fluxes at widely different wavelengths throughout the XUV range has been upgraded, but slowly, and much further work remains to be done in this direction (including the development of good techniques of recalibration in orbit).

In conclusion, one may speculate, that experimental and technological progress could conceivably lead to a situation where the proper selection among practicable experiments would present a problem of greater importance than that of developing further new techniques.

References

BOYD, R. L. F.: 1965, 'Techniques for the Measurement of Extraterrestrial Soft X-Radiation', *Space Sci. Rev.* **4**, 35.

FRIEDMAN, H.: 1961, 'X-Ray and Ultraviolet Radiation Measurement from Rockets', *Space Astrophysics* (ed. by W. LILLER), McGraw-Hill, New York.

FRIEDMAN, H.: 1963, 'Rocket Spectroscopy, in Space Science' (ed. by Donald P. LEGALLEY), John Wiley, New York.

HALL, L. A., SCHWEIZER, W., and HINTEREGGER, H. E.: 1965, 'Long-Term Variation of Solar Extreme Ultraviolet Fluxes', *J. Geophys. Res.* **70**, 2241.

HINTEREGGER, H. E.: 1961, 'Telemetering Monochromator Measurements of Extreme Ultraviolet Radiation', *Space Astrophysics* (ed. by W. LILLER), McGraw-Hill, New York, 1961.

HINTEREGGER, H. E.: 1965, 'Absolute Measurements in the Extreme Ultraviolet Spectrum of Solar Radiation', *Space Sci. Rev.* **4**, 461.

NEUPERT, W. M., BEHRING, E. W., and LINDSAY, J. C.: 1964, 'The Solar Spectrum from 50 Å to 400 Å', *Space Research* **IV**, North-Holland, Amsterdam.

RENSE, W. A.: 1953, 'Intensity of Lyman-α line in the Solar Spectrum', *Phys. Rev.* **91**, 299.

TOUSEY, R.: 1961, 'Ultraviolet Spectroscopy of the Sun', *Space Astrophysics* (ed. by W. LILLER), McGraw-Hill, New York.

TOUSEY, R.: 1963, 'The Extreme Ultraviolet Spectrum of the Sun', *Space Sci. Rev.* **2**, 3.

TOUSEY, R., AUSTIN, W. E., PURCELL, J. D., and WIDING, K. G.: 1965, 'The Extreme Ultraviolet Emission from the Sun between the Lyman-α line of H I and C VI', *Astronomical Observations from Space Vehicles* (ed. by J. L. STEINBERG), International Astronomical Union, CNRS.

SOLAR XUV OBSERVATIONS FROM STABILIZED SKYLARK ROCKETS

R. WILSON

Culham Laboratory, Abingdon, Berkshire, England

1. Introduction

To date all important spectroscopic observations of the Sun in the XUV have been made from stabilized platforms, for example, the bi-axial pointing control on Aerobee rockets [1–3], the OSO satellites [4], and most recently, stabilized Skylark rockets [5–7]. So far three prototype firings on the latter have been made, and the time scale of the project is perhaps of interest. In 1961 the original scientific proposals were made and feasibility studies carried out. The development of the Attitude Control Unit (ACU) and the scientific instruments began in 1962 and the first three launchings took place in August and December 1964 and April 1965 at Woomera, Australia. These flights included two experiments prepared at Culham Laboratory and the results of these will be described below.

2. The Attitude Control Unit (ACU)

The original specification for the ACU [8] was for the whole head to point at the sun with an accuracy of one solar diameter, i.e. 0.5 degree in the lateral axes and a few degrees in the roll axis. The actuator for the ACU is a gas jet system. The error signal for lateral control is provided by Sun-sensing photocells mounted on the side of the head; the error signal for roll control is provided by a magnetometer. The complete system is built into a single bay which is mounted at the bottom of the payload head (see Figure 1) and is jettisoned at the end of the observing period. The remaining equipment is then recovered by parachute return, the scientific data being recorded photographically.

3. The Experimental Payload

Two types of experiments have been carried out: ultraviolet spectroscopy of the solar chromosphere and corona from 1000 Å to 3000 Å and XUV spectroheliography of the solar disk in the range 50 Å to 350 Å.

3.1 UV SPECTROSCOPY

Because of the brightness of the photosphere at wavelengths above 1800 Å, the spectrograph slit is off-set from the limb of the sun. Since the separation of the photosphere and the corona is of the order of one minute of arc, it is necessary to have a pointing accuracy of the order of a few seconds of arc. For this purpose an optical fine alignment system has been developed [9]. This was achieved by tilting and rocking the primary collector mirror in order to maintain the image of the sun in the correct

J. G. Emming (ed.), Electromagnetic Radiation in Space.

Fig. 1. Solar alignment tests of stabilized Skylark head.

position with respect to the spectrograph slit. The error signals are taken from a static split-field detector consisting of a mosaic of four matched photocells upon which an image of the sun is focussed by a small mirror mounted on the same platform as the mirror.

The spectrograph used was a normal incidence type using a concave grating of 40 cm radius. The spectrum was recorded on a photographic film constrained to lie along a machined surface coincident with the Rowland circle. To obtain more recordings per flight, a revolving drum having six such surfaces was used. The photographic assembly is ruggedised to withstand re-entry, even in the event of parachute failure. In the first flight one spectrogram was taken with the slit positioned exactly on the limb of the sun. In the second flight, six spectrograms were taken with the slit positioned on the corona and in the third flight, six further spectrograms were taken, with the slit off-set 10 seconds of arc from the limb, i.e. positioned on the chromosphere.

3.2 XUV SPECTROHELIOSCOPY

The purpose of these experiments was to photograph the solar disk in the XUV region. Two types of instrument were flown.

(a) A pinhole camera with an aluminium filter to cut out all radiation except soft X-rays and the transmission band above 170 Å.

(b) A simple XUV spectroheliograph [10] using pinhole optics and a plane diffraction grating operated at grazing incidence in order to obtain photographs of the disk in the strong emission lines, e.g. He II at 304 Å.

As can be seen in Table I both the ACU and the Fine Optical Alignment System functioned successfully and achieved a performance much superior to that initially specified. The performance can be summarized by saying that the vehicle head was stabilized along the solar vector with an accuracy of a few minutes of arc in the lateral axes and a few degrees in roll while the fine alignment system controlled the solar image with an accuracy of a few seconds of arc.

TABLE I

Flight Technical Data

Flight Number	SL 301	SL 302	SL 303
Date of launch	11.8.64.	17.12.64.	9.4.65.
Local time (UT + 0930 hours)	15.10	13.05	10.30
Motor type	Raven 7A and Cuckoo Booster		
Motor jettison	H + 76 sec	+ 75 sec	+ 74 sec
ACU initiated	+ 85 sec	+ 86 sec	+ 88 sec
Attitude control achieved	+ 106 sec	+ 105 sec	+ 101 sec
Peak altitude	145 km	167 km	161 km
Attitude control ceased	+ 332 sec	+ 363 sec	+ 340 sec
Parachute recovery	no	yes	yes
Attitude control errors			
Lateral: peak to peak			
(max) P + Y	3 arc min	5 arc min	5 arc min
P − Y	8 arc min	3 arc min	5 arc min
Roll: peak to peak	10 deg	14 deg	4 deg
Fine alignment errors			
Lateral: peak to peak			
(max)	12 arc sec	13 arc sec	8 arc sec
RMS	± 2–3 arc sec	± 2–3 arc sec	± 2 arc sec

4. The Scientific Results

4.1 UV SPECTROSCOPY

Flight No. 1 – 11th August 1964

Only one exposure of 17 seconds at a rather low altitude (110–120 km) was obtained with the slit positioned on the solar limb. The short exposure time and the low altitude resulted in a spectrum of reduced intensity at wavelengths below 2000 Å and its elimination below 1000 Å. Several emission lines were recorded, including O VI 1032 Å and hydrogen Lyman-α 1216 Å, as well as a continuous spectrum with superimposed emission and absorption lines extending from 2000 Å to 2500 Å.

Flight No. 2 – 17th December 1964

Six exposures were obtained during the flight, in which the slit was off-set from the limb in the region of the inner corona. However, the surface on the collector mirror caused scattering of the intense photospheric light to the extent that it dominated the recorded spectra which were therefore characteristic for the total sun.

Flight No. 3 – 9th April 1965

In this flight the slit was off-set from the limb by 10 seconds of arc. Six good spectrograms were obtained with spectral resolution of about 0.3 Å. They revealed the chromospheric/coronal spectrum over the full wavelength range. About 300 emission lines were recorded together with a continuous spectrum.

Figure 2 shows a part of the spectrum recorded during a 95 second exposure at a mean altitude of 150 km. The central strip of the spectrum is produced by that region of the chromosphere and corona which was imaged on the centre of the slit. The length of spectral lines is increased by the effect of astigmatism in the spectrograph. The spectrum recorded along the outer borders of the film results from photospheric radiation scattered from the quartz collector mirror which enters the full length of the spectrograph slit. It therefore represents the spectrum of the total sun and can be compared at any wavelength with the central strip of the spectrum which represents the chromospheric and coronal emission. This is illustrated in Figure 3 which shows a detail of the spectrum at 2800 Å. The resonance doublet ($3s^2S$–$3p^2P$) of Mg II is seen in the photospheric spectrum as two strong absorption lines with faint emission cores, whereas the central spectrum shows the emission lines to be the dominant features. Many emission lines are recorded in the central spectrum which do not appear in the outer borders and which have not previously been observed in the solar spectrum. New identifications include intersystem transitions in C III, N IV, O V and new forbidden lines in Fe XI and Fe XII. The continuum appearing in the central spectrum is probably dominated by scattered visible light which reproduces the astigmatic pattern of the spectrograph.

4.2 SOLAR XUV PHOTOGRAPHY

The first two flights carried only the pinhole camera. The recorded image (see Figure 4) obtained in the second flight (17th December 1964) shows pronounced limb brightening, except in the regions near the poles, together with localized emission areas. This localized emission originates from active plage regions as can be seen from comparison with the spectroheliogram in Ca II light taken at approximately the same time at Kodaikanal, India. The bright plage region on the western limb in the X-ray photographs is not present in the Ca II photograph, but it was observed in Ca II light the previous day. This is due to the fact that X-ray emitting regions are at a higher level

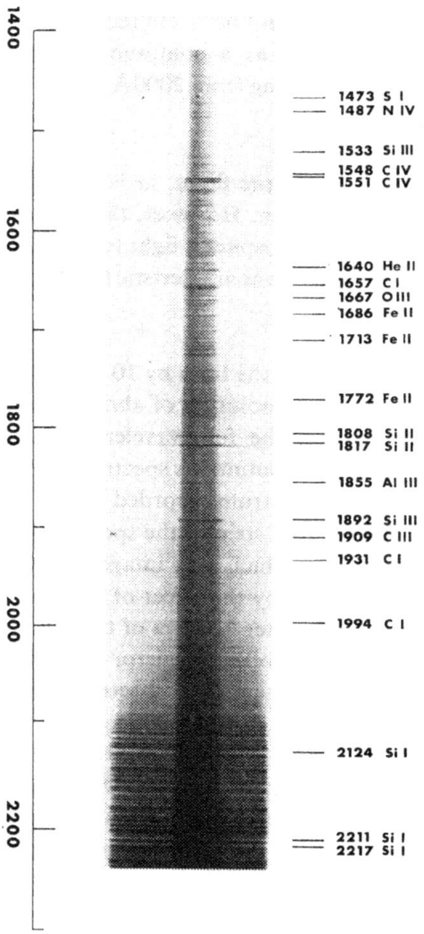

Fig. 2. Solar limb spectrum 1400–2250 Å (Skylark SL 303, 9th April 1965).

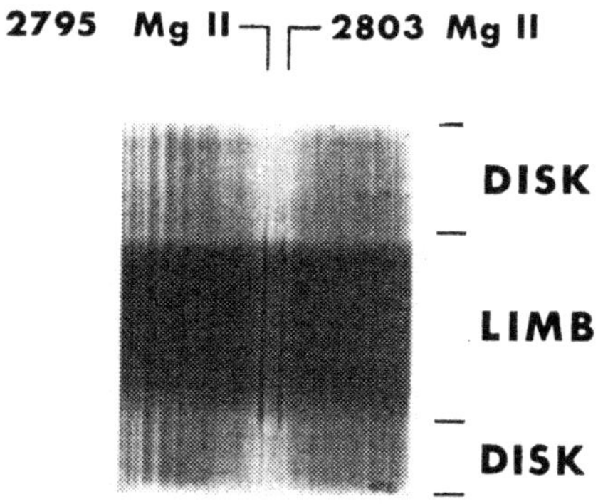

Fig. 3. Chromospheric emission of Mg II doublet (Skylark SL 303, 9th April 1965).

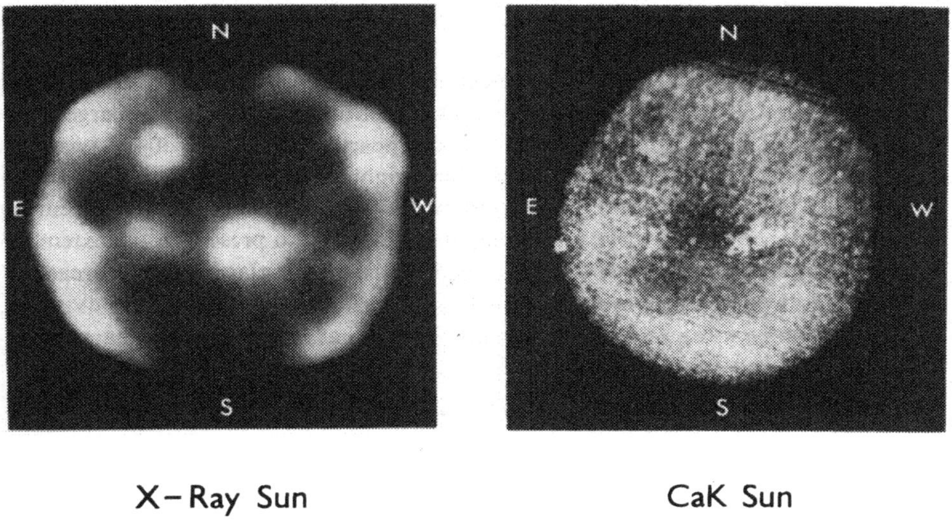

Fig. 4. X-ray pinhole camera photograph, recorded on Skylark SL 302 (17th December 1964) and the spectroheliogram in Ca II light taken at Kodaikanal, by Professor Vainu-Bappu.

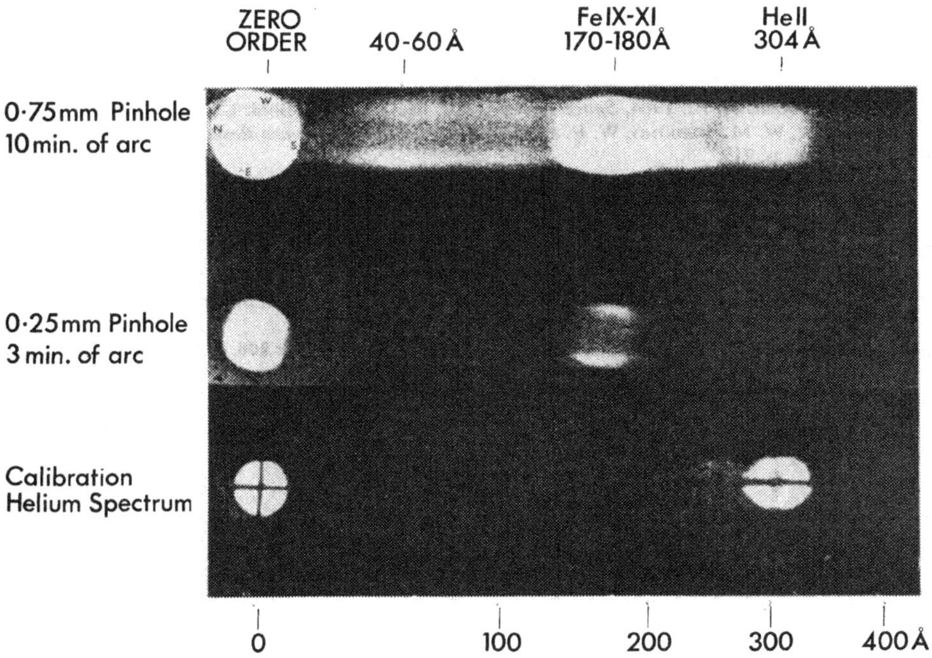

Fig. 5. XUV spectroheliogram (Skylark flight 9 April 1965, height 160 km). The image of a He II discharge plasma is shown for comparison.

than the Ca II regions and therefore can still be seen when the sun's rotation has moved the Ca II regions over the horizon.

In the third flight, the simple spectroheliograph was flown and produced the result shown in Figure 5 together with a pre-flight calibration using a He II discharge. The zero-order image in the upper spectrum is overexposed but the He II, 304 Å, image is correctly exposed and shows only slight limb brightening on a uniformly emitting disk image. The image by Fe IX, X, XI radiation at 170–180 Å is best seen in the central spectrum. This shows equatorial limb brightness which is also present in the extended image (top spectrum) from 40 Å to 150 Å together with bright plage emission areas.

The identification of the intense emission around 170 Å as being due to Fe VIII, IX, X, XI, XII was done using ZETA and thetatron plasmas at Harwell and Culham Laboratory [11].

5. Conclusion

The excellent performance of the stabilization systems involved indicate the great possibilities which exist for future research in solar physics and astrophysics in the XUV.

More accurate sun pointing systems and more powerful rocket motors are being developed. These will allow a wider range of more sophisticated experiments to be performed.

References

[1] Rense, W. A. and Violett, T.: 1959, *Astrophys. J.* **130**, 954.
[2] Tousey, R.: 1961, *J. Opt. Soc. Amer.* **51**, 384.
[3] Hinteregger, H. E.: 1964, *Space Astrophysics*, McGraw-Hill Book Co., p. 34.
[4] Neupert, W. M., Behring, W. E. and Lindsay, J. C.: 1964, *Space Research* **IV**, North-Holland Publ. Co., p. 719.
[5] Wilson, R.: 1964, *Ann. Astrophys.* **6**, 769.
[6] Black, W. S., Booker, D., Burton, W. M., Jones, B. B., Shenton, D. B., and Wilson, R. 1965, *Nature* **206**, 654.
[7] Burton, W. M. and Wilson, R.: 1965, *Nature* **207**, 61.
[8] Cope, P. E. G.: 1964, *J. Brit. Interpl. Soc.* **19**, 285.
[9] Black, W. S. and Shenton, D. B.: 1965, *Proc. IFAC Symposium*, Stavanger.
[10] Burton, W. M.: 1965, *J. Sci. Inst.* **42**, 477.
[11] Gabriel, A. H., Fawcett, B. C., and Jordan, C.: 1965, *Nature* **206**, 390.

OVERALL CONCEPT OF THE ESRO ASTRONOMICAL SATELLITE

L. A. HUSAIN

European Space Technology Centre, Delft, The Netherlands

Abstract. The most important scientific and technical specifications of the LAS are listed, including the scientific instrumentation specifications and the spacecraft design parameters. Possible technical solutions to the design problem of the LAS are reviewed. The conclusion is drawn that the design concept as proposed by ESTEC "presents an attractive line of development for the LAS and compares favourably with other overall concepts".

1. Introduction

Satellite concepts vary enormously depending on the mission which, in the most general terms, defines the overall requirements as regards complexity, size and shape. There are, however, over-riding factors which tend to influence the practical realisation of similar missions. These factors are usually economic but may well be also connected with peculiarities of the payload.

Satellites may be very broadly characterised as follows:

(1) "Small" Scientific: Fully integrated package and spacecraft. Size and shape largely defined by launch vehicle.

(2) Observatory: Too expensive to be fully integrated. Size and shape largely defined by payload.

(3) Manned: Satellite design almost wholly defined by payload, e.g. survival and safety.

We are concerned with the second category, i.e. Observatory type satellite. The problem is therefore to develop a spacecraft concept which satisfies the general payload physical characteristics of a stellar telescope, e.g. large "volume" for collecting power and long light paths, rigidity for alignment and focussing, plus a considerable degree of payload flexibility and development potential.

It is reasonable to suppose that a satisfactory solution for a stellar astronomical observatory would also be of interest to other fields, e.g. solar, where actual telescope dimensions are likely to be less important owing to the intensity of the source. This advantage may well be off-set, however, by the volume and weight required for thermal control equipment.

This paper very briefly surveys possible variations of the spacecraft concept, and suggests a possible solution for a general purpose spacecraft of the observatory type. No attempt is made to discuss design details since, in the first place, spacecraft at present are essentially integrated solutions resulting from costly development programmes thus making truly objective comparisons difficult; secondly, in the present case only one astronomical observatory concept, namely O.A.O., has been the subject of intensive development.

J. G. Emming (ed.), Electromagnetic Radiation in Space.

2. Spacecraft Requirements

The immediate requirement within ESRO is for a concept for a Large Astronomical Satellite (LAS). Considerable work by the Ad hoc Working Group resulted in a definition of desirable scientific objectives and probable vehicle requirements. Extracts from their final recommendations of particular interest in defining physical characteristics of the spacecraft are as follows:

A. PRELIMINARY SPECIFICATION FOR SCIENTIFIC INSTRUMENTATION

The first primary instrument will be expected to make stellar, and if possible, nebular spectrophotometric observations extending from 3300 Å to about 1100 Å (region A). If observations between this limit and 912 Å (region B) are feasible with the same instrument without appreciably reducing the efficiency over region A, the design should allow for their being made, even at low efficiency.

In consideration of the magnitude of the stars which it is hoped to observe, the primary mirror should exceed 50 cm in diameter. The possibility of two parallel mirror systems is not to be excluded. Greater size would be preferred but only if it did not cause loss of efficiency, or loss of resolution (as a result of the larger physical dimensions dictating a less good spectrograph design, or requiring the insertion of a plane mirror, etc.).

High spectral resolution is required throughout both regions A and B. It is felt that a few tenths of an Ångstrom is a reasonable target.

In the absence of relevant data it should be assumed that stellar observations must be made both in darkness and in sunlight.

B. BASIC VEHICLE PRELIMINARY RECOMMENDATIONS

The vehicle should contain a pointable platform which can be pointed in any direction with an accuracy of the order of 1 minute of arc whether in sunlight or in shadow.

The space available for the scientific instrumentation will be expected to house a mirror of at most 1 m diameter and should be as long as possible and not less than 2.5 m.

The weight available for scientific instrumentation should be about 200–300 kg.

C. SPACECRAFT DESIGN PARAMETERS

The above definition of the mission can be expanded into more specific design parameters for the spacecraft as follows. It should be noted that certain of the following parameters arise from arbitrary definitions like use of the ELDO launcher.

(1) *Point anywhere at anytime*

Optical axis must be gimballed about 2 mutually normal axes with respect to the reference axis.

(2) *Minimum systematic error between optical and reference axes*

Physical connection (primary structure) between telescope and space reference (star trackers) must be stiff and have maximum environmental protection. Minimum interaction between primary and any secondary structure.

(3) *Maximum payload volume*

The astronomical telescope case requires maximum diameter (for collecting power) and also maximum length due to manufacturing limitations on the *F*-number of the primary mirror.

(4) *Maximum environmental control of payload volume*

In addition to the primary structure, maximum protection must be afforded to Cassegrain mirror and spectrometer mountings if maximum resolution is to be attained in view of the long light paths.

(5) *Adequate space reference system*

A minimum of 2 reference directions (reference stars) are required at all times to fix a space reference, unless some form of inertial guidance is used. Thus minimum occultation of star trackers is required.

(6) *Adequate control and guidance system*

Fine pointing (order of $\pm$ 2 arc sec. to attain resolution of 1 Å for reasonable spectrometer and F-number for main mirror) must be based on observation star, unless systematic error in reference can be eliminated. Coarse pointing accuracy (probably $\pm$1 arc min achievable) must be maximised to reduce fine guidance sensor field of view.

Control system must be stable over long periods (e.g. for observation of faint stars).

(7) *Adequate Power Supplies*

Power required continuously for control and stabilisation and for long periods of observation which will almost certainly involve traversing of detectors. Solar paddles (if used) should preferably be always normal to earth-sun line and operate at maximum efficiency.

(8) *Structural Symmetry*

Vibration problem more difficult than usual owing to stringent alignment requirements: may involve in-flight focussing and alignment. Structural symmetry using materials with high internal damping preferable.

(9) *Weight*

Scientific package approximately 28% of all-up weight for observatory type satellites. Thus, 220 kg scientific package infers all-up weight of 800 kg.

(10) *Altitude*

Theoretically maximum altitude limited only by launch vehicle or sensor degradation due to ionising radiation belts. Minimum altitude limited by effect of aerodynamic forces on lifetime and minimisation of reaction wheel unloading gas. Probable nominal altitude for observatory type satellites for 1 year lifetime about 600 km.

(11) *Lifetime*

As long as possible, limited mainly by reliability. Aim: one year minimum.

(12) *Flexibility*

Provide services and accommodation for a reasonable variety of similar scientific payloads without imposing undue restrictions on payload design.

3. Launch Vehicle

The majority of items in the previous section are of general application. Some parameters, such as weight, size and altitude are included for convenience although they obviously require some knowledge of the intended launch vehicle. Studies for the ESRO Large Astronomical Satellite have centred around the use of the ELDO "A" Launcher as regards performance and nosecone geometry. These are as follows:

(1) *Performance Data*

Launch from Woomera. Sun-synchronous Polar Orbit: 600 kg at 600 km. Equatorial Orbit: not permissible.
Launch from Equatorial Site (no impact restrictions). Sun-synchronous Polar Orbit: 800 kg at 600 km. Equatorial Orbit: greater than 800 kg at 700 km.

(2) *Nosecone Dimensions*

These are shown in Figure 1, which is a copy of ELDO Sketch E.SK.23/2, Issue 2, dated 18/5/1965.

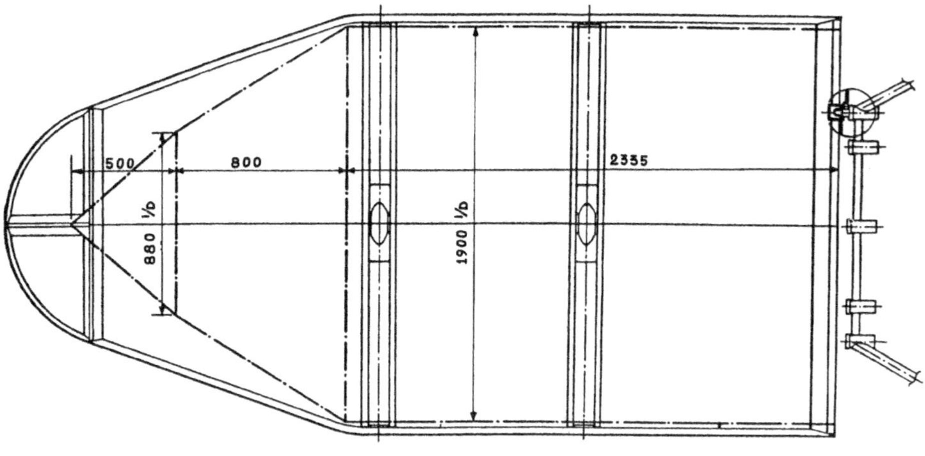

Fig. 1. ELDO-A launcher.

4. Possible Concepts

Figure 2 shows in diagrammatic form a range of possible concepts which seek to satisfy the requirements stated in Section 2. The concepts fall into two families based on the method by which the star-trackers (which lock onto chosen reference stars) are used to provide a space reference, viz. individual star trackers (OAO, DVL concepts) and "stable platforms" (RAE, CNES & ESRO concepts). Each family again breaks down into typical design concepts, depending on the actual mechanics of using the family characteristic.

In passing, a type not shown should be mentioned which could, in fact, fall into either family. This concept would incorporate an inertial platform periodically calibrated during the life of the satellite by one or more star trackers. Although attractive for a number of reasons, inertial platforms for this type of duty are still in the development stage and much work is required to establish the feasibility of such a system and its long-term reliability. This type of space reference has, therefore, not been adopted as the preferred method for the first version of the LAS since conversion of the present concept to an inertial reference would be easier than vice versa.

A. INDEPENDENT STAR TRACKER CONCEPTS

1. *OAO concept*

Figure 2a shows the basic features of the OAO concept. The space reference is obtained using six independent star trackers distributed over the outer surface of the spacecraft. Each star tracker is provided with a two-axis gimballed support and has access to only a limited area of the sky defined by the gimbal travel. The spacecraft is a hexagonal cylinder enclosing a concentric cylindrical scientific package compartment containing some form of telescope. Power is supplied by a pair of flat solar paddles aligned at 45 degrees to the spacecraft longitudinal axis.

In operation the whole spacecraft is aligned to the object star by means of two or more of the star trackers which are programmed to find and lock onto preselected suitable reference stars.

The obvious advantages of this concept are a clean symmetrical spacecraft with a well protected payload coupled with relative freedom in the choice of reference stars making problems connected with obtaining a space reference (e.g. occultation or failure of star trackers) less important. The major disadvantage however, which unfortunately is fundamental to the concept, is that the main spacecraft structure is inevitably part of the space reference system. Distortions due to varying environmental conditions resulting, for instance, from varying attitudes to the sun during the daylight portion of the orbit, may well limit the pointing accuracy obtainable. Much careful thought has been given to this problem and in-flight performance data must be awaited before any judgment can be passed. In addition, apart from the obvious complications in the on-board electronics which may be susceptible to radiation damage, a further undesirable feature is the proliferation of gimbal

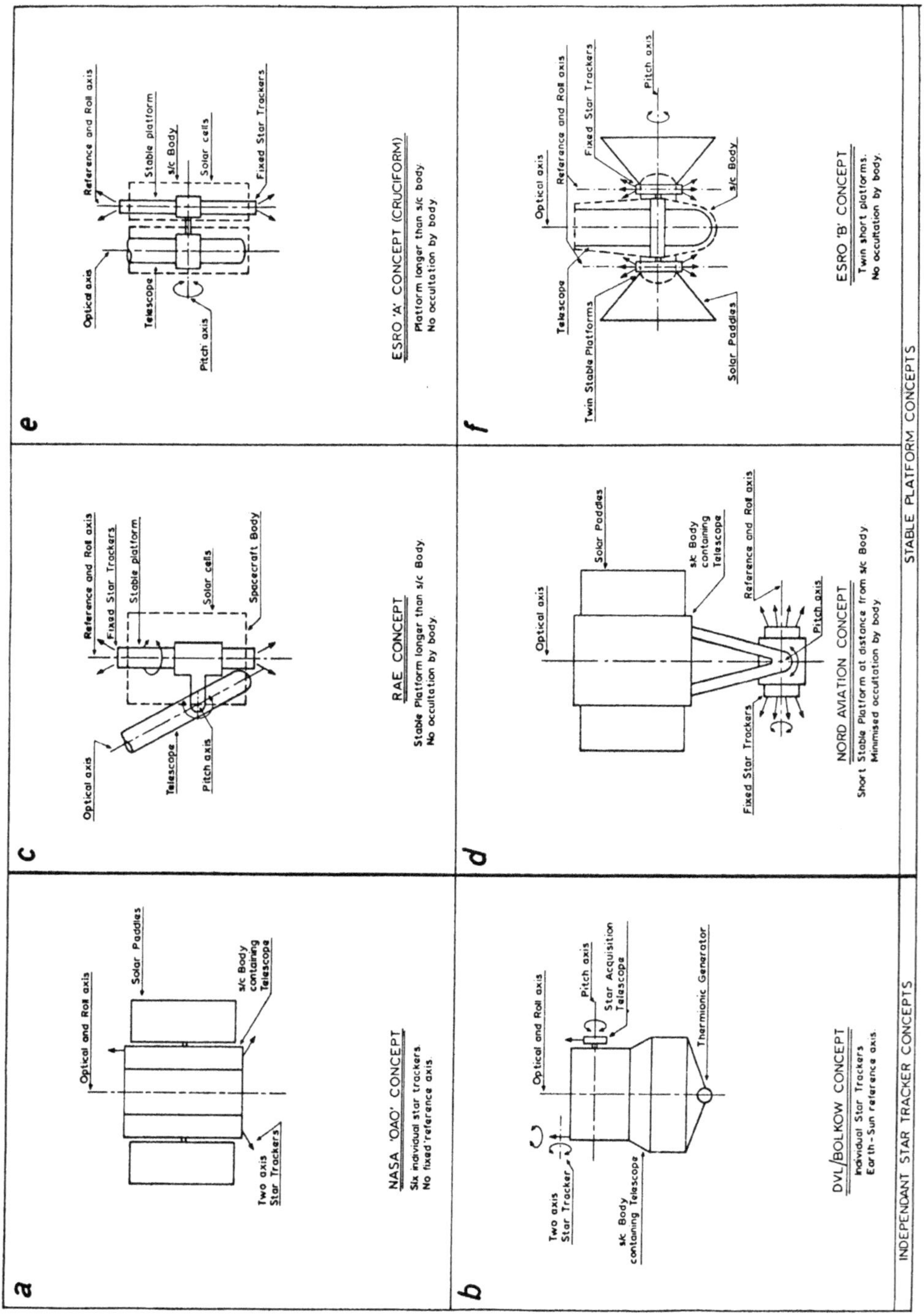

Fig. 2. Major astronomical satellite concepts.

bearings and angle transducers. This problem naturally became more acute when an initial proposal for a totally sealed system proved impractical and the whole star tracker assembly had to be redesigned for operation in vacuo.

2. *DVL/Bölkow Concept*

Figure 2b shows the basic features of the DVL/Bölkow feasibility study for the ESRO-Astro-Satellite. This concept is essentially similar to OAO but is an ingenious attempt to reduce the disadvantages of multiple star trackers and structural effects by using a single star reference, a sun reference, and the object star during its pointing manoeuvre. Although redundancy in the form of duplication is provided, only single elements are shown in the diagram for clarity. Figure 3 shows the arrangement of the proposal in the nosecone of the launch vehicle.

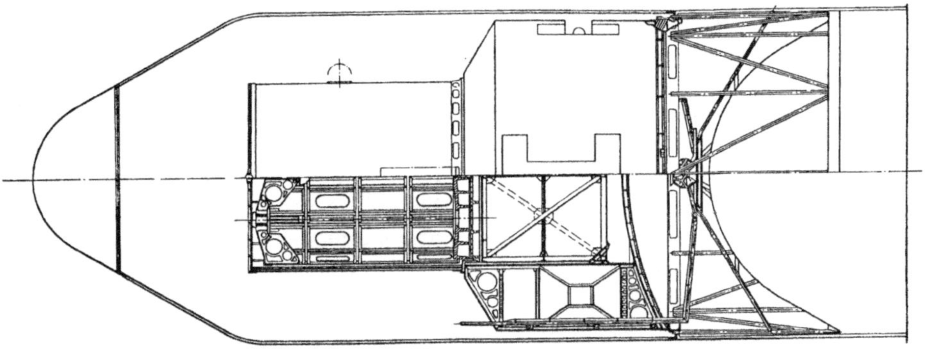

Fig. 3. DVL concept: Arrangement of satellite in nosecone.

Operation is somewhat complicated to visualise but is essentially as follows: during the daylight part of the orbit the pointing axis of the spacecraft is aligned with the sun (telescope pointing away) while the twin-gimballed star tracker locks onto a preselected reference star and establishes the "roll" angle. The single gimbal "star acquisition" telescope is then slewed with respect to the body to some predetermined angle based on the object star co-ordinates relative to the earth-sun line and the whole spacecraft is rolled about its longitudinal axis so that the object star is collected in the star acquisition telescope as soon as it appears over the horizon.

The spacecraft body is then slewed relative to the star acquisition telescope until their axes lie in the same plane and hence the object star appears in the main telescope. This performance is effectively repeated in reverse order to re-acquire the sun at the end of the observational part of the orbit. The advantages of this concept are similar to OAO in that symmetry is preserved and the package is well protected with reasonable flexibility and development potential. However, the choice of reference stars is limited and the complexity of the pointing manoeuvres, a large proportion of which must take place while the object star is in view, not only tends to restrict the observation time, but also demands an extremely accurate and reliable control system. Also, perhaps the most undesirable feature is that the importance of the object star in the pointing manoeuvre inevitably places some restriction on the magnitude of the stars to be included in the viewing programmes. It is also obvious that daylight observation is excluded since both pointing and power supplies depend on a specific sun orientation.

B. STABLE PLATFORM CONCEPTS

The main characteristic of this family of concepts lies in collecting together all the star trackers into a single unit in which they are rigidly aligned with each other and locate on a preselected pattern of reference stars. This "platform" which remains directionally fixed in space at all times is then connected by means of one double gimbal to the main telescope. The major family advantage of this philosophy compared to that of independent star trackers lies in the reduction of the number of gimbal bearings and angle transducers. The major family disadvantages are the restricted choice of reference stars and occultation problems. The family breaks down into three main categories, viz: "Long" platforms in which the platform is longer than the spacecraft body, i.e. reference stars are never occulted by the body, thus requiring the least number of star trackers. "Short" platforms which have very considerable advantages from the thermal design aspect, but where all reference stars might be occulted by the body and so requires the maximum number of star trackers. Finally, "medium" platforms which lie between the two in which only some of the reference stars may be occulted by the body.

1. *RAE Concept*

Figure 2c shows the basic concept of the RAE feasibility study for the ESRO Astro-Satellite.

This concept is an example of the "long" platform philosophy. Eight star trackers, four at either end giving 100% redundancy, are grouped in the platform which is mounted in, and protected by, the main spacecraft body. The platform is allowed to rotate about its longitudinal axis with respect to the body, thus providing one of the necessary gimbal axes. The telescope is located in a deep channel provided along one side of the body and is connected to the platform by means of a stiff "wishbone" structure. The telescope can rotate about an axis normal to the platform axis, thereby providing the second gimbal requirement. Since, in order to minimise the effects of scattered light from the sun in the trackers throughout the year, the preferred direction of the star reference axis is normal to the ecliptic plane, the platform and hence the spacecraft body remain normal to the earth-sun line at all times. Power is therefore provided by solar cells attached to the outer surface of the body. Figure 4 shows a diagrammatic view of the arrangement of the spacecraft in the launch vehicle nose-cone.

In operation the spacecraft is rotated about the platform until the telescope lies in the plane containing reference axis and object star. The telescope is then pitched to locate the star. The main advantages of this concept are simplicity of design and minimum power and weight required for establishing the space reference. Daylight observation is also possible. However, the need to protect the long platform environmentally has led to a somewhat asymmetric design and a relatively unprotected

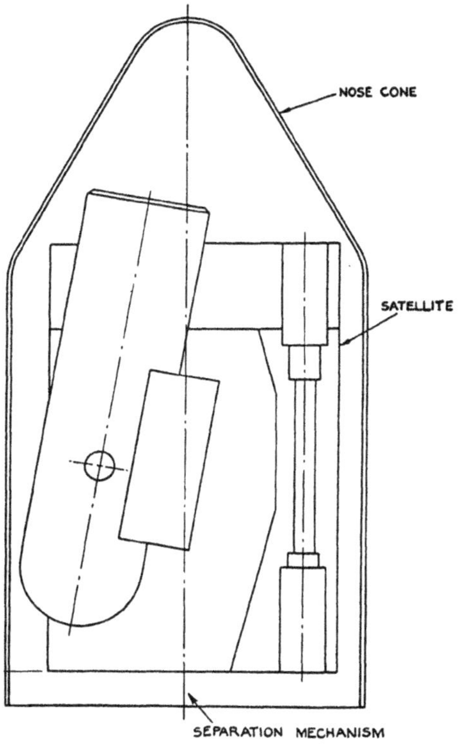

Fig. 4. RAE concept: Arrangement of satellite in nosecone.

telescope package. Problems of environmental control of the telescope are aggravated by the varying configuration factor with respect to the body. Finally, an undesirable feature is the reduced flexibility and development potential due to the need for close integration between telescope and package. For instance, the "wishbone" type suspension must inevitably place some restriction on telescope diameter and length of the spectrometer bay. Also the length of free portion (i.e. "forward" of the bearings) should not be such that it protrudes permanently outside the shroud in the interests of minimising environmental and occultation problems.

2. *CNES Concept*

Figure 2d shows the basic concept of the CNES feasibility study for the ESRO Astro-Satellite.

This concept is an example of the "short" platform philosophy. The main spacecraft body contains the scientific package and all other equipment while the platform is mounted on a frame structure as far from the body as possible in order to minimise occultation problems. The platform carries eight star trackers each having duplicate photo-multipliers for redundancy, i.e. eight reference stars are required in this case, compared to four in the case of the "long" platform. The star tracker assembly rotates within the platform body to provide one gimbal requirement while the connecting bearings between the platform body and support structure provide the second. Power is supplied by fixed solar paddles attached to the main body. Figure 5 shows a general arrangement of the spacecraft and the method of installation in the nosecone of the launch vehicle.

Operation is similar to the RAE concept except that the spacecraft body is pointed to the object star. Thus re-orientation with respect to the sun is necessary during the daylight part of the orbit in order to collect power, thus placing restrictions on any daylight observations.

The main advantages lie in symmetry and good protection for the scientific package. However, the attempt to minimise the occultation problems of a short platform has led to an even less satisfactory connection between the space reference and telescope axes than the OAO or DVL concepts where some environmental protection is

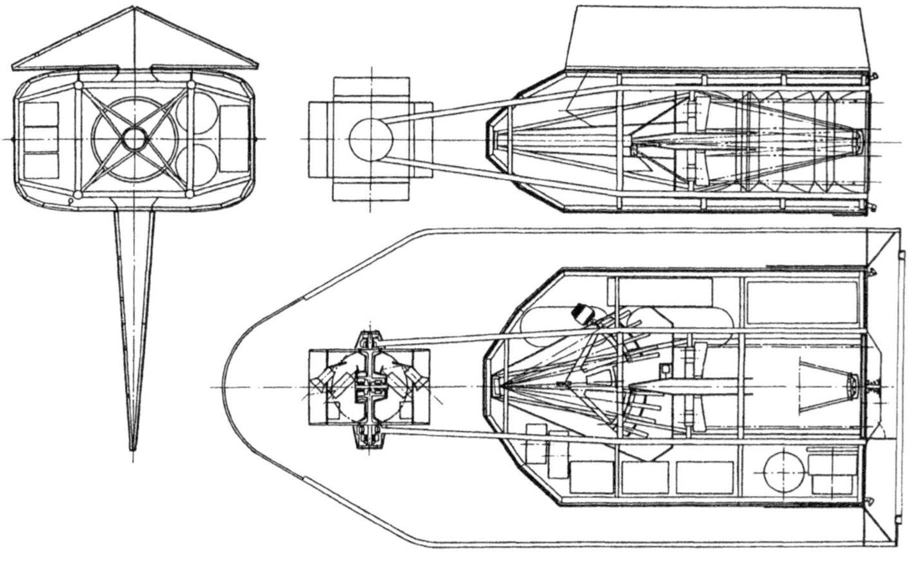

Fig. 5. CNES concept: Arrangement of satellite in nosecone.

possible. Thermal control of both the support frame and the platform would be difficult and might well limit the ultimate pointing accuracy achievable. A further undesirable consequence is an inevitable restriction on the maximum permissible telescope length within given nosecone dimensions.

5. The ESRO Concepts

It will be seen that none of the various concepts discussed in Section 4 completely satisfy all the requirements of Section 2. Basically the problem reduces to a spacecraft concept presenting the minimum imponderable engineering problems; minimum interference between spacecraft and the largest possible payload compartment; maximum environmental protection for both space reference, scientific package and their shortest, stiffest interconnection; and finally, maximum area and efficiency of solar cells (the most likely reliable source of power) independent of telescope pointing.

For completeness, two ESRO attempts to achieve these solar goals are shown:

ESRO "A" CONCEPT

Figure 2e shows the basis of a concept suitable for a relatively small, fully integrated satellite and is based on the "long" platform philosophy. The available nosecone diameter is divided into two parts, one containing the platform surrounded by some of the equipment, the other containing the telescope similarly protected. The two parts are connected by a simple journal bearing and thus rotate relative to each other. The second gimbal is supplied by rotation of the platform about its longitudinal axis within its casing. Power is supplied by solar cells attached to the platform half of the body, or a specially shaped solar paddle. Operation is similar to the RAE or CNES concepts.

While a short, stiff, well protected connection between platform and package has been achieved, the distribution of equipment between the two halves of the body must result in complete integration between spacecraft and payload, i.e. there is virtually no flexibility or development potential. Again, although superficially symmetrical in the "closed" position, if the l/d ratio is greater than 1 (as might well be the case, particularly if the concept were applied to large spacecraft), a varying cruciform configuration will exist during operation thus raising environmental control and other problems. Finally, perhaps, the most undesirable feature is the inevitable restriction on telescope diameter and length due to offsetting the optical axis from the centre-line of the launch vehicle nosecone. It is apparent, therefore, that this concept is really only of interest for a small compact spacecraft in which the various disadvantages might be minimised.

ESRO "B" CONCEPT

Figure 2f diagrammatically shows the features of a possible concept for the ESRO Large Astronomical Satellite which has evolved as a result of study of the various proposals. Figure 6 shows the basis of the proposed structural layout. The primary

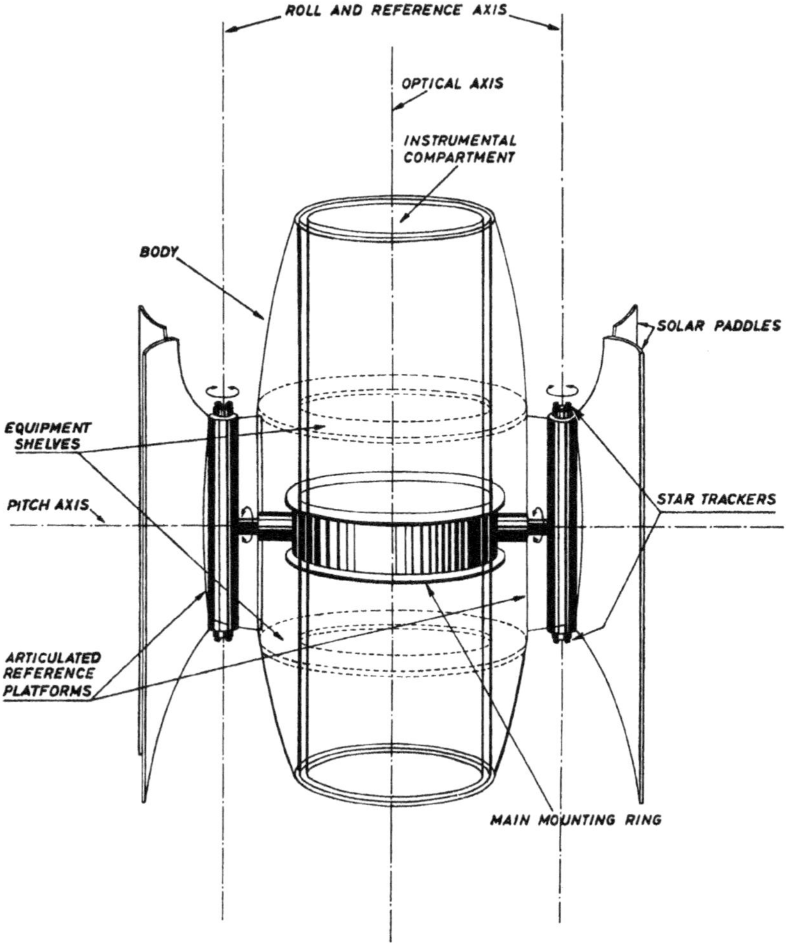

Fig. 6. ESRO-B concept: basic arrangement.

structure is shown heavily lined and consists of a stiff, well protected ring which forms a rigid platform for the attachment of the scientific package (not shown), the stable platforms and control mechanisms.

The whole is enclosed within a secondary structure comprising a thrust structure for supporting the primary structure during launch; shelves for the accommodation of equipment; and a light envelope for environmental protection whose shape can, to some extent, be tailored to that of the particular scientific package. The intention is that this secondary structure should have minimum interaction with the primary structure during operation. Thus the optical and reference axes have been brought as closely together as possible while at the same time making possible maximum dimensions, independence and flexibility for the payload within the boundary conditions imposed by the nosecone. Adequate equipment accommodation will be available since, although the "stiff ring" would be circular for reasons of symmetry, the spacecraft body could be elliptical in cross-section in this region. Figure 7 indicates how the spacecraft might be engineered while Figure 8 is a photograph of a mock-up model of the proposal.

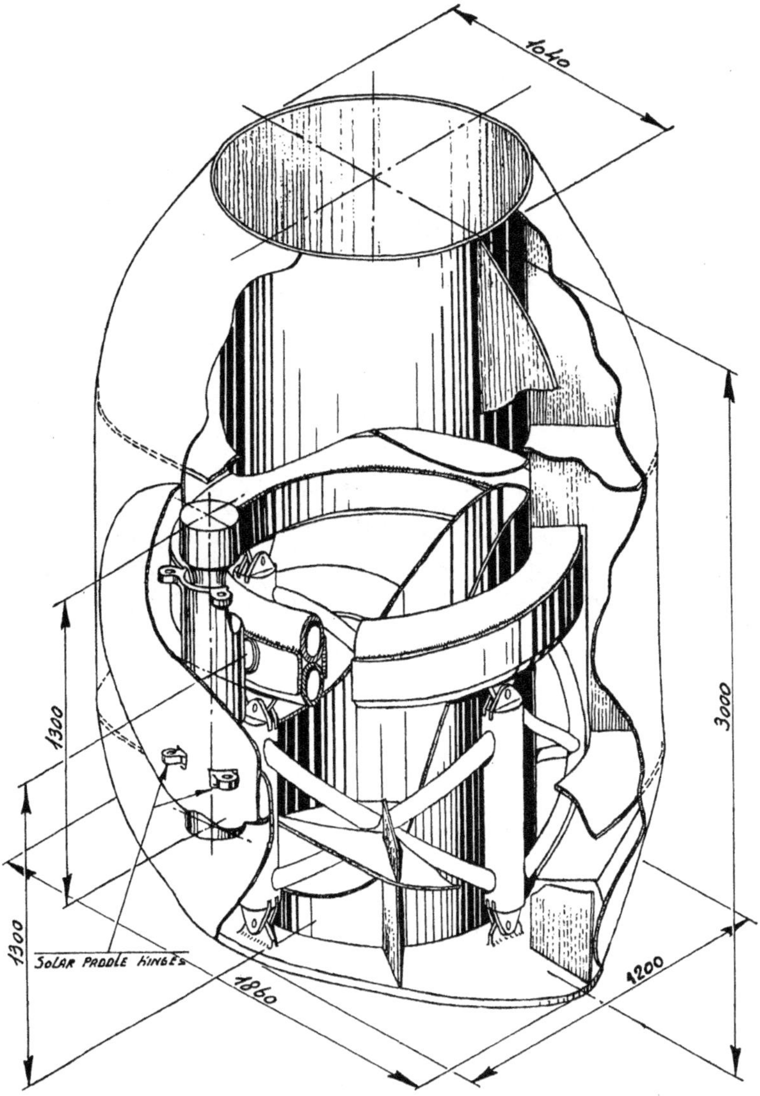

Fig. 7. ESRO-B perspective sketch.

Fig. 8. Mock-up of ESRO-B concept. (Courtesy of Culham Laboratories, England.)

The concept assumes two identical "medium" platforms located diametrically opposite to each other. This arrangement requires six reference stars, two of which (the axis reference stars), being roughly located at the north and south ecliptic poles, are therefore never occulted by the body, the remainder being chosen (two in each hemisphere) such that they always lie more than 30° from the sun. Investigation has shown that such reasonably bright star patterns exist, e.g. Polaris (α Ursae Minoris 2.12), Vega (α Lyrae 0.14), Mirfak (α Persei 1.9), Canopus (α Carinae −0.86), Rigel (β Orionis 0.34) and Hadar (β Centauri 0.86). Thus, if each platform carried six sensors aligned to the star pattern which should provide a space reference of at least two vectors at all times, the proposed arrangement should give 100% redundancy in both sensors and gimbals. It must be emphasised, however, that the above ideal can only be achieved providing the problem of accurate alignment of the two platforms can be solved. Practical considerations may require individual operation of the platforms on a master/slave basis, switching from one to the other if required.

It is important to note however that, although the platforms are exact duplicates of each other, operation on one platform is permissible. This fact can lead to the interesting possibility of eliminating one of the platforms in the interests of weight reduction and substituting the concept of a 180° slew about the reference direction to achieve "roll star" redundancy in terms of body occultation. In this case there can be no gimbal redundancy of course.

Power is supplied by means of four solar paddles at 90° to each other hinged to the platform casings. Thus, since the platform axis remains normal to the ecliptic plane, the solar paddle surface is always normal to the earth-sun line irrespective of the direction of the optical axis. Sufficient solar paddle area is provided such that sufficient projected area is presented to the sun in all positions about the "roll" axis. The platform environmental protection shrouds are likewise attached to the platform casings and rotate relative to the main body, thus affording protection in all pointing configurations.

TABLE I

Provisional Leading Particulars ESRO/LAS

Item	Proposed Specification	Remarks
Length Spacecraft	3.0 m	
Payload	2.7 m	Based on ELDO Europa
Diam. Spacecraft	2.0 m	Nosecone
Payload	1.0 m	
Weight Spacecraft	576 kg	Based on preliminary
Payload	224 kg	weight estimates
Altitude	650 ± 50 km	Based on estimated lifetime of 1 year
Orbit	Circular Polar or Equatorial	
Initial Mission	Stellar UV Spectrometry + Secondary Experiments	Payload axis will point in any direction at any time
Lifetime	1 year	
Pointing Accuracy		
Coarse	$\pm$ 1 arc min.	
Fine	± 2.5 arc sec.	(1 Å Resolution)
Power Supply	Solar cells	About 250 watts peak of which about 50 watts available for scientific package
Data Storage and Transmission	5×10^5 bits (Core store) 1.5×10^6 (Tape)	Approx. 3 orbit storage
Contact with Ground	Minimum: once/orbit about 3–8 minutes per contact	
Thermal Control of Payload Compartment	about ± 1 °C	Neglecting Payload inputs
Manoeuvre Time (180°)	about 9 minutes	

During launch, the optical and reference axes lie in the same plane and the solar paddles are "wrapped" around the main body and deployed just before separation from the final launch vehicle stage.

Operation is essentially similar to the RAE or CNES concepts. The spacecraft is "rolled" around the reference axis (the platforms contra-rotating within their casings) until the optical axis lies in a plane containing the object star. In-flight star-tracker calibration would be carried out by aligning the main telescope on some identifiable star, e.g. one of the axis reference stars.

Although the present paper is restricted to the evolution of an overall concept and hence a detailed examination of the feasibility of sub-systems would be out of place, it is nevertheless of interest to crystallise the discussion in terms of the probable physical characteristics of the proposed spacecraft. Table I summarises preliminary estimates of achievable objectives for the European Large Astronomical Satellite of the ESRO "B" configuration.

6. Conclusion

The ESRO "B" concept presents an attractive line of development for the European Large Astronomical Satellite, and compares favourably with other overall concepts in seeking to satisfy the basic requirements set out in Section 2. In particular, the attempt to define two major structural interfaces, viz: Package/Primary structure and Primary/Secondary structure, has considerable advantages regarding flexibility and adaptability.

It is emphasised, however, that a number of problems exist in all fields; perhaps the most important and difficult one peculiar to the concept being that of alignment and simultaneous operation of the two platforms. However, preliminary studies suggest that the concept is essentially feasible, and thus, if successfully developed, could form the basis for a multi-purpose spacecraft of the observatory category.

PART II

X-RADIATION

THEORETICAL ASPECTS OF SOLAR X-RADIATION

C. DE JAGER
University Observatory 'Sonnenborgh' and Space Research Laboratory, Utrecht, The Netherlands

1. Introduction

1.1 Classification of solar X-ray emissions

In physics X-radiation is defined as electromagnetic radiation originating from electronic transitions involving the lowest electron shells; gamma rays are of nuclear origin. Mostly gamma rays are harder than X-rays. This distinction has not rigorously been followed in the astronomical literature, and in actual fact, it has sometimes only been the *energy* of the solar quanta that was used to classify them. However, one had better avoid confusion and keep to the physical definition. Hence solar energetic radiation will only be called gamma radiation if it is certain that the radiation is of nuclear origin. Since such radiation has not yet been observed the conclusion must be that all energetic solar radiation observed so far must be classified as X-radiation. But in this class of radiation a certain distinction can certainly be made.

In this paper we suggest the following subdivision:

Soft X rays: $\lambda > 1$ Å or $E < 10$ keV;
Hard X rays: $\lambda < 1$ Å or $E > 10$ keV.

In the latter region we might, if desired have a finer subdivision as follows:

0.1 Å $< \lambda < 1$ Å (10 keV $< E < 100$ keV): deka-keV region,
0.01 Å $< \lambda < 0.1$ Å (100 keV $< E < 1$ MeV): deci-MeV region,
0.001 Å $< \lambda < 0.01$ Å (1 MeV $< E < 10$ MeV): MeV region.

The X-rays are characterized either by their production mechanism or by phenomena associated with their occurrence, as will be outlined in the following.

1.1.1 *Classification of X-rays according to the production process*

a) *Thermal and quasi-thermal X-ray emission.* Thermal radiation in the true sense of the word only exists inside an optically thick body in thermodynamic equilibrium. The sun does not emit X-rays of that nature. The X-radiation emitted by the coronal plasma is called quasi-thermal because the electrons have a Maxwellian temperature distribution, although the coronal emission is not comparable to that of a gas in thermal equilibrium: for X-rays the corona is optically thin and the outside radiation field is negligible. In order to calculate the intensity of this X-radiation it is sufficient to know the electron density and temperature as functions of height.

J. G. Emming (ed.), Electromagnetic Radiation in Space.

The coronal X-ray spectrum contains mainly emission lines; the continuous spectrum is ten times or more weaker than the integrated line emission spectrum. Still its smoothed shape can be described very roughly in terms of a Planckian distribution with a maximum wavelength around 30–50 Å. The position of this maximum is roughly in accordance with Planck's law: according to Wien's displacement law a plasma with a temperature of 10^6 °K would have its maximum intensity at 29 Å. However, the detailed shape of a spectrum and its absolute intensity do not at all agree with the Planckian curve: the intensity of the solar corona is about 10^{-17} times that of a black body with the same area and temperature as the corona.

b) *Non-thermal X-ray emission.* The non-thermal X-ray bursts are always related to energetic phenomena, e.g. particle acceleration. They are presumably due to the interaction of guided energetic streams of particles with an ambient gas (e.g. Bremsstrahlung-bursts); these particles may be confined by a magnetic field. For non-thermal X-bursts the notion "kinetic temperature of the gas" has hardly any sense; it is only the energy and flux of the particle stream that matters.

Obviously, if defined in this way, there is no sharply defined limiting wavelength between non-thermal and quasi-thermal X-radiation, and there is a range of energies where the two kinds of radiation can both occur. Furthermore, since the distinction between them is based on theoretical inferences, it will not always be possible immediately to classify a burst of energetic solar radiation. However, as a rule the lifetime of the event will be a useful parameter for classification: non-thermal bursts usually live fairly short.

1.1.2 *Classification of X-rays according to associated phenomena*

a) *Slowly varying component.* The radiation of the quiet corona has an eleven-year period. That of the active centres has an apparent characteristic time of 27 days due to the solar rotation, and an average real lifetime of around 3 months. Some faculae remain active for up to 6 months. Both the quiet corona and the active regions, including the flares, emit quasi-thermal X-radiation, with intensities defined by the temperature of the relevant plasma and the so-called *emission measure*, defined as the volume integral $\int N_e^2 \, dV$.

b) *Fast, flare-associated component.* Associated with solar flares the sun may emit radiation over an extremely wide energy range, from very long radio waves (radiation has been detected at frequencies down to 0.2 MHz, corresponding to 1.5 km) up to photons with energies near 1 MeV (corresponding to wavelengths of 0.01 Å). In the X-ray region solar flares may be associated with quasi-thermal enhancements of the soft X-ray flux in the wavelength range from about 1 Å upward, and with non-thermal enhancements at shorter wavelengths. There is no sharp limit between the wavelength ranges for the two kinds of solar X-bursts.

Figure 1 after Friedman, shows the intensity of soft X-ray emission during various conditions of solar activity. The results are average intensities over the spectrum of the emission lines. Figure 2 indicates the relation of the soft X-ray intensity to radiation at longer and shorter wavelengths.

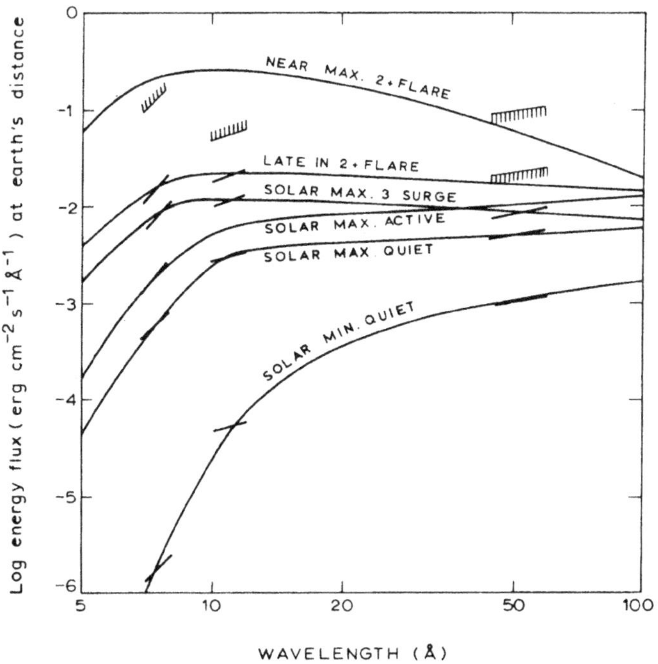

Fig. 1. Solar X-ray emission for various solar conditions. The curves drawn are based on measurements in three wavelength bands, as indicated by the heavy bar segments. The slopes of the bar segments are the slopes of the assumed X-ray emission functions used to reduce the photometer responses to the plotted energy fluxes. The energy fluxes refer to values observed at the earth's distance above the atmosphere (from H. FRIEDMAN: 1963, *Ann. Rev. Astron. Astrophys.* I, 59).

1.2 THE QUIET CORONA

All components of the quasi-thermal X-ray emission, whether due to free-free emission or free-bound emission are proportional to $N_e N_i$, and because in this highly ionized plasma the number of ions N_i is proportional to the number of electrons N_e, the intensity of the coronal emission is proportional to N_e^2. As in the corona N_e decreases outward with a scale height of about 5×10^4 km, the X-ray emission intensity should decrease outward with a scale height of about 2.5×10^4 km; i.e. the intensity would decrease to 0.01 of the limb value over a distance of a little more than 10^5 km.

In the last few years the detailed spectrum of the quiet corona has been observed, by the photographic technique down to 33 Å (by Tousey and coworkers), by the photo-electric scanning technique with diffraction gratings (Hinteregger; Manson), and photo-electrically by means of plane crystals as the dispersing medium (the Naval Research Laboratory group with CHUBB *et al.* (1961) and EVANS *et al.* (1967) from Leicester University). In the wavelength region above 30 Å the photographic observations of Tousey and the photo-electric observations of Hinteregger confirmed Elwert's prediction that the spectrum is virtually a line spectrum, with no appreciable

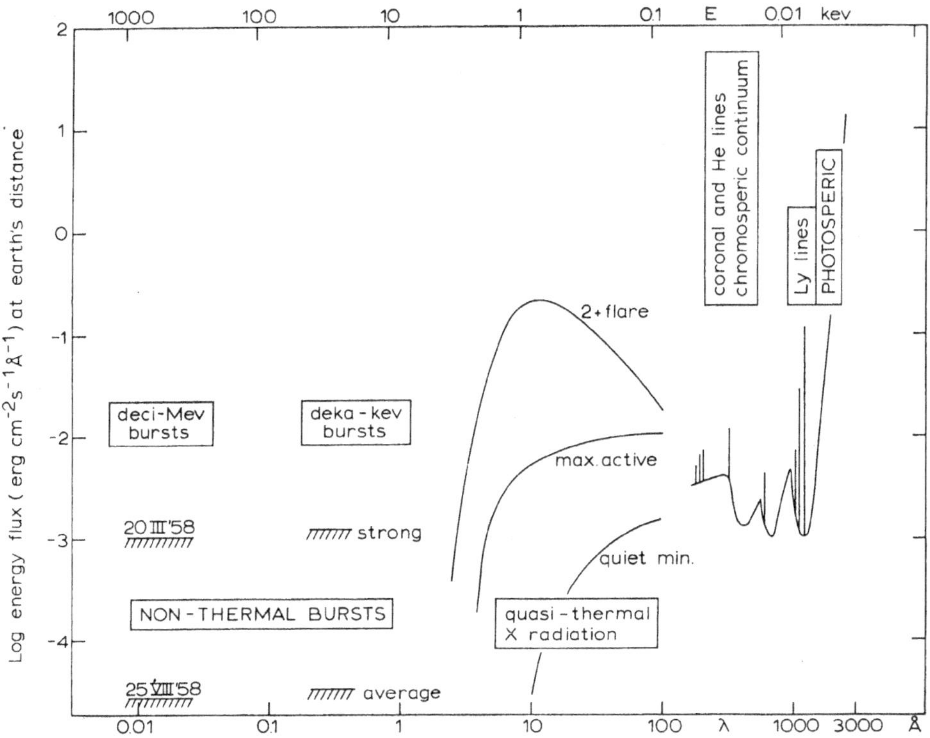

Fig. 2. Intensities determined in the ultraviolet and X-ray spectrum of the sun.

contribution of a continuous emission. Not in agreement with Elwert's predictions seem the observations in the shorter wavelength region. These observations show that also in this spectral range the main contribution to the emission is from spectral lines; the ratio between the integrated line intensity and the not yet detected continuum intensity is at least *six*. On the other hand, Elwert's theory for the quiet sun predicts a continuous spectrum with only a few weak lines.

However, the difference between Elwert's prediction and the observations may result from the fact that the sun was not wholly quiet during the observations. The spectral lines observed by the N.R.L. and Leicester groups were all emitted by *active regions*, and Elwert's computations show that such regions, with their higher temperatures, will emit mainly a line spectrum in the relevant region.

In the quiet sun equatorial limb brightening has been observed. It is a result of the transparency of the corona for X-rays, so that the emergent intensity is proportional to cos θ, where θ is the angle between the direction to the observer and the normal to the solar surface. Just beyond the limb the observed intensity should increase to twice the value inside the limb. No limb brightening has been observed at the poles; an effect that is due to an electron density that is smaller in the coronal polar regions than at the equator. This effect has been predicted by Elwert and observed by Pounds *et al.*, Giacconi *et al.*, and by Underwood and Muney.

1.3 Activity centres

The best information on these regions is given by monochromatic X-ray images of the sun. These show intense X-ray emission regions above faculae, of intensity 30 to 110, average 55 times that of the quiet sun. The intensity of the radiation emitted by active regions is proportional to $(N_e)^2_{ac} H_{ac}$ where H_{ac} is the scale height of the active region. The intensity of X-rays emitted by the quiet corona is proportional to $(N_e)^2_c H_c$ where H_c is the scale height of the quiet corona. Assuming $H_{ac} = \frac{3}{2} H_c$ it follows:

$$\overline{(N_e)^2_{ac}} = \tfrac{2}{3} \times 55 \times \overline{(N_e)^2_c}.$$

Hence,

$$(N_e)_{ac} \approx 6 \times (N_e)_c .$$

Similar values have been found from other observations. From a thorough discussion Newkirk (1967) finds that an active region during the states of its development when flares are present always contains enhanced densities of 2 to 5 times the local background to a distance of at least 4 $R_\odot$.

The only spectral line observations of coronal active regions made so far are the plane crystal spectrometer observations of the Naval Research Laboratory (Blake, 1965) and those made by Evans *et al.* (1967). The spectra shows lines of highly ionized ions, up to Fe XVII and Fe XVIII; there are also lines of Ne IX and X, and Ni XIX.

Furthermore we mention the earlier broad band spectral observations made by the Ariel I satellite where photon discrimination by means of proportional counters enabled the determination of the shape of the spectrum in the short wavelength region between 3 and 15 Å. Since the quiet corona hardly emits any radiation in the wavelength region below 10 Å (see Figure 1), this emission is virtually due to the coronal active regions. A comparison of the observed shape and intensity of the spectrum with theoretical computations by Elwert enables the observer to determine the electron temperature of these active regions. The Ariel observers obtained values of the order of 2×10^6 °K, thus confirming optical observations of the enhanced temperature of the coronal active regions.

The *conclusion* is that X-ray observations of the emission of coronal active regions lead to an average density increase with regard to the quiet corona by a factor 5 and a temperature of the order of 2×10^6 °K.

1.4 Flare-associated phenomena

A flare is a sudden and short-lived brightening of a small part of the disk, monochromatic in the visual spectral region. The optical radiation comes from lower layers of the corona and from the upper chromosphere.

However, a flare is an extremely complicated phenomenon and has at least three different aspects. The *optical flare* is produced by a relatively cool and dense plasma

($T \approx 10^4$ °K; $\log N \approx 13.5$), and emits a line spectrum of relatively low excitation emission lines.

The X-ray or radio flare, also called the *High Energy Flare*, is the hot counterpart of the optical flare. It consists of a plasma of very high temperature (T up to 10^7 °K or even higher), and moderate densities (Ne $\approx 10^{10}$ cm^{-3}), which emits quasi-thermal solar X-radiation of relatively short wavelengths (down to 1 or 0.1 Å) and which may produce at the same time a strong enhancement of radio radiation, mainly on centimeter and short decimeter waves.

The flares are also often related to the *acceleration* of energetic jets of charged particles. Type II radio bursts indicate proton beams or shock fronts moving at about 1500 km/sec. Type III radio bursts are due to electrons with average velocities of about 150000 km/sec. Riometer data show the arrival in the ionosphere of sub-relativistic particles, emitted during flares, with energies of several times 10^7 to more than 10^8 eV. Also solar cosmic rays of energies around and above 10^9 eV are detected during or after solar flares, but these are very rare. When such jets interact with the ambient coronal or chromospheric plasma or are forced to move in a magnetic field they lose energy in the form of radiation; this is what we define as a non-thermal X-ray burst. Of course, it is impossible to decide on observational grounds alone whether a burst belongs to the quasi-thermal or non-thermal X-ray phenomena as defined in this paper. However, in the main bursts are of the non-thermal kind if they do not last longer than a few minutes, or in any case if they live very much shorter than the optical or radio flare; quasi-thermal bursts may last longer, and may have a lifetime comparable to the optical or radio flares. The ground for these assertions is the fact that fast single jets of electrons are known to last for a short time or they occur in clusters which themselves do not last longer than one or two minutes (a result, mainly based on radio observations of type III bursts). Also magnetically trapped electrons may last only for a relatively short time in the magnetic field, as was shown by computations of SCHATZMAN (1965). On the other hand the hot plasma, which is thought responsible for the quasi-thermal X-ray bursts may last for at least a few minutes, usually as long as the main phase of the related solar flare.

1.5 QUASI-THERMAL X-RAYS EMITTED DURING FLARES

Quasi-thermal X-ray emissions associated with flares have been observed in the whole region below 100 Å by quite a number of observers: the N.R.L. observers recorded the short wavelength tail spectrum (20–60 keV) of flare-associated emission on 31 August 1959 and found it describable by a colour temperature of 1.2×10^8 °K. The emission was observable for at least six minutes (two rocket ascends). The N.R.L. observers thought this burst, with its rather long duration and its very high colour temperature, to be due to quasi-thermal emission by a hot plasma. However, Kawabata who re-examined the observations of this burst, and who compared them with simultaneous X-ray emissions in the 2 to 8 Å region and with cm and dm radio observations, concluded that the best description of the observations would be to assume the emission to be composed of two parts: a quasi-thermal component caused by a gas with a temperature of 2×10^7 °K and a non-thermal burst in the energy region above 20 keV.

Also other observations may sometimes show a combination of a longer lasting quasi-thermal burst, on which a short-lived non-thermal burst may be superimposed, often at shorter wavelengths.

Other observations are the rocket observations by Yefremev *et al.*, and the measurements with the N.R.L. series of Solar Radiation satellites, the U.K. Ariel satellite (Figure 3) and by the OSO satellites. All these observations confirm that

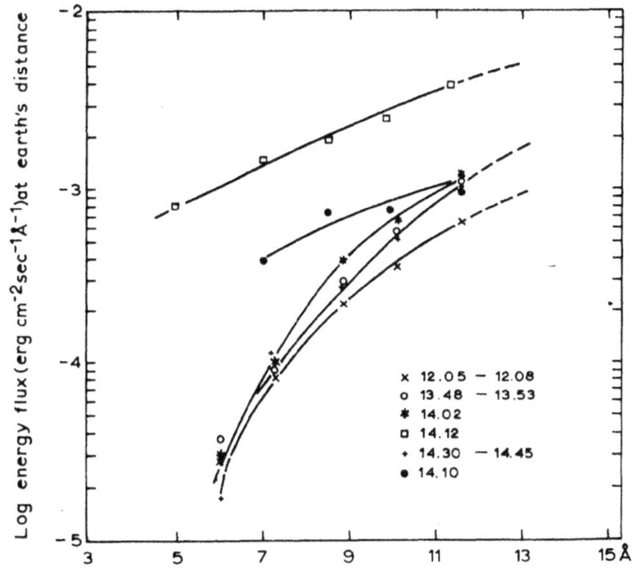

Fig. 3. Development of solar X-ray intensities for a class-2 flare on 3 May 1962: compared with the quiet sun spectrum observed the same day; Ariel satellite observations (from CULHANE *et al.*: 1964, in *Space Research* IV, p. 741).

together with solar flares the sun may emit enhanced quasi-thermal X-radiation in the whole wavelength region below 100 Å, with particular enhancements below about 10 Å, where normally hardly any solar radiation is emitted. The colour temperature of the emitted radiation may be far above 10^6 °K. The general character of the bursts, as observed in particular and in detail in the 1–18 Å band by the U.S. OSO III satellite is that of a gradual hardening and intensification of the spectrum during the growth phase of the flare.

These observations confirm a result for the first time brought forward by radio observers, namely that solar flares may quite often be accompanied by a hot component: the radio or X-ray flare, or: the High Energy Flare. The location and structure of this component is still unknown. The temperature seems to be well above 10^6 °K and may rise to above 10^7 °K. Theoretically one may expect a gas cloud with a temperature of several 10^6 °K to expand and to leave the sun nearly explosively, shooting a blast wave into interplanetary space, unless being held near the sun by a magnetic field.

1.6 Flare-associated X-ray emission in the deka-keV and deci-MeV regions

Mainly through the work of Winckler *et al.* (1967), who monitored the solar X-radiation in the OGO I and III satellites in the energy range between about 20 and 150 keV a fairly great number of solar hard X-bursts has been detected in that range. Furthermore, a few bursts have been observed in the deci-MeV region. Although it is mostly thought that the hard X-ray bursts are of non-thermal origin this has not yet been proved to be definitely true for the deka-keV bursts. It may perhaps be safe to assume that the higher the photon energy, the greater the chance that the burst is not thermal.

The hard X-ray bursts show the following properties.

1. They occur mostly during the flash phases or the rising phases of flares. We recall that the acceleration of particles to high speeds occurs often in the beginning phases of flares (Radio bursts of type III, the start of II radio bursts, other optical phenomena showing the emission of high velocity gas clouds).

2. The hard X-ray bursts correlate also with centimeter radio bursts. Such radio bursts tend to occur in large activity centres in particular in those where the flare "touches" or partly covers the penumbras of sunspots.

3. The energies of these bursts are of the order 10^4 to 10^6 eV so that they may be due to Bremsstrahlung by electrons moving with velocities of 60000 to 280000 km/sec.

The interpretation of the observed phenomena will be discussed further in Section 3.

TABLE I

X-rays associated with solar flares

Class	I: soft	II: hard
Character	not impulsive	short-lived or impulsive
Lifetime	about as long as flare	mostly shorter than flare
Spectral range	$\lambda > 0.1 - 1.0$ Å	$\lambda < 1$ Å
Associated radiowave emission	microwave, sometimes metric	mainly microwave; sometimes type III
Associated ionospheric phenomena	S.I.D.	not always S.I.D.'s
Interpretation	Quasi-thermal radiation from plasma with $T \leqslant 10^7$ °K, $N \approx 10^{10}$ cm^{-3}	mostly non-thermal radiation
$N_e V$	10^{29}	10^{27} (deka-keV) to 10^{25} (deci-MeV)
	Subdivision of type II X-ray bursts	
Class	II – deka-keV	II – deci-MeV
Associated radio and optical phenomena	clearly correlated with microwave bursts	not always associated with radio or optical events

2. Quasi-thermal X-Ray Emission

2.1 The definition of temperature

Temperature is often used merely as a mathematical parameter in astrophysics. Its definition depends on the observational circumstances. Some common definitions are listed below.

i) *Kinetic* (real) *temperature.* This is the T-value to be substituted in Maxwell's velocity distribution law, in order to have the best description of the observed velocity distribution.

ii) *Electron* and *ion temperatures* are defined by approximating the velocity distribution of these particles by a Maxwellian distribution. Normally in outer layers of stellar atmospheres, one has $T_{el} \approx T_{ion}$.

iii) *Bolometric temperature* of a body is defined as the temperature of a black body of the same size and distance having the same total energy emission.

iv) *Radiation temperature.* Intensity measurement is made over a small wavelength range in the visible region and equated to the radiation of a black body of the same size and distance in the same wavelength range. If the relevant spectral region is extremely narrow one speaks of Monochromatic Temperature.

v) *Colour temperature* is obtained by approximating the shape of the observed radiation spectrum or of a part of it to a Planck curve. Some X-ray astrophysicists call the colour temperature the *Gray Body temperature.*

Gradient temperature is obtained by comparing the slope of the spectral distribution (preferably over a band not too broad) with that of a Planck curve. There is no essential difference between the colour temperature and the gradient temperature.

vi) *Ionisation temperature* is obtained from the ratio of two ion populations, using Saha's law. Normally each pair of ions gives a different value.

vii) *Excitation temperature* is obtained by approximating the populations of excited states of an ion, as given by Boltzmann's law; normally each pair of levels gives a different value.

In the X-ray region ($\lambda < 100$ Å) the values obtained by these methods for the solar corona are roughly

$$T_{kin} \approx T_{el} \approx T_{ion} = 2 \times 10^6 \, {}^\circ K$$

$$T_{col} = 3 \times 10^6 \, {}^\circ K$$

However, the faintness of the absolute intensity of the coronal radiation reflects in the bolometric and radiation temperatures: T_{rad} would be found from the observed quiet solar X-ray flux (0.1 erg cm^{-2} sec^{-1} at the earth's distance) from

$$\sigma T_{rad}^4 = 0.1 \times \left(\frac{R}{r}\right)^2 \text{erg cm}^{-2} \text{sec}^{-1},$$

where R is the Astronomical Unit, and r the solar radius. One finds

$$T_{rad} \approx 100 \, {}^\circ K.$$

2.2 Ionisation and recombination in the corona

As the corona is not in local thermal equilibrium Saha's law does not apply. Therefore we must investigate the physical processes that determine the degree of ionisation of the atoms in the corona.

There are six processes that determine, in principle, the ionisation equilibrium:

i) Photoionisation,	$R_1 = N_r \times Q_r^{r+1}$
ii) Radiative recombination,	$R_2 = N_{r+1} \times N_e \times Q_{r+1}^r$
iii) Collisional ionisation,	$R_3 = N_r \times N_e \times S_r^{r+1}$
iv) Three-body recombination,	$R_4 = N_{r+1} \times N_e^2 \times S_{r+1}^r$
v) Auto-ionisation,	$R_5 = N_r \times A_r^{r+1}$
vi) Dielectronic recombination,	$R_6 = N_{r+1} \times N_e \times A_{r+1}$

where R denotes the rate at which a process occurs, and N_r, N_{r+1} represent the number densities of ions in the r-th, $(r+1)$-th degree of ionisation, respectively. Q, S and A are probability functions depending on temperature, radiation field, and the atom concerned.

Elwert has shown that R_4 is always very much smaller than R_2, and in the quiet corona R_1 is much smaller than R_3. So, if auto-ionisation and the reverse process, dielectronic recombination could be neglected the steady state requires for the quiet corona:

$$R_2 = R_3, \qquad \text{i.e.,} \qquad \frac{N_{r+1}}{N_r} = \frac{S_r^{r+1}}{Q_{r+1}^r} \tag{1}$$

In order to arrive at the ionisation equilibrium the functions S and Q must be known. Many attempts to compute them have been made, notably by Elwert.

First the function Q_r^{r+1} is found by making the following two assumptions:

i) the atoms are hydrogen-like, but with appropriate Gaunt factor.

ii) the radiation field is given by Wien's law, an acceptable approximation.

Then, using Kramer's law, an expression for Q_r^{r+1} is obtained. Q_r^{r+1} may be derived from Q_r^{r+1} by assuming local thermal equilibrium; in such conditions

$$N_r \times Q_r^{r+1} = N_{r+1} \times Q_{r+1}^r N_e;$$

$$\frac{N_e N_{r+1}}{N_r} \quad \text{being determined by Saha's law.}$$

Since the function Q_{r+1}^r is a general function it is also applicable to non-L.T.E. conditions.

Elwert has obtained a value for S_r^{r+1} from a treatment of collisional ionisation. From the observed line intensities, and the fractions of ions in various stages of ionisation as computed with Equation (1), he deduced $T_{\text{kin}} \approx 8 \times 10^5$ °K. On the other hand, the widths of forbidden visible emission lines indicate $T_{\text{kin}} = 2$ to 3×10^6 °K.

The fundamental difficulty encountered here could be solved when Unsöld suggested that dielectronic recombination should be taken into account. Calculations have been done by BURGESS and SEATON (1964). The process of auto-ionisation was discussed by VAN RENSBERGEN (1967a, b). Auto-ionisation may occur when two electrons of the same atom (or ion) are excited at the same time. In such a case it may occur that one of these electrons is excited to an energy exceeding the ionisation energy of the singly excited atom (or ion). Then the electron may move to the continuum levels of the atom by a radiationless transition. Such transitions occur very easily, or stated otherwise: the lifetime of the level occupied by such an electron is very short. Hence, once a second electron has been excited the ionisation cross section is large.

Dielectronic recombination is the inverse process of auto-ionisation, it is a recombination via nearly-unstable states of an ion with two electrons in excited states.

If these two new processes are incorporated in the theory we have to consider that

$$R_1 + R_3 + R_5 = R_2 + R_4 + R_6,$$

but since $R_2 \gg R_4$ and $R_1 \ll R_3$,

$$R_2\left(1 + \frac{R_6}{R_2}\right) = R_3\left(1 + \frac{R_5}{R_3}\right),$$

or

$$\frac{N_{r+1}}{N_r} = \frac{S_r^{r+1}}{Q_{r+1}^r}\,\varphi(T, N_e). \tag{2}$$

Here

$$\varphi(T, N_e) = \frac{1 + \dfrac{A_r^{r+1}}{N_e S_r^{r+1}}}{1 + \dfrac{A_{r+1}^r}{Q_{r+1}^r}}$$

has been computed for several ions, for different temperatures by VAN RENSBERGEN (1967).

In most cases

$$A_r^{r+1} \ll N_e S_r^{r+1},$$

so that effectively $\varphi(T, N_e)$ is only a function of T.

Figure 4 shows some results of Burgess and Seaton computed with this theory, and compared to the earlier work of Elwert. For Fe X and Fe XIV the results show that the dielectronic recombination theory gives higher temperatures for the same degrees of ionisation.

The temperature of the quiet corona obtained in this way is

$$T_{\text{kin}} = (1.2 \text{ to } 1.5) \times 10^6 \,^\circ K.$$

The slight discrepancy still remaining between the values from line widths and those from line intensities has been interpreted by Seaton to be a result of turbulence. At

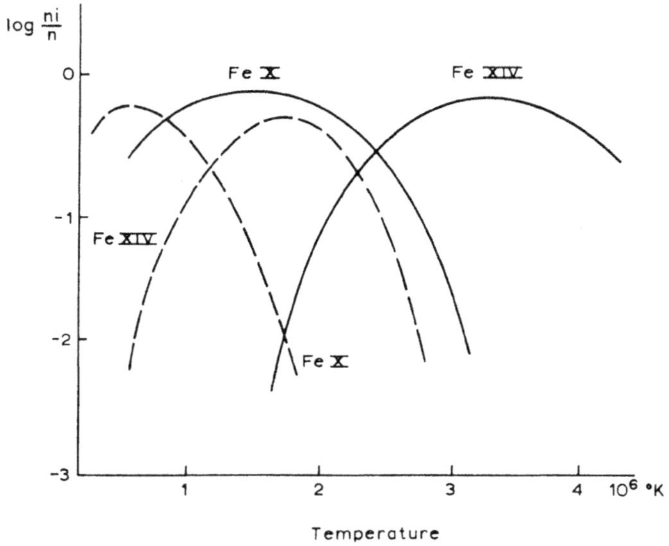

Fig. 4. Ionisation in the solar corona, as calculated by Burgess and Seaton (full line), and by Elwert (dotted line). The curves represent the logarithm of the fraction of atoms in the respective ionisation stage as a function of temperature.

$T_{kin} = 10^6$ °K the electron velocity is ≈ 6000 km/sec while that of the Fe ions is only about 50 km/sec. Thus a turbulent velocity of 50 km/sec would double the temperature derived from line widths, without influencing the derived degree of ionisation.

2.3 Theory of the soft quasi-thermal X-radiation

2.3.1 *Continuum*

Free-bound transitions of electrons in the field of the lightest nuclei, H and He give radiation in the extreme ultraviolet (eg. H 912 Å, He I 504 Å, He II 227 Å ionisation limits). Most of the ionisation limits for highly ionised heavier ions like O, C, N, Ne, and also of ions like Fe and Si occur in the range of soft X-rays.

These spectra, and those of free-free transitions, which are not important, have been calculated by Elwert and Kawabata for temperatures between 7×10^5 and 10^8 °K.

2.3.2 *Line emission*

The calculation of line emission involves the following stages:

a) Estimation of the wavelength of the electronic transition. To give a very simple example, the energy levels of nearly completely stripped oxygen ions can be predicted by comparison with atoms with well-known energy levels:

O VIII is hydrogen-like
O VII is like He I
O VI is like Li I

For O VIII, with charge eight times the H nucleus, the Lyman-alpha line should occur at $1216/8^2 = 19$ Å. Measurements by the N.R.L. group show the wavelengths of the lines predicted by this method in the region around 20 Å to deviate by less than 0.2 Å from that of the actual lines. This method of using the so-called *isoelectronic sequences* for predicting the wavelengths of spectral lines of heavily ionized ions works well for simple ions, but the application demands more and more precautions the greater the number of electrons around the nucleus.

b) Determination of the mechanism of excitation. Collisional excitation is the most important process.

c) Determination of the f-value (oscillator strength), from quantum-mechanical computations.

d) Assumption of a coronal model (i.e. $N_e(h)$, and $T_e(h)$).

e) For strong lines self-absorption must be considered.

Elwert has made calculations of intensities of the most important lines for $T = 0.6$ to 6×10^6 °K, which were extended by Kawabata up to 10^8 °K. Recent spectral line observations by various authors have shown the solar soft X-ray spectrum to be more complicated and to contain more lines than was assumed in Elwert and Kawabata's simplified theories. Table II shows Elwert's predictions and Friedman's observations of the total X-ray flux of the sun.

TABLE II

Computed and observed X-ray flux in erg cm^{-2} sec^{-1}. Q is an uncertainty factor of order unity (between 2 and 4)

	Elwert (computed)	Friedman (observed)
Wavelength	100 Å	44–60 Å
Solar minimum flux	$0.03 \times Q$	0.13 (1953)
Solar maximum flux	$0.46 \times Q$	1 (1959)

2.4 The structure of quasi-thermal flare-associated X-ray sources

The structure of quasi-thermal X-ray sources can be determined from a simultaneous investigation of the microwave radio emission and the soft X-ray flux and spectrum of the burst. Tacitly it is assumed that the emission comes from the same volume. Such investigations have only been performed a few times in different degrees of completeness. The most complete discussions are those of Bowen *et al.* (1964, see also De Jager, 1964, page 32) who discussed the X-ray flare of 3 May 1962. Furthermore Kawabata (1966) made a very detailed discussion of the soft X-ray burst of 4 September 1960 of which good recordings were available both in the soft X-ray region (SR I satellite) and on microwave radio frequencies. The results are shown and compared in Table III.

Kawabata suggests, on the basis of other visible observations that the 4 Sept. 1960 source may have had the structure of a loop with a projected length of 2×10^{10} cm and a width of 5×10^8 cm. It also seems, on the basis of Sudden Cosmic Noise

TABLE III

Integral number of excited electrons in X-ray bursts

class	type of radiation	date	quantity	value	reference
I	quasi-thermal	4 IX 1960	$\int N_e^2 \, dV$	10^{49}	KAWABATA (1966)
		3 V 1962	$\int N_e^2 \, dV$	6×10^{47}	BOWEN *et al.* (1964)
II	non-thermal	31 VIII 1959	$\int N_e \, dV$	4×10^{36}	CHUBB *et al.* (1960)
	(deka-keV)				SEVERNY (1960)
	non-thermal	20 III 1958	$\int N_e \, dV$	10^{35}	Table V
	(deci-MeV)	18 IX 1963	$\int N_e \, dV$	4×10^{35}	Table V

Absorption observations that during the peak of the X-ray phase the temperature of the source may have been higher than the value of 6×10^6 °K which applies to the peak of the associated microwave burst. Kawabata suggests that the temperature of the source may have been as high as 10^7 °K.

Clearly these observations can not yet be used for a general theory on the formation of quasi-thermal X-ray sources; they show that we are just at the beginning and that more and complete observations of soft X-ray bursts must be secured and discussed in conjunction with their simultaneous radio and optical observations.

3. Non-Thermal X-Ray Events

3.1 The Interaction of Energetic Particles with Fields and Matter

We have to assume that in most cases the non-thermal X-ray events are due to accelerated clouds of electrons confined by a magnetic field. If the particles were not confined, the accelerated particles would run away in an extremely short interval of time and the X-ray burst would have a very short lifetime. This may have happened in one case, the X-ray event of 18 March 1958 which had a duration of at most some twenty seconds. In the other observed cases the X-ray burst had a duration of at least a few minutes. With a view to the great energies involved which were often of the order of a few hundreds of keV's, the particles must have been confined. So we have to examine what happens to a cloud of accelerated particles moving in helical orbits in a magnetic field. The electrons may collide with the heavier particles of the cloud and lose part of their energy by the emission of Bremsstrahlung. Surely the particles will also lose energy by the emission of synchrotron radiation, mainly in the microwave frequency range.

We assume that the radiative energy loss may be written as

$$-\left(\frac{dE}{dt}\right)_{\rm rad} = P(H, E, N).$$

Next we define an energy exchange time t_e such that the energy loss due to collisions with the ambient gas may be written as

$$-\left(\frac{dE}{dt}\right)_{\rm col} = \frac{E}{t_e}.$$

Then the total energy loss is

$$-\frac{\mathrm{d}E}{\mathrm{d}t} = P + \frac{E}{t_e}.$$

By integration one then obtains

$$E(t) = E(0) - \int P\,\mathrm{d}t - \int \frac{E}{t_e}\,\mathrm{d}t. \tag{3}$$

However, the electrons do not only lose their energy by collisions and radiation, but part of the electrons may escape from the cloud. A characteristic escape time may be found from the deflection time which we define as $t_a(E, N)$. A good approximation for t_a may be obtained by defining it as the average time for a particle to obtain a deflection of more than 90°. It seems reasonable that after such a deflection the particles may move out of their helical spiral orbits and leave the cloud of confined electrons.

The energy loss due to these deflections is then

$$N(t) = N(0)\exp\left(-\int_0^t t_a^{-1}\,\mathrm{d}t\right). \tag{4}$$

From a combination of Equations (1) and (2) we obtain the number of particles with an energy E_1 at a time t after the initial acceleration of the cloud

$$N(E_1, t)\,\mathrm{d}E = N\left\{E_1 + \int P\,\mathrm{d}t + \int E t_e^{-1}\,\mathrm{d}t\right\}_{t=0} \cdot \left\{\exp\left(-\int_0^t t_a^{-1}\,\mathrm{d}t\right)\right\}\frac{\mathrm{d}E(0)}{\mathrm{d}E(t)}\,\mathrm{d}E. \tag{5}$$

Expressions for the stopping time t_e have been summarized by Schatzman and may be written as

$$t_e = \frac{\sqrt{2}}{3}\frac{y^{3/2}}{A_{el}c}.$$

Here $y = E/m_0c^2$ and

$$A_{el} = \frac{1.14 \times 10^{-24}}{y}\left[2N_e^*\left(13.67 + \tfrac{1}{2}\log\frac{T}{N_e} + \log y\right) + N_0(4.84 + \log y)\right].$$

In this equation N_e^* is the electron density of ambient electrons; and N_0 is the number density of heavy particles.

The deflection time t_d may be written as

$$t_d = 2 \times 10^{-3}\,\frac{m_0^2 v^3}{e^4 N_e^*},$$

where v is the velocity of the electrons, and m_0 their rest-mass. The expression has been used by Takakura and Kai (1966) and is due to Spitzer.

TABLE IV

Range and stopping time for electrons and protons in the solar outer layers

Height above photosphere	− 400	0	2000	6000	km
$\log N_e$	14.8	12.0	10.4	9.6	
$\log N_0$	17.2	15.7	11.4	0.4	
$X_{\text{electrons}}$ 51 keV	0.074	2.35	3×10^4	5.6×10^5	km
$t_{\text{electrons}}$ 51 keV	7×10^{-7}	2×10^{-5}	3×10^{-1}	–	sec
$X_{\text{electrons}}$ 5 keV	2.1×10^{-3}	2×10^{-2}	420	6300	km
$t_{\text{electrons}}$ 5 keV	6×10^{-8}	2×10^{-6}	0.01	0.2	sec
X_{protons} 100 MeV	0.68	2×10^3	4×10^7	1.6×10^9	km
t_{protons} 100 MeV	7×10^{-4}	2×10^{-2}	380	1.5×10^3	sec
X_{protons} 11 MeV	9×10^{-3}	0.3	5000	20 000	km
t_{protons} 11 MeV	9×10^{-7}	3×10^{-5}	0.5	23	sec

Schatzman has given numerical values for various characteristic quantities for different heights above the photosphere. Part of these results are summarized in Table IV. Here X is the range of the particles in km, t is their stopping time t_e.

On the basis of the above given Equation (5) Takakura and Kai have computed the energy distribution functions of energetic electrons as a function of time for different assumed initial distribution functions of the electrons. Figure 5 gives such a function for an initial distribution assumed to be the power law

$$N(E, 0) = E^{-\gamma}.$$

The computations described in Figure 5 apply to the case of a magnetic field of 500 gauss, a particle density of 10^9 cm^{-3} and $\gamma = 5$.

As should be expected the energy spectrum of the accelerated particles hardens in the course of time: the particles with the smallest energies disappear first. Furthermore the general number of accelerated particles decreases rapidly in the course of time. In the next sections we shall describe some observed non-thermal X-ray events.

3.2 The X-ray event of 20 March 1958

This event was the first observed energetic X-ray burst; it was detected by Peterson and Winckler (1959) by means of a Geiger-Müller counter and an ion chamber floating at balloon altitude over Cuba during an importance 2 explosive type flare. The X-ray and radio observations (at 21 cm and 3 cm) are shown in Figure 6. The X-ray and radio observations occurred at precisely the same time and were all of extremely short duration. It was shown that approximately 10^{35} electrons were involved in this event; their energy was around $\frac{1}{2}$ MeV. The radio burst was supposed to be due to synchrotron radiation of electrons guided along the magnetic field lines into the dense parts near a sunspot. The computed radio intensity should be about 10^4 higher than the observed flux, but this difference may be reconciled by the assumption that the greater part of the radio emission was produced in those parts of the solar atmosphere from where the radiation could not escape.

3.3 The X-ray event of 25 August 1958

On August 25, 1958 a short-lived burst of photons of at least 0.4 MeV energy has been recorded by the satellite 1958 epsilon between 17.54 and 17.55 UT. The origin of this burst has been obscure for quite some time. However, Křivský (1964) could attribute it to the very first beginning phase of a solar flare of importance 1^+. This flare was visible on Hα filtergrams taken by the McMath and Sacramento Peak Observatories. It started before 17.58 UT and ended 18.44 UT. Later it started again and pulsated

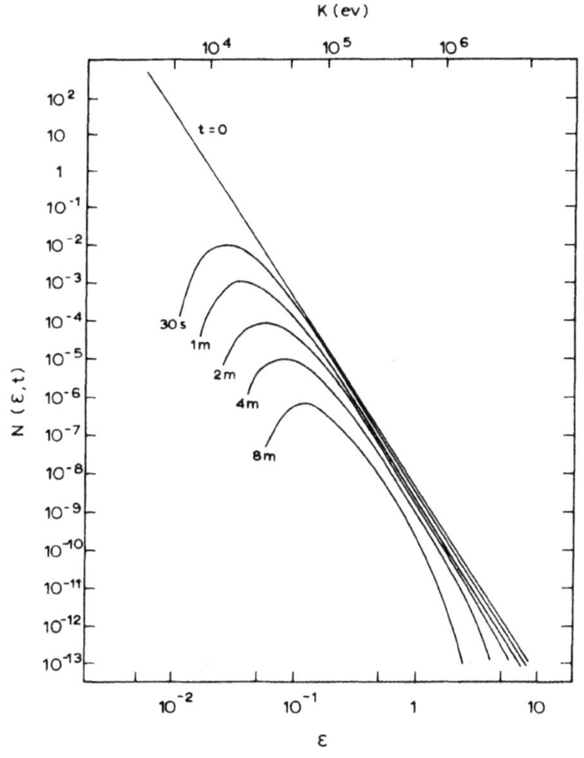

Fig. 5. The time variation of the energy distribution functions of energetic electrons as computed by Takakura and Kai. The initial distribution function is given by the power law: $N(E, 0) = E^{5}$. The curves are labeled with the time. The kinetic energies of the electrons are expressed both in units of the rest energy (m_0c^2) and in electron volts (eV).

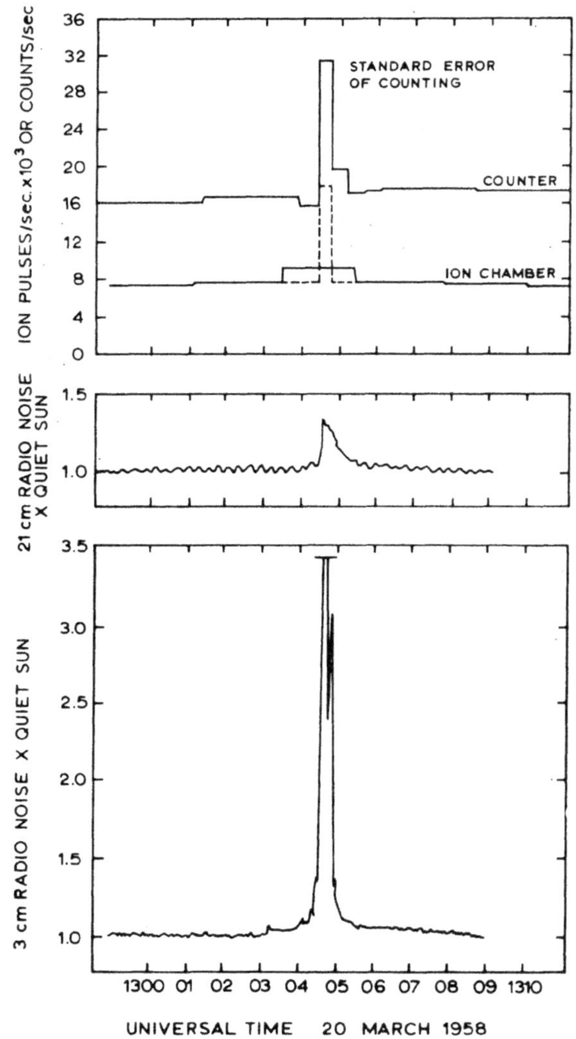

Fig. 6. Hard X-ray observations and radio observations during the solar event of 20 March 1958.

between 19.04 UT till later than 20.01 UT. The flare occurred in a rather complicated E type spot group on the western hemisphere of the sun. No radio emission concurrently with the flare has been observed; however, the next day, August 26, an extraordinarily intense flux of radio emission was observed on decimeter waves, originating from the same activity region. During the occurrence of the hard X-ray burst a SEA effect occurred which started as early as 17.00 UT on August 25.

The peculiarity of this apparently non-thermal burst is that it occurred with a flare without the simultaneous occurrence of phenomena that indicate the existence of a

X or radio flare. This means that the existence of a hot plasma is not a necessary condition for the occurrence of an acceleration process which produces energetic jets of electrons. Neither should the optical flare always be of the rapidly developing or of the explosive type: the relevant flare had its maximum at about 18.18 UT, i.e. 25 minutes after the hard X-ray burst and did not show any explosive phenomena. Apparently the burst occurred at the very beginning of the flare so that this observation shows that impulsive acceleration processes may occur in the very first phase of rather unimportant optical flares without the simultaneous occurrence of the hot plasma which manifests itself sometimes in the radio or X-ray flares.

3.4 The X-ray event of 11 August 1960

This event was observed by Winckler and coworkers with balloons at two different latitudes (Minneapolis and Fort Churchill). Both balloons gave essentially the same radiation profile so that the burst is of solar origin and consists of photons. The X-ray burst occurred in time with an explosive flare of importance 2; radio observations show a cluster of type III bursts all the time the X-ray photons were observed. This feature may be explained in such a way that the accelerated cloud produced at the same time two different kinds of electrons. Part of the electrons moved upward into the solar corona. Such particles, which move into an extremely tenuous part of the sun cannot be stopped, but they excite the corona to the emission of type III radio bursts. Other particles that are at the same time accelerated downward are stopped in the dense photospheric layers and produce Bremsstrahlung X-rays. (De Jager, 1960). This burst was of the deka-keV type.

3.5 The X-ray event of 28 September 1961

This event was observed by Anderson in a high altitude balloon flight over southern Canada. Simultaneous radio observations made at Fort Davis Texas (Maxwell) showed the burst to coincide in time with a microwave radio burst and with type III bursts. There was also a prolonged type IV continuum. Simultaneous X-ray observations in Minneapolis showed the same time structure of the X-ray event so that, again, the burst was photonic and of solar origin. Since the balloon observations were made with a scintillation counter, the energy of the X-ray event could be determined; it turned out that the photonic energy spectrum remained more or less constant throughout the event. It was suggested by Anderson and Winckler (1962) that the time history of the event may be split into three separate events. The beginning of each of them may be associated with fast type III bursts seen on a radio spectrogram. The attractive part of this hypothesis is that type III events are clearly due to electrons accelerated to energies of the order of a few hundreds of keV's, which shows that the beginning of each of the parts of the X-event is associated with the occurrence of accelerating electromagnetic fields in the low solar corona. Analogously, as in the case of the burst of 11 August 1960, parts of these fields may be orientated in such a way that electrons are shot upward into the corona and produce type III fast drift bursts; other electrons are shot downward and kept in a magnetic field where they emit the microwave radio burst by the synchrotron mechanism and produce X-ray photons in stopping collisions with heavier particles. Also this burst was of the deka-keV type.

3.6 The X-ray event of 18 September 1963

This burst was observed by J. P. Legrand by means of a set of Geiger-Müller counters at balloon altitude. The observations were made at high geographic latitude near Kiruna, north Sweden, but from the associated radio and optical phenomena it is clear that the burst was of solar origin; furthermore it was photonic since a particle telescope did not show any response (Figure 7). The burst had its maximum around 13.56 UT and lasted a few minutes; a steep increase was followed by an exponential decay. During its declining phase a radio burst was observed on 3 and 10 cm. Maximum

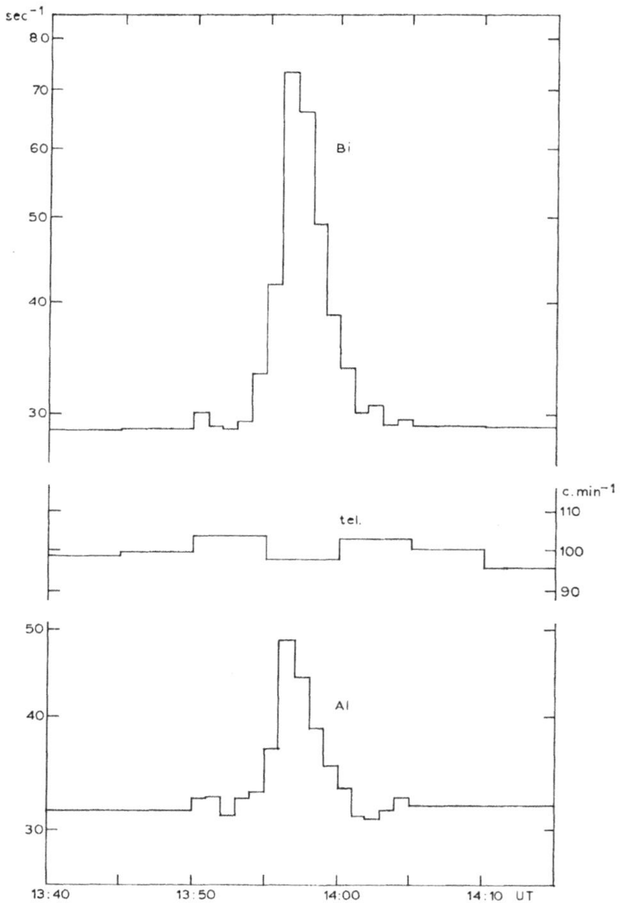

Fig. 7. X-ray observations of 18 September 1963.

radio intensity occurred two minutes after that of the X-ray burst. The X-ray and radio bursts ended almost simultaneously. Optically a small shortlived (≈ 1 minute) flare point occurred simultaneously with the X-ray burst in a magnetically interesting part of the active region of September 1963. The X-ray burst photons seem to have had an energy of about 0.5 MeV. The burst has been discussed theoretically by DE JAGER (1967). The discussion shows that a mass of gas magnetically confined to a volume of about 8×10^{25} cm^3 in the low corona, containing about 4×10^{35} electrons was accelerated to energies of about 0.5 MeV. The radioburst is perhaps due to synchrotron radiation.

3.7 SUMMARY; MODEL OF THE HIGH-ENERGY FLARE, AND OF NON-THERMAL X-RAY EVENTS

The characteristic data of the three X-ray bursts of the deci-MeV type observed so far are summarized in Table V.

From this table we derive the following results. Deci-MeV X-ray bursts have photon energies of the order of 0.5 MeV. They are of short duration, not longer than a few minutes. They are not necessarily associated with very important flares; although the observational evidence is still small it may eventually turn out that even very

TABLE V

Hard X-ray observations in the 0.5 MeV region

event of	20 March 1958	25 Aug. 1958	18 Sept. 1963
U.T.	13.04.27	17.54	13.56
appr. duration	$\lesssim$ 20 sec	1 min	2 min
appr. photon energy	$\approx$ 0.5 MeV	$\gtrsim$ 0.4 MeV	$\approx$ 0.5 MeV
max. flux rate (erg cm^{-2} sec^{-1} Å^{-1})	10^{-3}	2×10^{-5}	1.2×10^{-1}
suggested emission mechanism	Bremsstrahlung	?	Bremsstrahlung
$\int N_e \, dV$	10^{35}	?	10^{35}
ref. X-ray observations	PETERSON and WINCKLER, 1959	MIYAZAKI and TAKEUCHI, 1960	DE JAGER and LEGRAND, 1965
ref. X-ray interpretations	ibid	–	DE JAGER, 1967
radio events at	3 cm (extremely strong) 21 cm (25 times weaker) not at $\lambda \geq 60$ cm	none	3 and 10 cm (weak) not at $\lambda \geq 15$ cm
suggested radio mechanism	synchrotron (DENISSE, 1959) inverse Compton (ŽELEZNYAKOV, 1965)	–	synchrotron (DE JAGER, 1967)
optical flare importance	2 (explosive)	1+	1−
general reference, review	WINCKLER, 1964	KŘIVSKÝ, 1964 DE JAGER and LEGRAND, 1965	DE JAGER, 1967

unimportant optical flares may be associated with bursts of very high energy photons. The total number of electrons involved is fairly large and ranges between 10^{33} and 10^{36} particles. The electron density could be determined for the burst of 18 September 1963 and is slightly less than 10^{10} electrons cm^{-3}. Hence, the volume involved is fairly large. If we assume the same value of N_e for all types of hard X-ray bursts, the consequent volumes range between 10^{25} and 10^{28} cm^3. If one would assume a spherical cloud of accelerated particles this volume would correspond to a radius of 1000 to 10000 km. This radius is the minimum radius calculated on the basis of the assumption that *all* the electrons in this volume are accelerated. The size of this volume makes it clear at the same time why the importance of the optical flare is but loosely related to the importance of the X-ray burst: the phenomenon may be for the greater part of coronal origin; what one observes in the photosphere may simply appear to be a loose reflection of the more energetic phenomena occurring in the higher optically transparent layers of the corona.

It is clear that the problem of the acceleration of electrons to energies to hundreds of keV offers certain problems. The velocities involved are of the order of 0.5 c to 0.9 c. The acceleration of the huge clouds of electrons described above to such energies is extremely hard to explain. If expressed in temperatures these energies would correspond to temperatures of more than 10^9 °K. It is clear that the notion "temperature" has no sense for these bursts, at least not in the beginning phase. After one or two minutes collisions may have defined a certain kinetic equilibrium so that from that moment onward the notion temperature may have sense; at that time temperatures of the order of more than 10^8 °K may occur in the solar corona. This observation may be reconciled with the earlier described rocket observation of Friedman *et al.*, who found that the rocket spectrum of a burst observed in the softer X-ray region (20–60 keV) on 31 August 1959 could be explained by thermal radiation with a temperature of 10^8 °K.

Comparison of the source characteristics for the quasi-thermal and non-thermal X-ray bursts

A comparison of the summarizing data given in Tables III and V shows the important result that an *essential* difference between the physical structure of soft, quasi-thermal X-ray sources and of hard X-ray sources does perhaps not exist. The main difference seems to be one of the energy of the excited electrons. For soft X-ray sources temperatures have been found of the order of 2 to 6×10^6 °K, while the energies of the hard X-ray emitting electrons are larger than these and would correspond to higher temperatures. A comparison of $\int N_e \, dv$ found for the hard X-ray sources with $\int N_e^2 \, dv$ found for the soft X-ray sources shows that these numbers correspond to physical data which are of the same order of magnitude. Only the volume involved seems smaller for the more energetic types of bursts (see Table I).

Another fundamental difference is that the kinetic energy of the electrons in the soft X-ray sources corresponds to temperatures of 2 to 6×10^6 °K while the energy of the electrons in the hard X-ray sources is larger.

Model of the source

Finally we describe a model of the source introduced by Takakura and Kai (1966). These authors suggest that the electrons are confined in the magnetic field between two sunspots. The source is situated reasonably low in the corona, yet above the

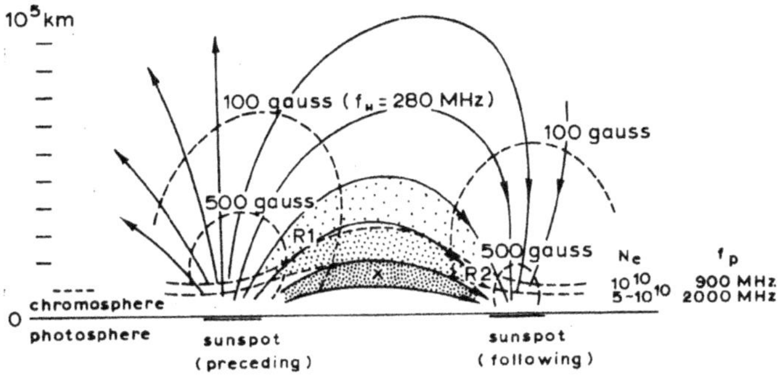

Fig. 8. Schematic model of the high energy flare after Takakura and Kai.

photosphere, so that in photospheric emissions the source is invisible. The magnetic field at the place of the source is certainly much weaker than in the spots and may be of the order of 50 to 100 gauss (see Figure 8). The points R_1 and R_2 in the figure may be the sources for microwave impulsive bursts. The source of the X-ray burst is found in the central parts of the dotted region, marked by X.

Although this model was introduced to explain one particular observation it is sufficiently general to be applicable to most or all of the hard X-ray bursts described here. This picture is also not in disagreement with other images of the general configuration in the vicinity of solar flares as described earlier by other authors (see e.g. De Jager, 1963).

Altogether we may conclude that the existing observations of hard X-ray bursts are already sufficient in number to give us an approximate picture of the events, and lead us to a provisional model of the configuration. Further investigations, both observationally and theoretically, may improve this picture.

References

Anderson, K. A. and Winckler, J. R.: 1962, *J. Geophys. Res.* **67,** 4103.

Blake, R. L.: 1965, *Astrophys. J.* **142,** 1.

Blake, R. L., Chubb, T. A., Friedman, H. and Unzicker, A.: 1963, *Astrophys. J.* **137,** 3.

Bowen, P. J., Norman, K., Pounds, K. A., Sanford, P. W., and Willmore, A. P.: 1964, *Proc. Roy. Soc.* **A 281,** 538.

Burgess, A.: 1964, *Astrophys. J.* **139,** 776.

Burgess, A. and Seaton, M.J.: 1964, *Monthly Notices Roy. Astron. Soc.* **127,** 355

Chubb, T. A., Friedman, H. and Kreplin, R. W.: 1960, *J. Geophys. Res.* **65,** 1831.

Chubb, T. A., Friedman, H. and Kreplin, R. W.: 1961, *Space Res.* **I,** 695.
Denisse, J. F.: 1959, in *Paris Symposium on Radio Astronomy* (R. N. Bracewell, ed.), Stanford Univ. Press, p. 237.
Elwert, G.: 1958, *J. Atmos. Terrest. Phys.* **12,** 187.
Elwert, G.: 1961, *J. Geophys. Res.* **66,** 391.
Evans, K., Pounds, K. A., and Culhane, J. L.: 1967, *Nature*, Febr.
Jager, C. de: 1959, 'Structure and Dynamics of the Solar Atmosphere', *Handbuch der Physik* **52,** 80.
Jager, C. de: 1960, *Space Res.* **1,** 628.
Jager, C. de: 1963, *Space Sci. Rev.* **1,** 487.
Jager, C. de: 1964, 'Solar Ultraviolet and X-ray Radiation', *Research in Geophysics* (H. Odishaw, ed.), MIT Press.
Jager, C. de: 1967, *Solar Phys.* **2,** 327.
Jager, C. de and Kundu, M. R.: 1963, *Space Res.* **3,** 836.
Jager, C. de and Legrand, J. P.: 1965, *SPARMO Bulletin* **3,** 1–3.
Kawabata, K.: 1960, *Rep. Ion. Space Res. Japan* **14,** 405.
Kawabata, K.: 1966, *Rep. Ion. Space Res. Japan,* **20,** 118.
Křivský, L.: 1964, *Bull. Astron. Inst. Czechoslov.* **15,** 131.
Manson, B. E.: 1967, *Astrophys. J.* **147,** 703.
Maxwell, A.: 1963, *Plan. Space Sc.* **11,** 897.
Miyazaki, Y. and Takeuchi, H.: 1960, in *Space Research* **1** (H. K. Bijl, ed.), North Holland Publishing Co., 869.
Newkirk, G.: 1967, *Ann. Rev. Astron. Astrophys.* **5,** 213.
Peterson, L. E. and Winckler, J. R.: 1959, *J. Geophys. Res.* **64,** 697.
Pounds, K. A. and Russell, P. C.: 1966, *Nature* **209,** 490.
Rensbergen, W. van: 1967, *Bull. Astron. Inst. Neth.* **19,** 6.
Rensbergen, W. van: 1967, *Solar Phys.* **1,** 354.
Schatzman, E.: 1965, 'Particle and Radio Emission from the Sun', in *The Solar Spectrum* (C. de Jager, ed.), Reidel, Dordrecht.
Severny. A. B.: 1960, *Izv. Krymsk. Astrophys. Obs.* **22,** 67.
Takakura, T. and Kai, K.: 1966, *Publ. Astron. Soc., Japan* **18,** 57.
Unsöld, A.: 1955, *Physik der Sternatmosphären*, 2 Aufl., Springer-Verlag, Berlin.
Winckler, J. R.: 1964, in *The Physics of Solar Flares* (W. N. Hess, ed.) NASA, **117**.
Winckler, J.R., Arnoldy, R.L., and Kane, S.R.: 1967, *Solar Physics,* **2,** 171.
Winckler, J. R., Mai, T. C., and Masley, A. J.: 1961, *J. Geophys. Res.* **66,** 316.
Železnyakov, V. V.: 1965, *Astron. Žurn.* **42,** 96.

OBSERVATIONS OF SOLAR X-RAYS

J. LABEYRIE

Centre d'Etudes Nucléaires, Saclay, France

Abstract. A comprehensive review of the measurements of X-rays emitted by the quiet as well as the active sun is presented.

Starting with the historic sounding rocket flight by Burnight in 1948 these observations were carried out by means of balloons, sounding rockets and satellites, covering more than one solar cycle. The instruments used as well as the experimental results are discussed in detail.

Some experimental and theoretical aspects of solar X-ray observations are considered in general, including absorption of X-rays, production in the solar atmosphere and correlation with geophysical phenomena.

1. Introduction

These lectures intend to give a survey of the observational results obtained in the field of solar X-rays, and a description of instruments used aboard balloons, rockets and satellites for measuring the fluxes and spectra of X-rays and for determining the solar regions responsible for this emission.

The theoretical aspects of the emission of X-rays by the sun and the mechanisms occurring in the solar corona are discussed by Prof. C. DE JAGER [1] in his lectures.

1.1 REFERENCES

Many data given here were taken from the review article on instrumental techniques by BOYD [2], from several papers of H. Friedman and his coworkers, and from MANDEL'SHTAM *et al.* [3]. In addition some unpublished data of the X-ray group of the Service d'Electronique Physique, C.E.A. SACLAY [4] have been used. In addition many data come from different papers mostly from *Space Research* and the *Journal of Geophysical Research.*

1.2 DEFINITIONS

The X-rays to be considered are photons coming from the sun with wavelengths between 100 Å and about 0.02 Å. At the long wavelength end of this range they pertain to the world of optics, where the unit is the Ångström. At the short wavelength end the X-ray photons belong to the world of nuclear physics where the common unit is the electron-volt. These two units are related as

$$\lambda(\text{Å}) \times E(\text{keV}) = 12.34$$

Thus the X-ray photons considered here have energies between about 0.1 and 620 keV.

The fluxes discussed in these lectures are the fluxes of solar X-rays as they arrive at the top of the earth's atmosphere.

J. G. Emming (ed.), Electromagnetic Radiation in Space.

The fluxes are expressed either in erg $sec^{-1} cm^{-2} Å^{-1}$ (spectral distribution) or in erg $sec^{-1} cm^{-2}$ (within a given wavelength band). Occasionally, number of photons is used instead of energy.

2. Absorption of X-Rays in the Terrestrial Atmosphere

The terrestrial atmosphere shields almost completely the earth from solar X-radiation. For example, at the Chacaltaya laboratory (5200 m) which is the highest cosmic and astronomical observatory in the world, X-rays at 8 Å are reduced in intensity by a factor 100 in 1 cm of air.

However, X-rays of higher energies may traverse a certain thickness of the upper atmosphere. The K-absorption edges of nitrogen, oxygen, and argon are at about 30, 20, and 4 Å respectively. The mean absorption of X-rays of different wavelengths in the atmosphere can easily be determined from laboratory experiments. Figure 1 shows the transparency of the atmosphere as function of height for wavelengths between 100 Å and about 0.1 Å. Several conclusions can be drawn from these curves:

i) At 40 km, which is the highest practical altitude for balloons, the intensity ratio P of vertically incident X-rays with respect to the value at the top of the atmosphere is found to be:

E (keV)	100	50	30	20	15
λ (Å)	0.123	0.25	0.41	0.61	0.82
P (%)	60	50	30	6	1

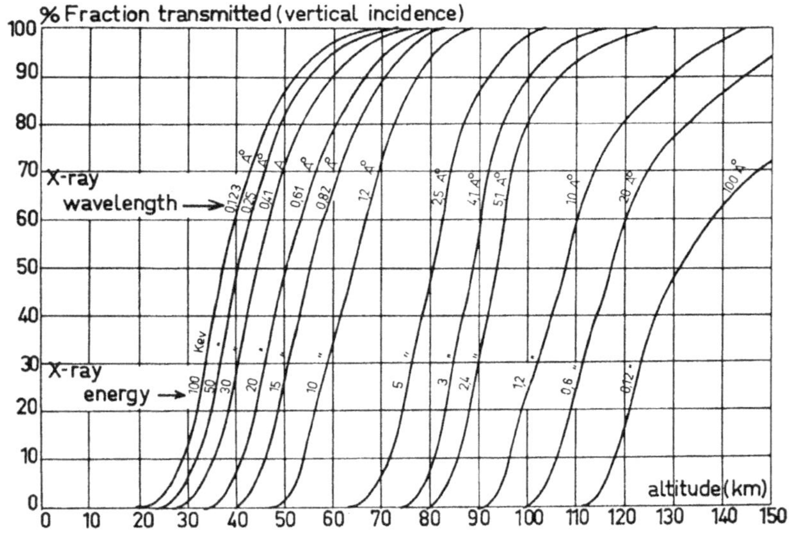

Fig. 1. Transparency of atmosphere to X-rays.

At lower altitude, the absorption of X-rays is very strong, e.g. at 20 km less than 1% of the 100 keV radiation is arriving.

ii) The soft X-ray component (below about 15 keV) is detectable only at altitudes which rockets or satellites can reach. At 80 km, which is the middle of the so-called D-region of the ionosphere, about 10% of the 3 keV radiation arrives, and about 1% of the 2.5 keV radiation. As will be seen below, the ionisation in this layer under quiet solar conditions is essentially due to the absorption of Lyman-α (1216 Å) which is only capable of ionising the NO molecule. The number density of this molecule is only about 10^{-4} times that of the other constituents of the atmosphere at this altitude.

iii) At 110 km (central part of the day-time E-region) only about 1% of the 100 Å radiation and 20% of the 20 Å radiation arrives, but nearly 100% at 4 Å. Space vehicles must reach at least 125 km in order to detect more than 30% of the 100 Å radiation.

3. Theoretical Predictions of Solar X-Rays

When Lyot [5] constructed the first coronagraph in 1930, he observed that the coronal lines of visible light were so broad* that temperatures of the order of one million degrees were necessary to explain their width in terms of Doppler broadening. Such a temperature corresponds to a mean thermal kinetic energy of the order of 100 eV in the coronal plasma. A gas of this temperature if radiating as a black body must emit X-rays with maximum intensity around 30 Å.** In 1938 Hubert and Vegard suggested that solar X-rays are responsible for the ionisation in the *E*-layer. The well-known properties of X-ray emission imply that, in a hot plasma, accelerated electrons radiate energy in a broad spectrum. X-ray photons may come from at least three processes:

i) *free-free* transitions of electrons in the field of protons, helium ions and heavier ions. This is called "bremsstrahlung" emission and contributes to the so-called "continuum" of X-rays.

ii) *free-bound* transitions of electrons recombining with ionised atoms. This process also contributes to the continuous X-rays.

iii) *excitation* of levels of atoms or ions by electron impact. This last process produces spectral lines.

In 1941 Edlen following a suggestion made by Grotrian in 1939 identified the coronal lines in the visible light as forbidden transitions to the ground levels of Fe X, Fe XI, Fe XIII, Fe XIV, Ca XII, Ca XIII, Ni XII, Ni XVI etc. The fact that the ionisation potentials of these ions range from 233 eV (Fe X) to 814 eV (Ca XV) and even higher for the very highly ionised iron or nickel atoms confirmed the existence of very high

* Particularly the strong coronal line at 5303 Å.

** If the plasma forming the solar corona is considered as a black body of temperature T, the wavelength λ_M of the maximum of the emission is given by Wien's law:

$$\lambda_M = \frac{0.3}{T}$$

with λ_M in cm and T in °K.

temperatures in the corona and suggested strongly that X-ray photons resulting from de-excitation of these ions were incident at all times upon the upper layers of the earth's atmosphere.

On the other hand HOYLE and BATES [6] showed in 1948, as it had been suggested 10 years before, that the flux of soft X-rays from such a corona would be adequate to account for the permanent ionisation of the E-region of the terrestrial ionosphere around an altitude of 110 km. They estimated that between 10^{-2} and 10^{-1} erg cm^{-2}sec^{-1} would be required for a certain wavelength band centered around 40 Å.

Hence, X-rays appeared as a very interesting component of the solar radiation, both because they seemed to offer the possibility of investigating the mysterious corona and because they might affect our atmosphere.

4. Absorption of X-Rays by Foils

As we have seen, in order to observe solar X-rays, we must send detectors on top of the atmosphere. These detectors are either devices based on the ionisation in a gas (ionisation chambers, proportional counters and Geiger-Müller counters), or photographic emulsions, or scintillation counters.

All of these detectors have to be equipped with thin-foil windows either to prevent the escape of gas or to prevent visible and UV light from entering the sensitive volume. The foils are as thin as possible and of as low an atomic number as can be because the absorption increases rapidly with atomic number.

The lightest element which can be used for foils is beryllium ($Z=4$). Figure 2 shows the transmission of X-rays through Be foils of various thicknesses.

It will be noted that at 111 Å there is an abrupt decrease in transmission which is due to the K-edge. The transmission increases again towards shorter wavelengths. The transmission of aluminium foils is also plotted in Figure 2.

For other currently-used foils the edges are listed in Table I.

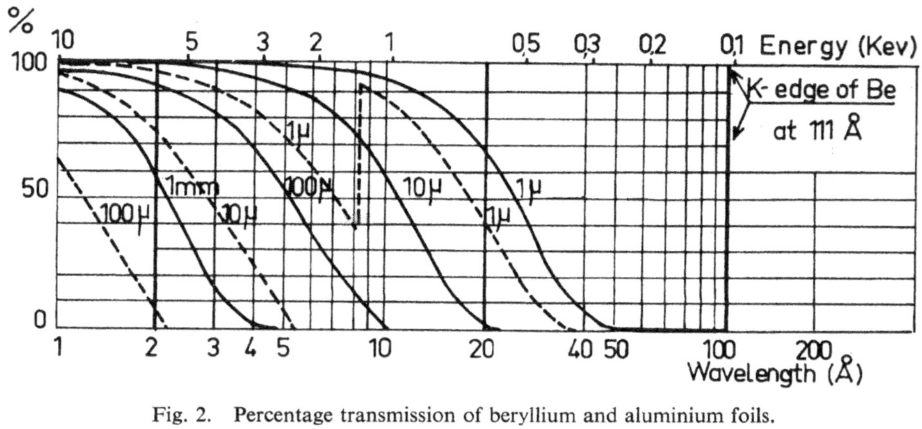

Fig. 2. Percentage transmission of beryllium and aluminium foils.

TABLE I
K-edges of various elements

Atomic number	Element	K-edge	
		in Å	in eV
4	beryllium	111	111
6	carbon (mylar)	44	270
13	aluminium	8	1540
28	copper	1.4	8500

5. First Observations of Solar X-Rays

The first observational evidence of solar X-rays comes from the rocket flight on 5 August 1948 by BURNIGHT [7] who used Schumann plates as detector. The plates were covered by a beryllium foil 0.76 mm thick which had a transmission of about 10% at 3 Å increasing rapidly towards shorter wavelengths.

A blackening of the photographic plate was found which was attributed to X-rays shorter than 4 Å. In 1949 FRIEDMAN *et al.* [8] showed however, that it was most likely due to the soft component with wavelengths longer than 111 Å. The historical picture is shown in Figure 3.

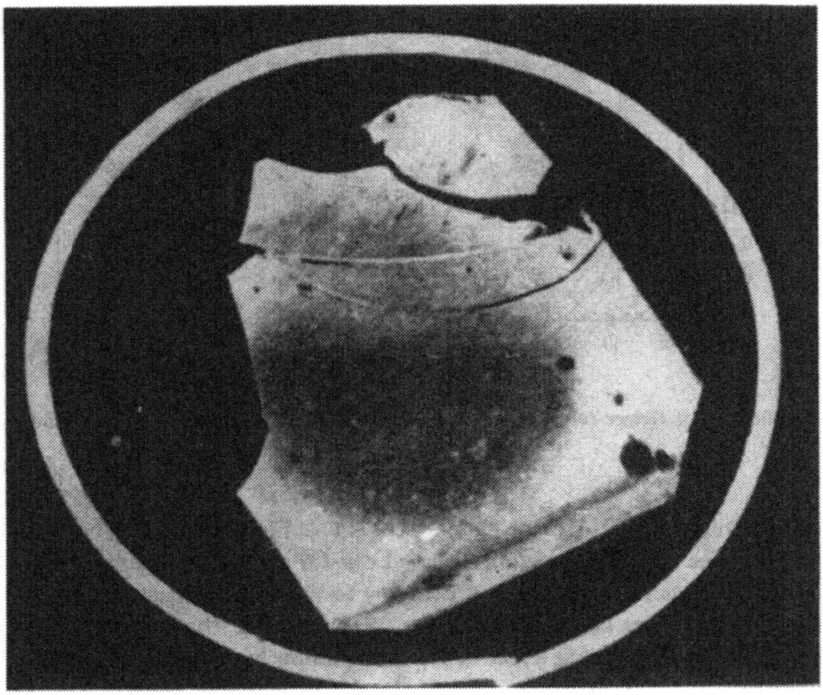

Fig. 3. Burnight's discovery of solar X-rays.

During the same year Burnight flew another payload containing a photographic plate covered with an aluminium foil 0.0076 mm thick which transmitted about 50% at the K-edge (8 Å) and about 2% at 16 Å. The blackening indicated that there were also X-rays in this range.

The following September, FRIEDMAN *et al.* [8] flew two G.M. counters on a V-2 rocket. The counters were sensitive to radiation shorter than 10 Å. During the ascent of the rocket X-rays were first observed at 87 km. By comparing the increase of flux with altitude with the computed atmospheric absorption of X-rays at different wavelengths (see Figure 1) it was possible to show that the X-ray flux was negligible below 7 Å.

6. Variation of X-Ray Flux Within the Solar Cycle

During the following 15 years several rockets carrying similar counters or ionisation chambers were launched by Friedman and his co-workers from the U.S. Naval Research Laboratory in order to measure solar X-rays in several wavelength bands.

By the skilful use of various foils and gas fillings, it was possible to distinguish 4 wavelength bands, i.e. 2–8 Å, 8–20 Å, 44–60 Å and 44–100 Å, and to determine the

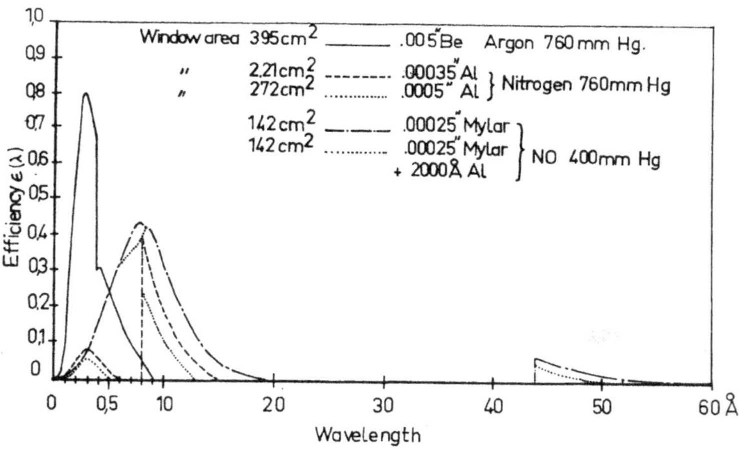

Fig. 4. Efficiency of Geiger tubes used aboard Naval Research Laboratory rockets and satellites for solar X-ray flux measurement.

mean flux value in each of these bands [9, 10, 11]. Figure 4 shows the wavelength response of three of these detectors. Carried out for more than a full sunspot cycle, these measurements have shown that, contrary to what takes place with visible light, the X-ray flux changes considerably along this cycle, and that the variations differ markedly for the four bands. Near sunspot minimum in 1953 and 1954, the rocket measurements indicated a great reduction in X-ray emission below 20 Å. In some experiments no emission at all was detected below 10 Å. When approaching the solar maximum the X-ray flux generally increased, but especially at shorter wavelengths. In the 2–8 Å band the maximum-to-minimum ratio was several hundred: in the 8–20 Å band this ratio was at least 60 and in the 44–60 Å band it was 7.

The variation of total flux between 2 and 100 Å corresponds to a factor of about ten (see Figure 5).

However, these results are rather inaccurate owing to the uncertainty involved in the calibration of the detectors which depends on the assumed spectral distribution

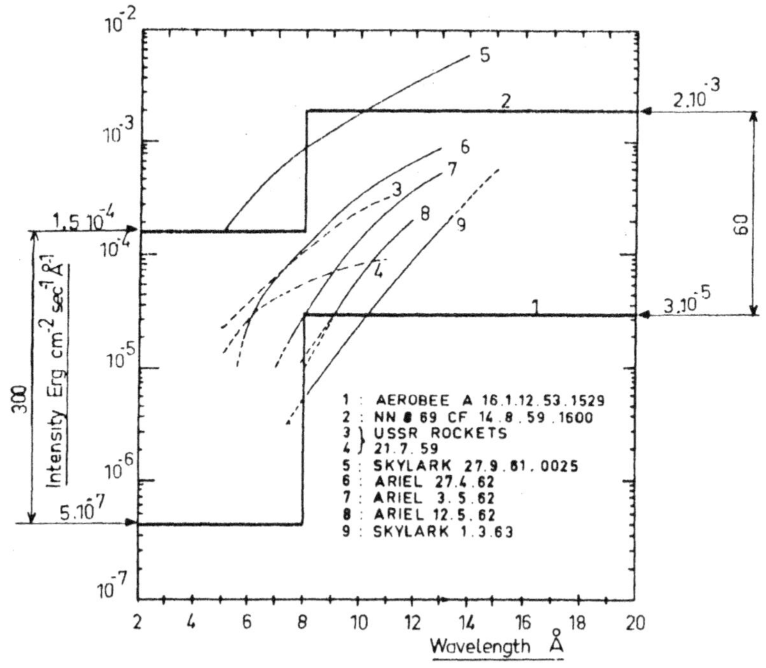

Fig. 5. A comparison of quiet sun X-ray measurements.

of the detected X-rays. To illustrate this effect, let us assume that the sun radiates as a "gray body" (see Figure 6).* Then if the temperature of the corona was 10^6 °K at solar minimum (1953) the total flux in the band 44–100 Å would have been 0.14 $\mathrm{erg\,cm^{-2}\,sec^{-1}}$ (or 2.5×10^{-3} $\mathrm{erg\ cm^{-2}\,sec^{-1}\,Å^{-1}}$ as a mean value). Doubling the temperature would increase the total flux to 0.35 $\mathrm{erg\,cm^{-2}\,sec^{-1}}$.

As an example, let us consider a G.M. counter with Xe filling and Al window.

* *Editor's footnote:* a "gray body" is a fictitious emitter of radiation whose spectral distribution is identical to that of a black body but with an absolute intensity which is reduced by an arbitrary factor. This concept has been widely used in the interpretation of measurements of extra-terrestrial X-ray fluxes in spite of the fact that it is meaningless from a thermodynamic point of view. It bears no relationship to the "gray atmosphere" used in stellar models.

Its efficiency is high between 8 and 20 Å and low between 20 and 100 Å. If the sun radiates as a gray body at $T=0.8\times10^6\,^\circ$K its main emission is between 12 and 400 Å (see Figure 6), a region where the G.M. counter has a very low efficiency.

If the temperature of the gray body is $4\times10^6\,^\circ$K it emits predominantly in a wavelength region where the counter has great sensitivity (40 times greater than in the previous case). Thus, a much smaller total X-ray flux would produce the same counting rate.

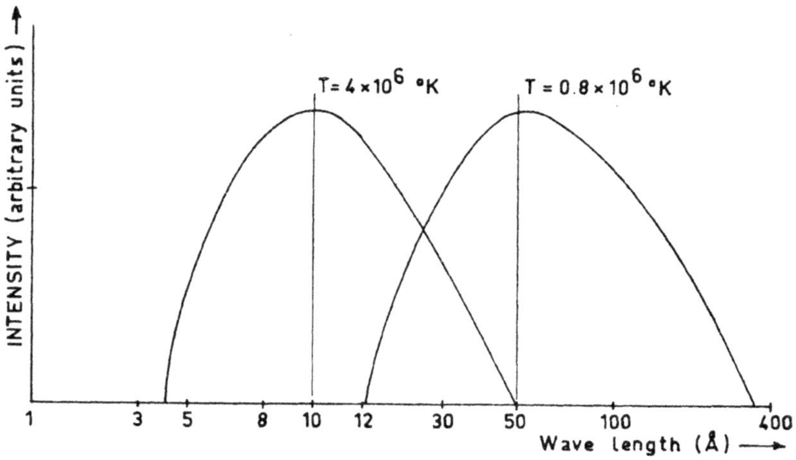

Fig. 6. Normalized gray body spectral distribution.

7. Influence of Solar X-Rays on the Ionosphere

As early as 1955 by measuring the variation of the flux of X-rays with altitude [12], it was shown that the X-rays in the band 10–100 Å were absorbed in the E-region of the ionosphere (between 100 and 140 km). The X-ray energy absorbed in this layer appeared adequate to account for a major portion of the ionisation. As can be seen from Table II large ionisation occurs also in the D-, F_1- and F_2-layers but according to Figure 1 this cannot be due to absorption of solar X-rays.

TABLE II

Height, electron density and corresponding critical frequency of ionospheric layers (from Smith and Smith [13].

Region	Height in km	Electron density in cm^{-3}	Critical frequency in MHz
D	70–80	10^2 to 10^4	0.1 to 1
E	90–150	1 to 20×10^4	1 to 4.5
F_1	150–250	1 to 6×10^5	3 to 7.5
F_2	250–400	1 to 25×10^5	3 to 15

On the other hand, monthly values of electron density N in the E-layer (deduced from the critical frequency* at which reflection occurs) show detailed agreement with the pattern followed by monthly values of sunspot numbers, indicating that at least part of the ionising flux emanates from the sunspot regions. Prior to 1958, however, no direct identification of localised sources of X-ray emission in the corona had been made.

8. X-Ray Observations During the 1958 Eclipse

The total eclipse of October 12, 1958 offered an opportunity to launch rockets carrying UV and X-ray detectors to observe the distribution of emission sources over the disk and to determine whether any residual emission of X-rays or ultra-violet radiation was detectable at totality. The U.S. Naval Research Laboratory launched two sea-borne rockets during totality through the eclipse shadow at E-region altitude. Each rocket was equipped with a Lyman-α ionisation chamber and with X-ray detectors sensitive in the two wavelength bands 8–18 Å and 44–60 Å.

About 0.05 percent residual Lyman-α flux and from 10 to 13 percent residual X-ray flux at 44–60 Å was observed [14].

Hence under the solar conditions prevailing during the eclipse 10 to 13 percent of the X-rays in the 44–60 Å band were emitted by the corona above the limb. Consequently, the X-ray emitting regions may be located very high in the solar atmosphere.

Since during the totality phase of an eclipse the E-region does not disappear completely, as would be expected if the source of ionising radiation were totally obscured and recombination were fast, this experiment tends to support the idea that X-rays are more responsible than Lyman-α for the ionisation in the E-layer.

The eclipse of 15 February 1961 which passed over the Crimea was observed in a similar way by Russian rocket astronomers [15] with essentially the same results.

The first objective of the U.S. Naval Research Laboratory experiment during the 1958 eclipse was to identify localised sources of X-ray emission on the disk. On the day of the eclipse, the area of the disk near the east limb contained a number of active regions identified by intense plages in Ca-K light whereas an equivalent area bordering the west limb was almost free of activity. Rockets were fired so as to observe the exposed crescent on the east and west limbs before the second contact and after the third one (Figure 7).

The east limb crescent which contained the plages was observed to be six times as bright in X-rays than the west limb crescent, demonstrating for the first time that the sources of X-rays are situated in the vicinity of plages.

* The critical frequency of a plasma is the frequency for which the index μ of refraction of this plasma is equal to zero:

$$\mu = 1 - (e^2/\pi m)\,(N/f^2)$$

where N is the density of particles having charge e and mass m. The critical frequency f, in Megacycles, is therefore given by:

$$N(\text{cm}^{-3}) = \pi m f^2/e^2 = 1.24 \times 10^4 f^2$$

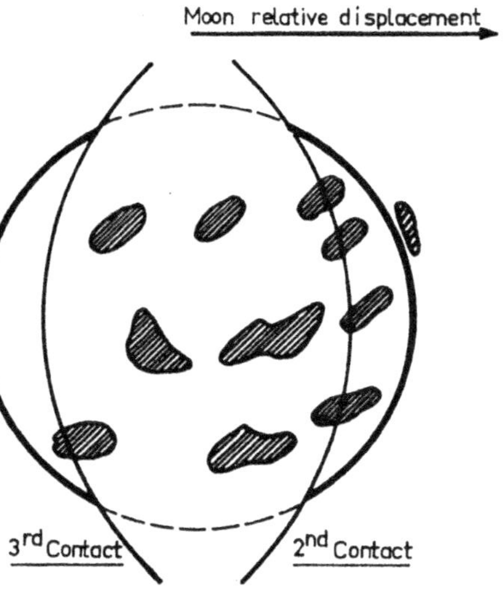

Fig. 7. Active regions, as observed in Ca-K light during a solar eclipse.

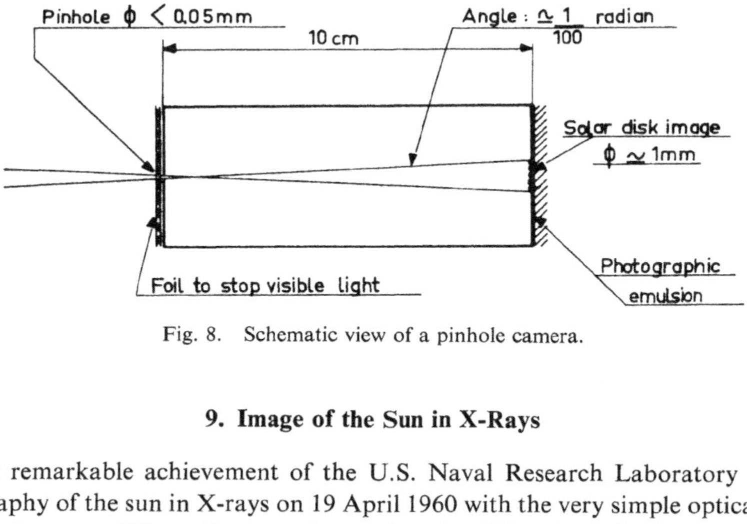

Fig. 8. Schematic view of a pinhole camera.

9. Image of the Sun in X-Rays

Another remarkable achievement of the U.S. Naval Research Laboratory was the photography of the sun in X-rays on 19 April 1960 with the very simple optical device of a pinhole camera (Figure 8) mounted on an Aerobee-Hi rocket [16]. The camera length was about 15 cm resulting in an image of about 0.15 cm in diameter: the pinhole was 0.125 mm in diameter giving a resolution a little better than a tenth of a solar diameter. The pinhole was covered by a plastic film of Parlodion, of surface density $2.5 \times 10^{-4} \mathrm{g\,cm^{-2}}$, overcoated by an evaporated film of aluminium $5.9 \times 10^{-5} \mathrm{g\,cm^{-2}}$. The transmission of the foil is shown in Figure 9. The duration of exposure was 286 seconds. Peak altitude was 220 km and the emulsion used was an industrial Ilford X-G film (see Figure 10).

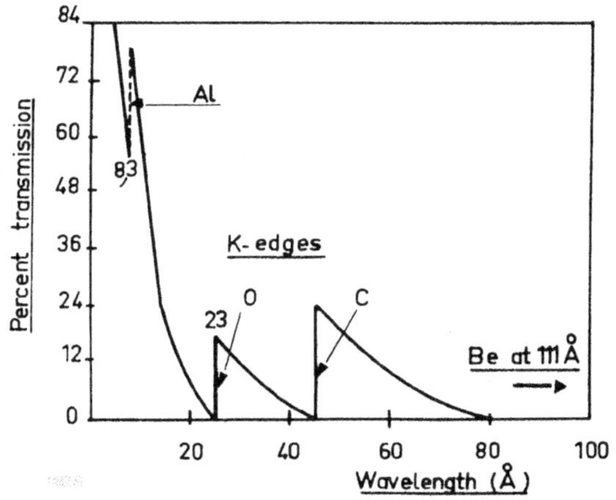

Fig. 9. Transmission properties of filter used to obtain first X-ray photograph of the sun.

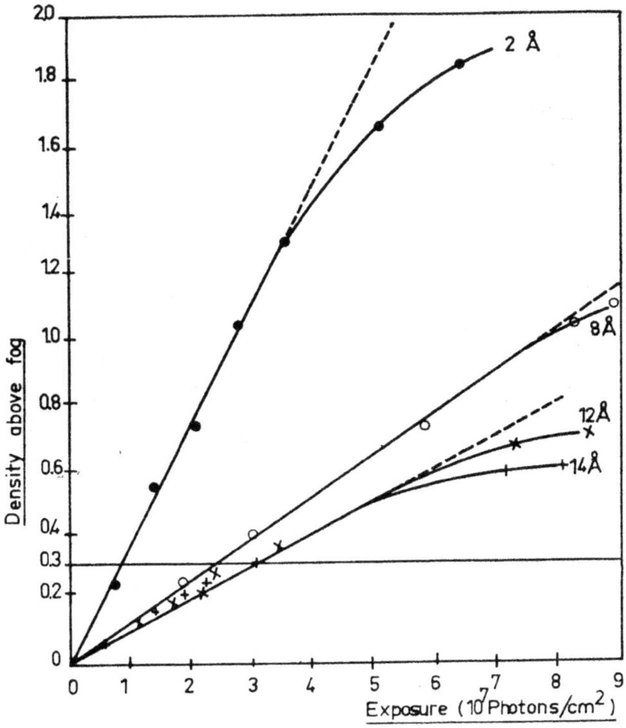

Fig. 10. Response of industrial Ilford-G emulsion.

The camera was partially stabilised by biaxial attitude control which kept the camera pointed towards the sun but left it free to rotate around its axis. During the flight the image was first rotated 20° clockwise and then returned counter-clockwise to complete the full arc of 160° extent. The axis of the camera was kept pointed at the sun within 1 minute of arc. The image obtained is shown in Figure 11.

The similarity of the emissive X-ray centres with the centres of isophotes on a map of radio decimeter-wave emission was particularly striking. Furthermore, correlation

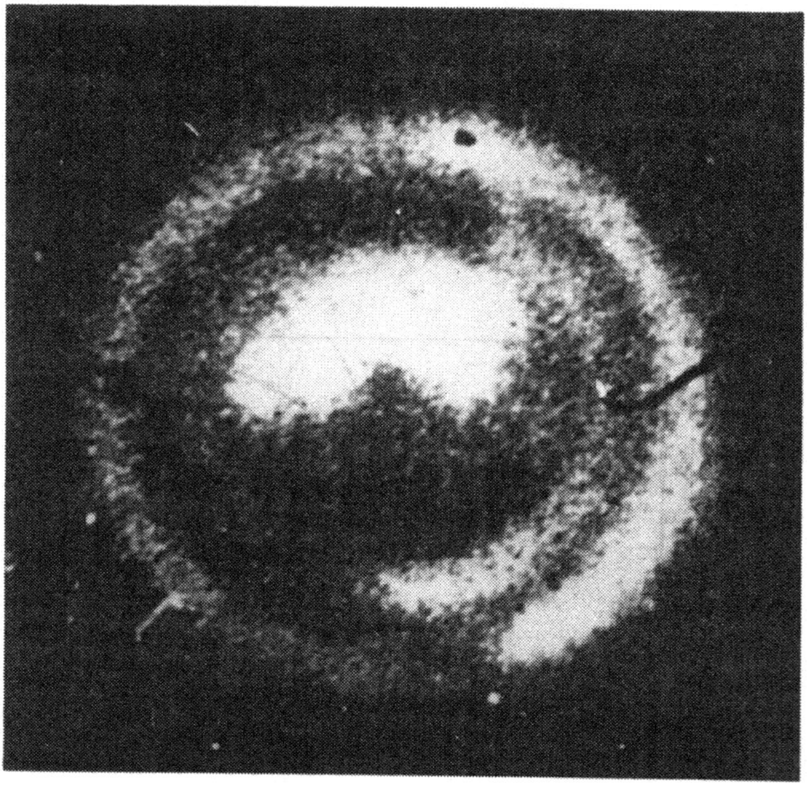

Fig. 11. First X-ray photograph of the sun.

with the more active plage regions in Ca-K was established, 75% of the X-emission being produced by plages accounting for less than 6% of the solar disk.

Moreover, the size of the X-ray sun was found 6% greater than the optical image; thus the X-ray emission extended to about 0.06 solar radius (43000 km) above the visible limb.

Very similar results were obtained on 11 August 1964 and on 17 December 1964 by Russell [17–18] and the group of Leicester using a triaxial attitude control unit on a Skylark rocket. The characteristics of their camera was as follows: pinhole diameter 125 μ; focal length 19 cm; exposure time 160 seconds. On Figures 12 and 13, one can see the good correlation between X-ray emission centres and the plages and flares observed optically.

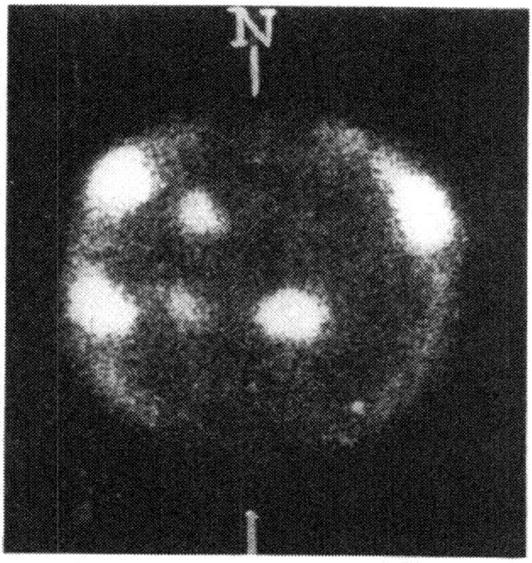

Fig. 12. Soft X-ray image of the sun.

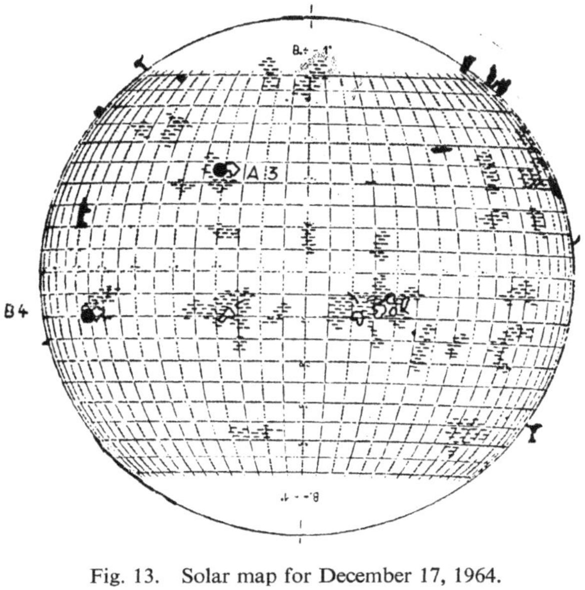

Fig. 13. Solar map for December 17, 1964.

These results have been confirmed by a rocket flight on 17 March 1965 by GIACCONI, LINDSAY and co-workers [19], who obtained the first photograph of the sun taken by a new instrument i.e. the optical X-ray telescope.

10. X-Ray Telescopes

Based on studies of WOLTJER [20] and described by GIACCONI and ROSSI [21] as early as 1960 this telescope is composed of two coaxial mirrors, one being parabolic the other hyperbolic. The X-ray paraxial beam is focussed on the axis (see Figure 14). The telescope produces a stigmatic image within a certain field of view.

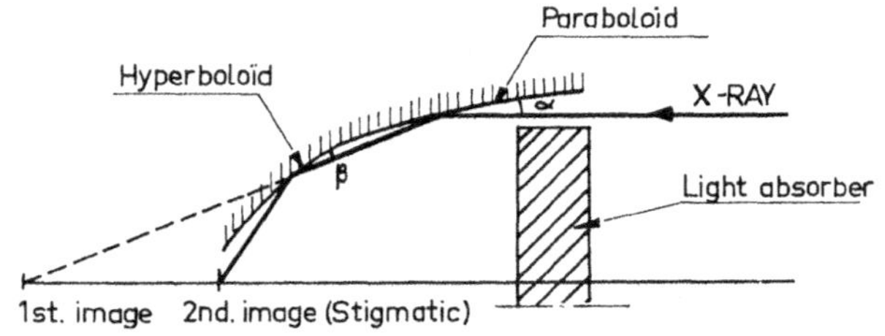

Fig. 14. Principle of X-ray telescope.

For very small angles of incidence X-rays are totally reflected. The maximum angle for which total reflection occurs is roughly proportional to the wavelength as indicated in Table III which pertains to a one-mirror telescope (hence astigmatic) [22].

TABLE III

Properties of X-ray mirrors for different wavelengths (from BOYD [22])

Max. angle of grazing incidence	0.0075	0.03	0.12	radian
Minimum wavelength reflected	2.5	10	40	Å
Possible mirror diameter	6	25	102	cm
Resulting collecting area	16	256	4130	cm^2

The parameters of the instrument flown by Giacconi *et al.* were as follows:

Diameter of telescope	7.6 cm
Focal length	83.6 cm
Collecting area	1.6 cm^2
1st reflecting angle	40 arc min
2nd reflecting angle	120 arc min
Diameter of solar image	0.75 cm
Efficiency of the double reflection (at 8 Å)	15%
Overall effective collecting area	0.24 cm^2

The exposure time was 50 sec, and a 0.15 mil mylar foil overcoated by 2200 Å aluminium was covering an Ilford G commercial X-ray film.

B. Rossi has presented at the Alpbach ESRO Summer School this first telescopic image of the sun on which can be clearly seen a flare-associated X-ray emission centre, a diffuse emission at the limb and several diffuse plages on the disk.

Once efficient electronic devices for recording and transmission of X-ray images are developed, X-ray telescopes equipped with these devices will probably become a very promising tool for the study of flares and other transient phenomena on the surface of the sun and in the solar corona.

Many photographs of the sun X-ray emission have been taken during 1965 and 1966, showing generally very strong correlation with McMath Calcium plages.

A new technique, which combines aboard the same rocket photographs and spectrographic X-ray measurements seems very interesting. A recent achievement of this type [44] is described later.

11. X-Ray Bursts, Solar Flares and S.I.D. Events

In 1937 Dellinger [23] had discovered that solar flares produced abrupt fade-out of short-wave radio communications. He attributed this fade-out to ionisation produced deep in the ionosphere by a burst of UV radiation. The name Sudden Ionospheric Disturbance (S.I.D.) was given to the phenomenon. Whereas ionisation at E- and F-region levels, 90–150 km and 150–400 km, respectively, increase ionospheric reflectivity, ionisation in the D-region (at 70–90 km) leads to absorption of the waves travelling between the earth and the E and F mirrors which may result in communication fade-out. At E and F layers reflection occurs because electrons oscillating under the action of the E.M. radio field can re-emit their energy before losing it by collision. But in the D-layer the pressure is between one hundred times and several million times higher than in the E- and F-layers, with the result that electrons lose their energy by collision with ions without having time to re-emit it. Under normal conditions electron density in the D-region is low (10^2–10^3 electrons/cm^3), hence the region is transparent to short waves. But any increase of ionisation enlarges the absorption effect. In 1937 solar X-rays were unknown and all ionospheric ionisation including the increase during flares was attributed to UV radiation.

The discovery that there was also a marked decrease (10 to 20 km) of the height of the base of the D-layer during flares and S.I.D. events suggested that UV radiation was not responsible for the increase in ionisation, since it would require an increase in the Lyman-α line of about 100000 times the normal value because of abrupt increase of absorption of UV with decreasing height in this region of the atmosphere. Since the flare area on the sun is less than one-thousandth of the area of the solar disk, the flare region would have to increase its UV output (mainly in the line Lyman-α) by a factor of 10^8 which seems very improbable.

On the other hand, the arrival of only 10^{-5} erg cm^{-2} sec^{-1} of X-rays in the 1–2 Å range would produce the observed lowering of the base of the D-region down to 65 km.

Therefore, it became highly desirable to fly experiments in order to see if during flares X-rays of high energy were emitted (we have seen that in quiet conditions X-rays below 2 Å are absent).

In order to ascertain that the extreme (very improbable) enhancement of UV light during flares was absent Lyman-α detectors should be included. Figure 15 shows the sensitivity of a typical detector, flown later by the Naval Research Laboratory aboard the solar monitoring satellite SR-1.

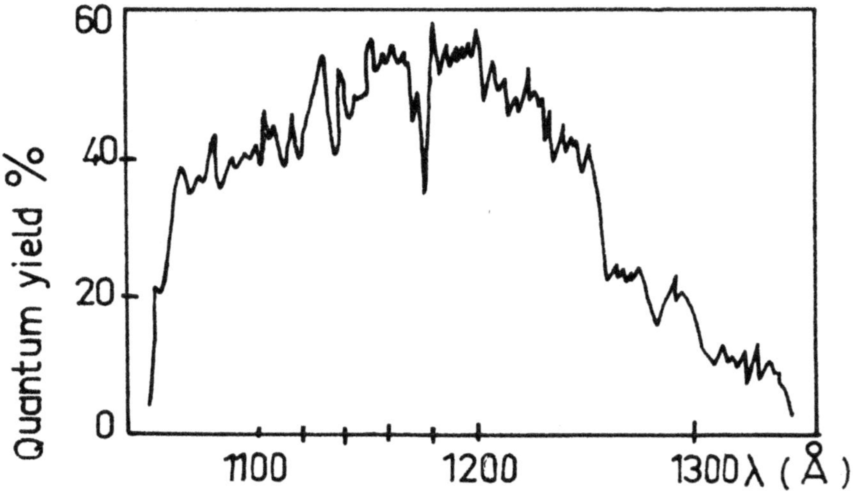

Fig. 15. Efficiency of Ly-α ion chamber on SR-1.

11.1 Rocket experiments

The earliest attempts to detect high energy X-rays during a flare were made by the Naval Research Laboratory in summer 1956 with the "Rockoon", a combination of a small, solid-propellant rocket carried aloft on a "Skyhook" balloon. The procedure was to launch a rockoon balloon in the morning from a ship and to keep it floating at 24 km altitude. When a solar flare was detected optically or indirectly by observation of a S.I.D., the rocket was fired by radio-command. A Geiger-Müller counter sensitive to the wavelength band between 1 and 12 Å was used together with a small detector for monitoring the UV flux.

Although this procedure was not very efficient it succeeded in observing one small flare (class 1^-). Data obtained during this flight indicated that during the flare there was

no increase in Lyman-α emission

an increase of X-ray emission in the wavelength region below 12 Å.

Further progress was made in 1959 when the Nike-Asp became available, a two-stage, rail-launched, solid-propellant rocket capable of transporting payloads of 25 kg to altitudes of 300 km. The Naval Research Laboratory flew rockets whose payloads included four X-ray detectors covering the wavelength range from 0.06 Å to 60 Å and,

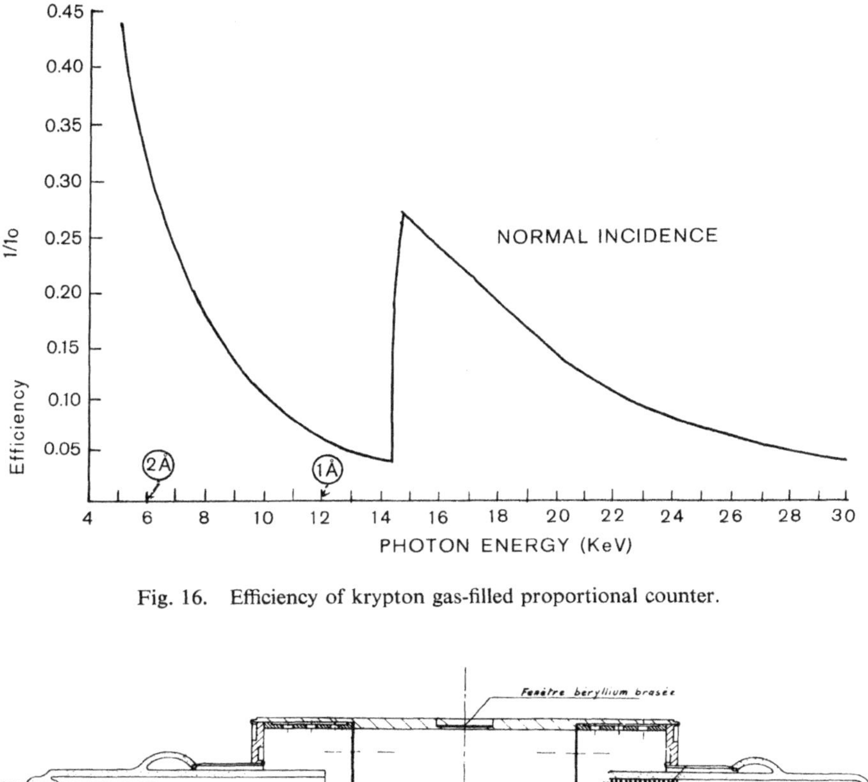

Fig. 16. Efficiency of krypton gas-filled proportional counter.

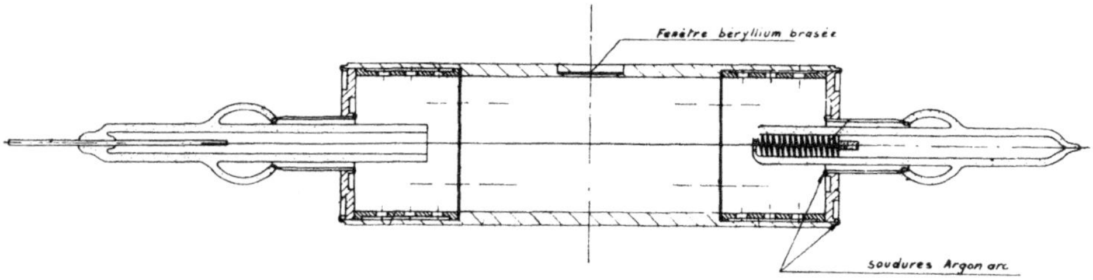

Fig. 17. Solar X-ray proportional counter.

for the first time, a proportional counter covering the range 2–20 keV and a scintillation counter for the detection of very hard X-rays (20–150 keV). The rockets were launched on alert, requiring only the push of a button to launch them when a flare was observed [24]. The efficiency of the proportional counter as function of wavelength is shown in Figure 16. Figure 18 shows the efficiency of a scintillation counter used for somewhat higher energies by Saclay workers. In Figures 17 and 19 one can see the

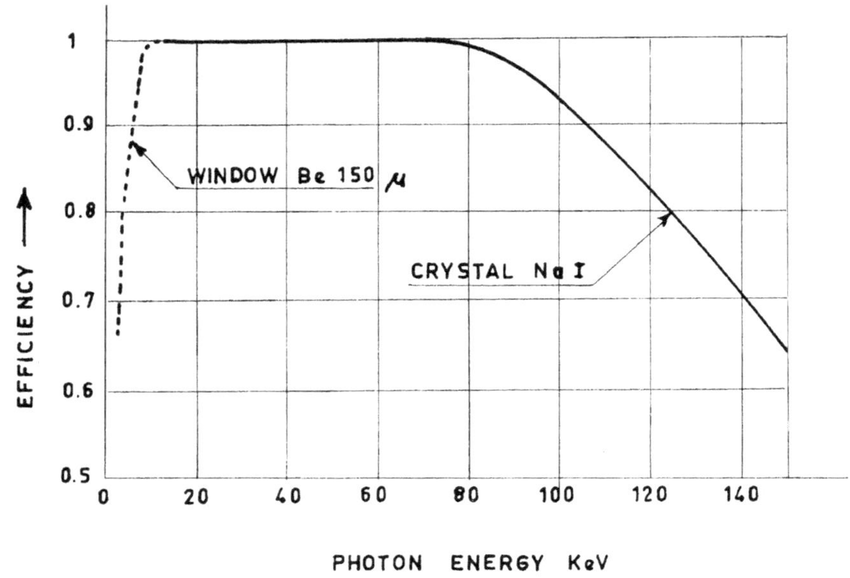

Fig. 18. Efficiency of the scintillation counter shown in Figure 19.

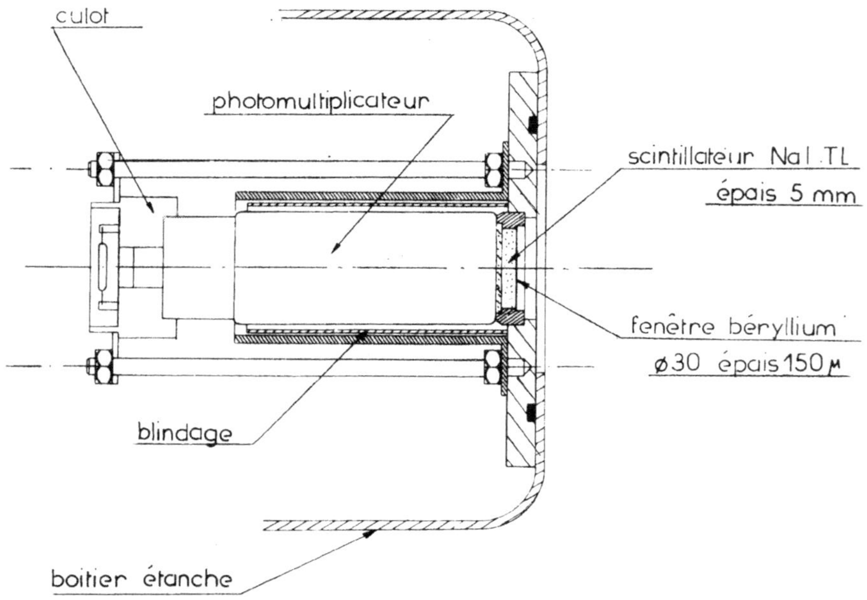

Fig. 19. X-ray scintillation counter.

proportional counter and scintillation detector which are used in France for similar experiments (see also Figures 20 and 21).

Three rockets were launched during flares of class 2^+ by the Naval Research Laboratory group.

The rocket launched on 31 August 1959 reached peak altitude only four minutes after the maximum of a class 2^+ flare. In the band 44–60 Å the recorded flux was 1.7 $\mathrm{erg\,cm^{-2}\,sec^{-1}}$. Moreover, X-ray quanta with energies between 20 keV and 70 keV

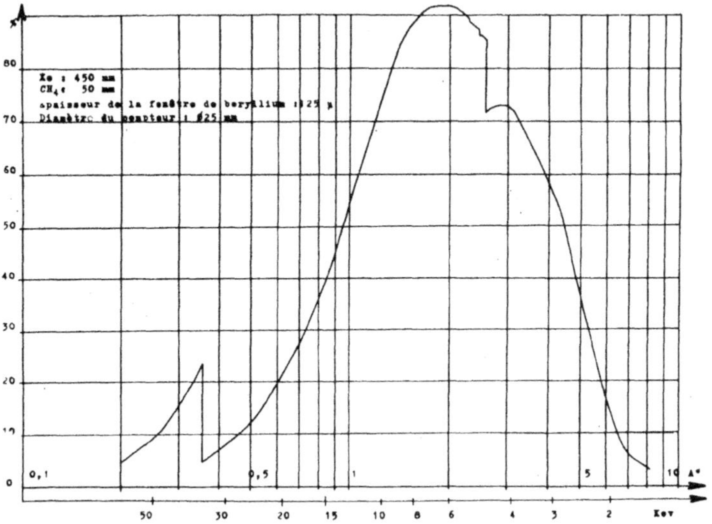

Fig. 20. Efficiency of proportional counter shown in Figure 17.

were observed with the scintillation counter (see Figure 22). These high-energy solar X-rays began to be observed at altitudes below 43 km during the ascent of the rocket, only two minutes after the optical flare maximum. During that flare the evolution of the X-ray emission was very fast: in a 3-minute interval beginning 2 minutes after the optical flare maximum the flux at a given wavelength had decreased by a factor of two. The energy radiated above 20 keV was about $4.5 \times 10^{-6}\,\mathrm{erg\,cm^{-2}\,sec^{-1}}$.

In addition, four rockets were launched during solar maximum but in absence of flares. At the time of the 1959 series of these Nike-Asp firings the average flux in the 44–60 Å band was found to be about 0.14 $\mathrm{erg\,cm^{-2}\,sec^{-1}}$ in the absence of solar flares.

At lower photon energies the proportional counter also yielded interesting results (see Figure 23). During quiet sun the flux is continuously decreasing towards high energies. During the flare of class 1^- the proportional counter spectra exhibit a large maximum at about 12 keV.

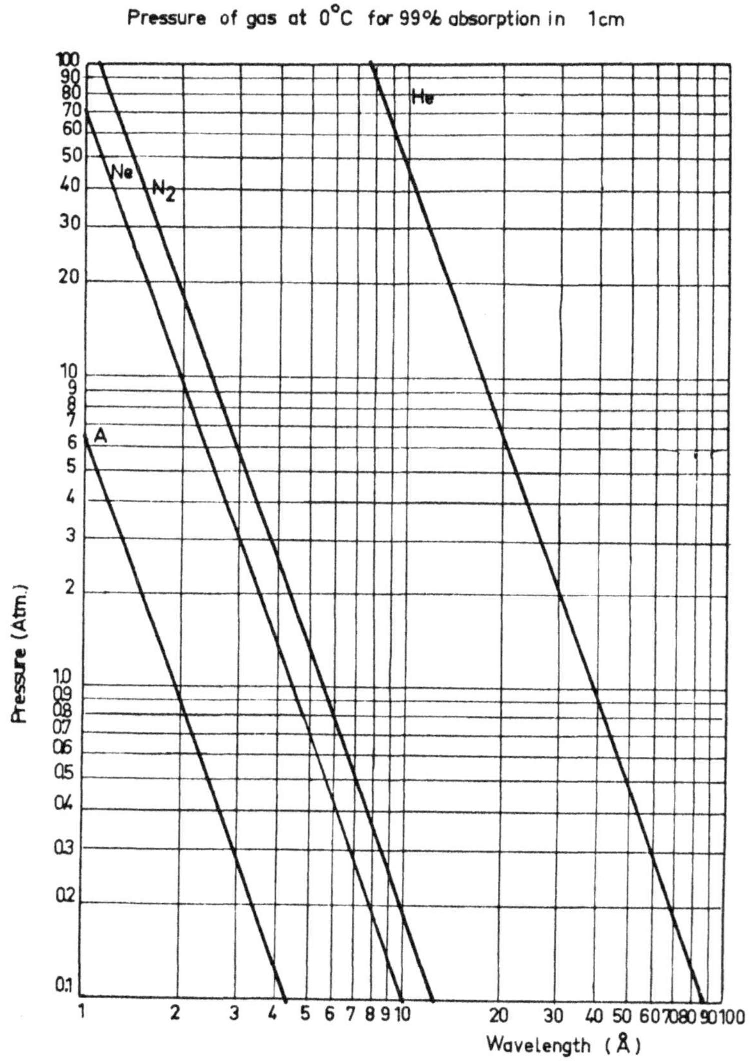

Fig. 21. Absorption of soft X-rays in gases.

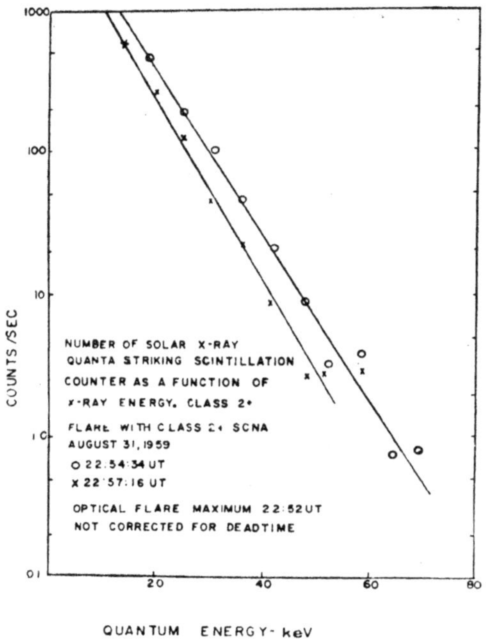

Fig. 22. Spectra of X-rays emitted during 2+ flare, 31 August 1959.

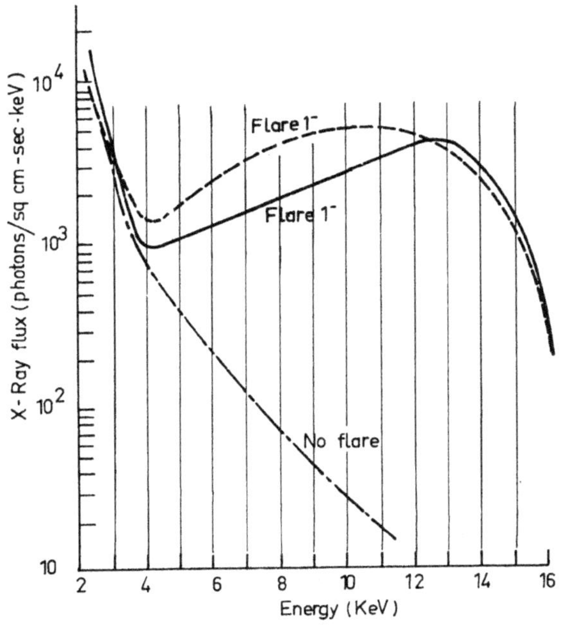

Fig. 23. Solar X-ray spectra obtained from proportional counters on three different flights.

The flux values in four bands covering the whole wavelength range analysed during typical quiet and flare conditions are given in Table IV (see also Figure 24).

Table V gives a summary of the results obtained from rocket flights of the Naval Research Laboratory [26].

TABLE IV

Solar X-ray flux (in erg cm^{-2} sec^{-1}) (from Chubb *et al.* [25]).

Wavelength Å	20–100	8–20	2–8	below 0.6
A) Quiet maximum				
14 August 1959 16.00 UT	0.6	0.018	0.00055	
B) Flare 2^{+}				
31 August 1959 22.53 UT	4.0	0.9	0.03	0.0000045

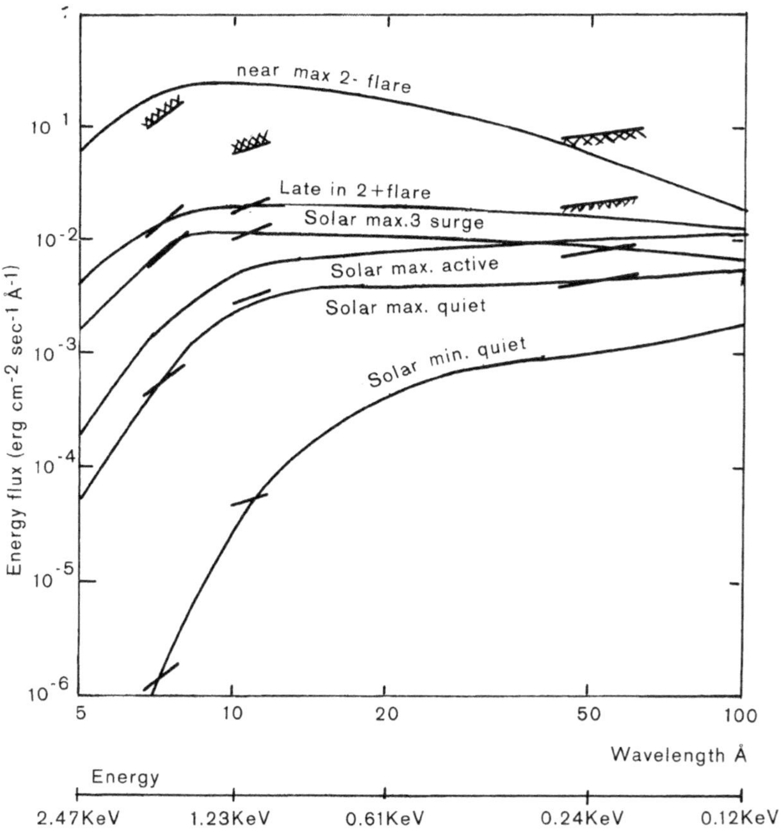

Fig. 24. Results of solar X-ray flux measurements obtained by Friedman *et al.* during the period 1950–1961.

TABLE V

Results obtained by Naval Research Laboratory rocket flights (from FRIEDMAN [26]).

Rocket	Time		Counts $cm^{-2} sec^{-1}$		Sun condition
			2–8 Å	8–20 Å	
V-2 49	1730	29/9/49	1.0×10^4		160 min after class 1
A-9	1459	1/5/52	495		Quiet solar minimum
A-10	1344	5/5/52	< 125		Quiet solar minimum
A-14	2240	15/11/53	< 40		Quiet solar minimum
A-15	1546	25/11/53	332		Quiet solar minimum
A-16	1529	1/12/53		4.5×10^4	Quiet solar minimum
A-34	2250	18/10/55		1.4×10^5	Quiet
D-8	1915	20/7/56	1.2×10^5		Late in class 1
8-69	1600	14/8/59	1.3×10^4	2.0×10^6	Quiet solar maximum
8-73	2253	31/8/59	$> 7 \times 10^5$	$> 1 \times 10^7$	Class 2^+

An important feature of these bursts of X-rays is that during a flare the increase of X-ray flux is small in the long wavelength region but becomes more and more pronounced as higher energies are considered: the X-ray spectrum becomes "harder" during flares.

A beautiful result on a class 2^+ flare was obtained by the NRL group on Sept. 1, 1959, flare [43]. For the first time the rocket flight began 2 minutes before the maximum of the Hα flare. X-ray energy spectra from 20 keV to 100 keV were analysed at different times during the flare evolution: rapid changes occurred in the spectral energy distributions and are interpreted as a rapid cooling of a thermalized electron plasma. The data indicate initial local plasma temperatures of about 285 million degrees falling to about 160 million degrees in 1.5 minute and decaying more gradually thereafter.

The energy flux at earth integrated over the whole flight corresponds for $h\nu > 20$ keV to 1.2×10^{-4} erg cm^{-2} sec^{-1}.

The maximum of equivalent plasma temperature was observed during the first spectrum (73 to 92 sec after the beginning of the flight) and, therefore, this maximum occurred at least half a minute before the Hα-maximum. The last spectrum was taken during the time interval 395–399 sec. During the period between 168 and 399 sec the X-ray spectrum and intensity remained the same (and probably would have been observed during many tens of minutes after), showing a very slow cooling of the gas.

The mean output of X-ray energy from the flare above 20 keV during the period 73–399 sec was about 3.4×10^{23} ergs sec^{-1}. If the flare X-ray emission decay has followed the flare optical light decay (the optical-flare disappeared 130 minutes after the maximum), the total output of X-ray (>20 keV) energy integrated over the whole flare is about 10^{27} ergs.

11.2 BALLOON EXPERIMENTS

The process of inflating balloons of large volumes (50000 m^3) takes about one hour. Another hour is needed to reach top altitude. Hence large balloons cannot be launched on alert, as rockets can. On the other hand they can float at an altitude of about 40 km for many hours. Therefore, in periods of high solar activity there is a reasonable probability of observing X-ray events of high photon energy (>20 keV) which, as we now know are coincident with large optical Hα-flares.

PETERSON and WINCKLER [27] of the University of Minnesota have observed a burst of very high energy radiation (γ-rays) during a cosmic-ray balloon flight at 10 gm/cm^2 residual pressure, coincident in time with a class 2 solar flare. The flare occurred on 20 March 1958. The γ-ray burst occurred in less than 18 seconds (which was the time resolution of the Geiger counter and ionisation chamber). From the ratio of the outputs of the two instruments they deduced a photon energy of 0.2 to 0.5 MeV, which was much higher than any energy previously observed.

Moreover the γ-ray burst was coincident with a very strong 3-cm radio-burst; several other terrestrial events occurred at the same moment, i.e. S.I.D., an earth current disturbance and a magnetic crochet.

The γ-ray flux was estimated to have been

$$2.8 \times 10^{-5}\ \mathrm{erg\,cm^{-2}\,sec^{-1}} \quad \text{if the energy was 500 keV}$$

or

$$1.6 \times 10^{-5}\ \mathrm{erg\,cm^{-2}\,sec^{-1}} \quad \text{if the energy was 200 keV.}$$

The burst was attributed to "bremsstrahlung produced in the solar photosphere by electrons of 0.5 to 1 MeV energy". The authors also assumed that the electrons which spiral in a 1000 gauss field in the flare region can produce the observed radio burst by betatron radiation (of which only 0.01% escaped from the flare region toward the earth because of absorption in the solar plasma).

On 12 October 1960, VETTE and CASAL [28] from the Convair Research Laboratory at San Diego, on a balloon flight at 4.8 g/cm^2 over Arizona, observed two X-ray bursts with energies exceeding 80 keV during a class 2 and a class 1 flare; rough spectral information as well as the complete history of these events and their correlations with terrestrial phenomena were obtained*. The detector was a NaI scintillation counter pointed at the sun. It was mounted in a box of lead with small holes giving a field of view of 6° half-angle. For energy discrimination there were three pulse-height channels corresponding to 20, 80 and 150 keV lost in crystal. As the process by which photons lose energy in a crystal is essentially photoelectric in nature these figures represent the energy of the photons which had entered the crystal.

In Figure 25 are shown the values of the flux received in the three wavelength bands as function of time.

On 28 September 1961 two balloons (one flown over Minneapolis, Minn., by Winckler and the other over Flin Flon, Manitoba by Anderson from a Berkeley Cosmic Ray Group), both equipped with scintillation counters and three channel pulse-height analysers covering the 20 to 150 keV energy region, recorded a beautiful burst of X-rays [29] at precisely the same time (see Figures 26 and 27).

* Other terrestrial events observed during these flares: SEA (Sudden enhancement of atmospherics, due to the increase of reflectivity of ionospheric layers for radio noise generated by lightnings), SCNA (Sudden cosmic noise absorption) and radio-storms of type III, V, I, II and IV.

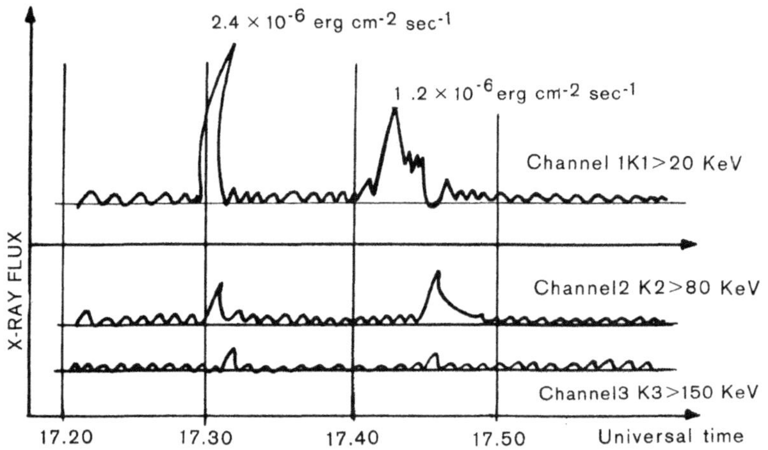

Fig. 25. Event of 12 October 1960.

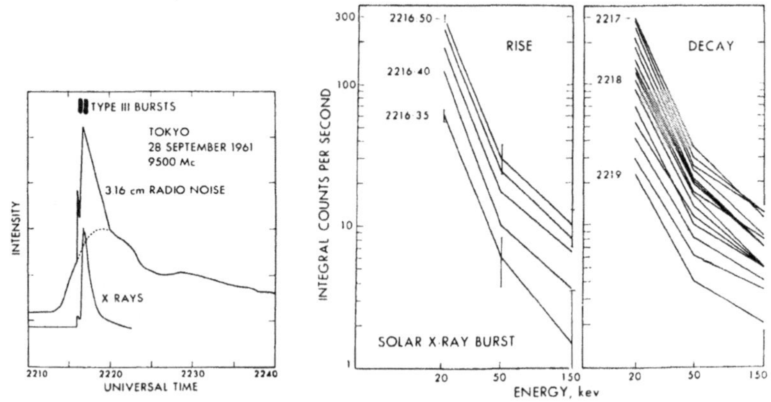

Fig. 26. Solar flare X-ray burst, 28 September 1961.

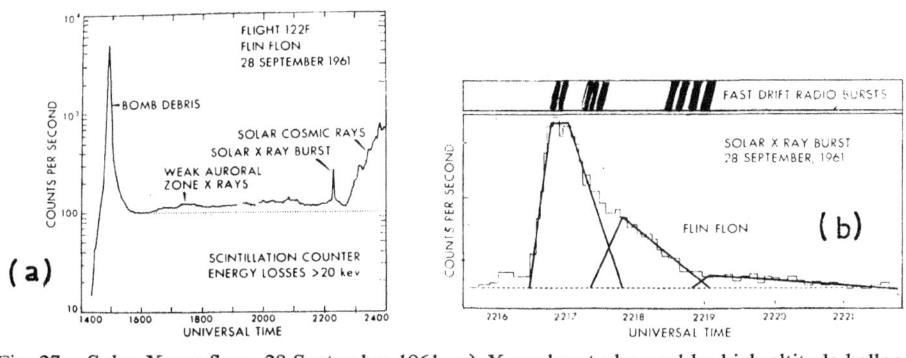

Fig. 27. Solar X-ray flare, 28 September 1961. a) X-ray burst observed by high-altitude balloon. b) Detailed intensity variation of X-rayburst and comparison with simultaneous type III radio bursts.

Within 1 minute this burst was in coincidence with a large 3-cm radio-burst recorded by the Tokyo radio-telescope (Figure 26). In Figure 26 we can also see that the beginning of this radio-burst occurred at 22.16 hours, about 10 min. before the maximum brightness of the class 3 or 3^+ optical flare, whose beginning (in Hα light was at 22.02 hours.

An interesting feature of the X-ray burst is that the spectral shape between 20 and 150 keV is almost entirely independent of the intensity of the X-rays (Figure 26b). This observation is of great importance because it shows that the bremsstrahlung is not of a "thermal" nature, contrary to the conclusions drawn from Friedman's results which seemed to be compatible with a kind of gray body spectral distribution (see Figure 23).

The two bursts of X-rays and 3-cm radio waves were attributed by Anderson and Winckler to bremsstrahlung and synchrotron radiation, respectively, produced in the low (10000 km) and in the high (100000 km) corona.

Anderson and Winckler observed that the X-ray and 3-cm radio bursts were coincident with short metre-wavelengths bursts, which are seen on top of Figure 27b. These radio bursts are known as Class III bursts. They are characterized by short duration (< 1 min.) and a drift of their frequency towards lower values. They are due to plasma oscillations in the upper corona which in turn are generated by streams of relativistic electrons. The two authors suggested that the same electrons which had produced the X-ray and 3-cm radio bursts in the lower corona could also be ejected into the high corona at relativistic speeds, thereby generating the plasma oscillations which produce the type III radio bursts. But this interpretation meets with difficulties. One of the main difficulties, as pointed out by De Jager and Kundu [30], is that the same electrons which have produced the bremsstrahlung and synchrotron radiation would have to preserve their energy in order to excite type III oscillations at higher altitudes in the corona.

Measurements with sounding rockets above 40 km are limited to about 5 minutes. On the other hand, balloon observations, although restricted to high-energy X-rays, have long observation times (hours), which makes it possible to observe the entire course of an X-ray event and to correlate it to radio and optical observations with relatively good accuracy.

12. X-Ray Measurements on Board of Satellites

12.1 The Naval Research Laboratory Satellite SR-1

a) *Instrumentation*

The first satellite equipped with an X-ray detector was SR-1 (Greb 1 or 1960 Eta 2), launched on 22 June 1960. It supplied useful data for about 5 months [31].

The X-ray detector was a simple d.c. ionisation chamber with a Be window having an area of 2.33 cm^2 and a thickness of 0.125 mm (25 mg/cm^2) filled with argon at atmospheric pressure, whose depth (2.54 cm) was equivalent to 4.5 mg/cm^2. The efficiency as a function of wavelength is shown in Figure 28. It was greater than 10% between 1 and 7 Å.

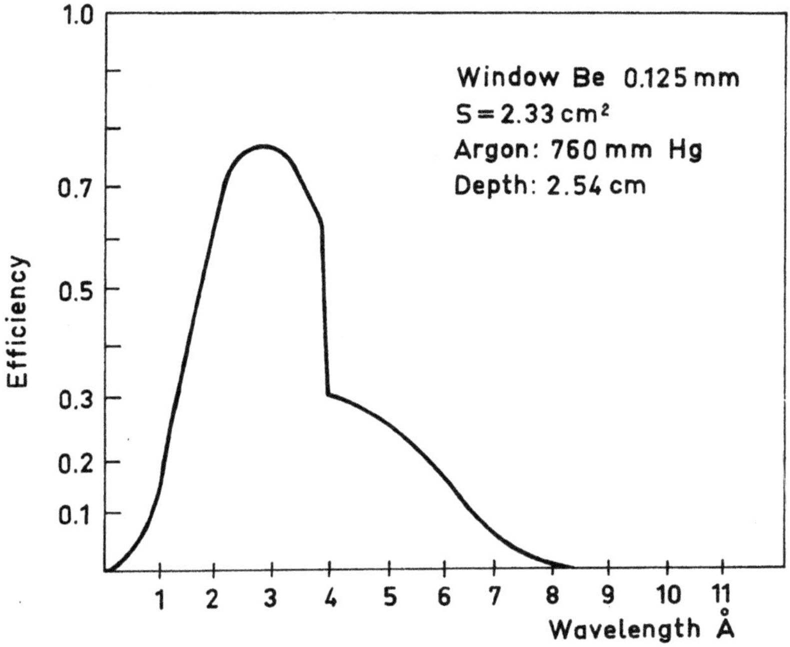

Fig. 28. Efficiency of the SR-1 ionization chamber.

A Lyman-α ionisation chamber and an aspect photocell were also mounted in the equatorial plane of the satellite. The spectral sensitivity of the Lyman-α chamber which had a LiF window of area $2 \times 9.4 \times 10^{-5}$ cm^2, 1 mm thick, and which was filled with NO at a pressure of 15 mm Hg is shown in Figure 15.

An Alnico magnet (2400 gauss) was used to deflect Van Allen belt electrons. It served to bend away from the Be window of the X-ray counter (which otherwise would have been sensitive to electrons of energy greater than 0.15 MeV) the electrons with energy less than 1 MeV.

The axis of the magnet was in the equatorial plane of the satellite. The magnet interacted with the terrestrial field causing the spin axis of the satellite to precess in a large cone with a period of 8 days. Owing to this precession and to other effects, useful data were transmitted (in real time) only 1.2% of the total flight time.

b) *Experimental results*

The spectrum of the X-rays was unknown. The flux received in the band 2–8 Å was computed on the assumption that the spectrum was that of a gray body with characteristic temperature of 2×10^6 °K.

The results show that the normal X-ray flux between 0 and 8 Å from 22 June to 1 November 1960 was smaller than 0.6×10^{-3} erg cm^{-2} sec^{-1}.

In addition, about 100 short-lived flux increases were observed. All of them occurred during solar optical flares observed on the disk or during unusual limb events. In 12 of these events the flux was greater than 1.5×10^{-3} erg cm^{-2} sec^{-1}.

The Lyman-α flux was found to be constant at 6 erg cm^{-2} sec^{-1} within 11% even during flares. This result refutes the hypothesis of any Lyman-α flare playing a role in S.I.D. phenomena, thus confirming the earlier rocket measurements. Correlation with terrestrial observations show that a S.I.D. appears when the 0–8 Å flux exceeds 2.0×10^{-3} erg cm^{-2} sec^{-1}.

The results may be summarised as follows:

i) From 22 June 1960 to 1 November 1960 the flux in the wavelength band 0 to 8 Å was less than 0.6×10^{-3} erg cm^{-2} sec^{-1} (assuming that the X-rays are emitted by a gray body at 2×10^{6} °K).

ii) Most of the time when the flux (0 to 8 Å) was larger than 0.6×10^{-3} erg sec^{-1} cm^{-2}, one or more distinct solar events were visible on the sun (during one or two days, there was only a generally high level of activity without noticeable event).

iii) When the flux (0 to 8 Å) was larger than 2×10^{-3} erg sec^{-1} cm^{-2}, short-wave fadeout and other ionospheric effects were observed on earth.

iv) Significant changes in X-ray flux were found to take place on a time scale of about 1 minute.

v) Long-duration X-ray events of moderate intensity sometimes accompanied rising limb prominences.

12.2 The satellite Ariel (UK-1)

a) *Instrumentation*

This satellite was equipped for X-ray measurements by workers of the University of Leicester and of the University College, London. Data obtained in the period 26 April to 15 May 1962 have been analysed [32].

For detecting solar X-rays Ariel carried a proportional counter filled with a mixture of Ne and CH_4. The window was covered by a foil of 25 μ beryllium [33] whose area was about 10^{-3} cm^2. For spectral analysis there were 5 pulseheight channels (4.7 to 13.8 Å).

The aims were to study the time variations in both normal and flare-enhanced emission. In order to resolve rapid variations, the time resolution of the electronic circuits was 25 sec and 1 min. However, even this relatively short time proved too long.

The lowest detectable flux (limited essentially by spurious counts due to cosmic ray particles) was:

$$3 \times 10^{-4} \text{ erg cm}^{-2} \text{ sec}^{-1} \quad \text{below 14 Å}$$
$$1 \times 10^{-5} \text{ erg cm}^{-2} \text{ sec}^{-1} \quad \text{below 8 Å}.$$

b) *Experimental results*

About 15 spectra per day (under non-flare conditions) were obtained during the three weeks mentioned above. The results can be summarised as follows (see Figures 29 and 30):

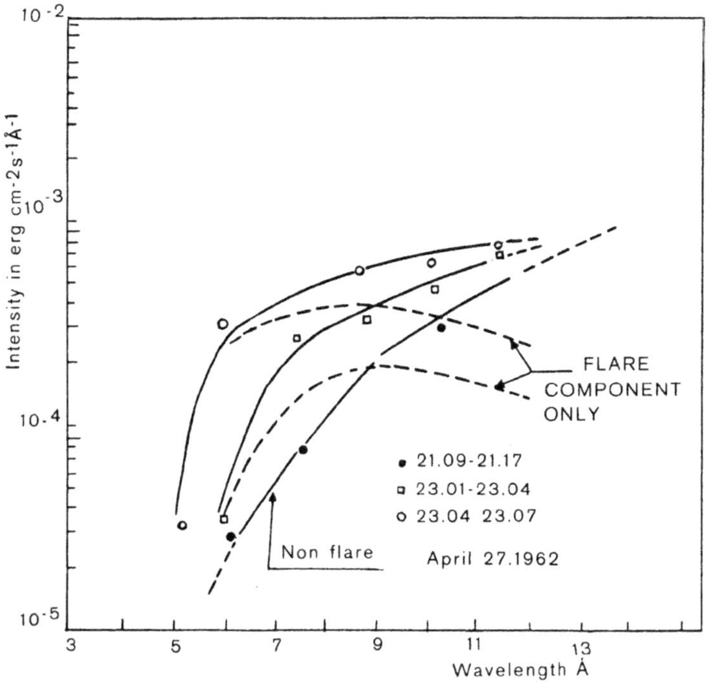

Fig. 29. Typical hardening of X-ray spectrum during a flare 1 or 1^- (Ariel).

i) *Non flare conditions.* Typical values of the equivalent gray body temperature were:

13–11 Å	1.4×10^6 °K
11– 9 Å	1.6×10^6 °K
9– 7 Å	1.7×10^6 °K

while the integrated intensity was found to be

3.0×10^{-3} erg cm^{-2} sec^{-1} in the 8–14 Å band

1.3×10^{-4} erg cm^{-2} sec^{-1} below 8 Å

ii) *Flare conditions.* Intensification and hardening of the pre-flare spectrum. The degree of flare enhancement varied much from one flare to another.

X-ray emission as function of time was not smooth, there were many peaks.

Both thermal and non-thermal processes seem to occur in the X-ray production of some flares.

A progressive hardening is visible, for example on Figure 29 where the X-ray flux of a small flare 1^- without Class III radio emission [34] is shown. In Figure 30 a similar hardening is seen occurring during a Class II flare.

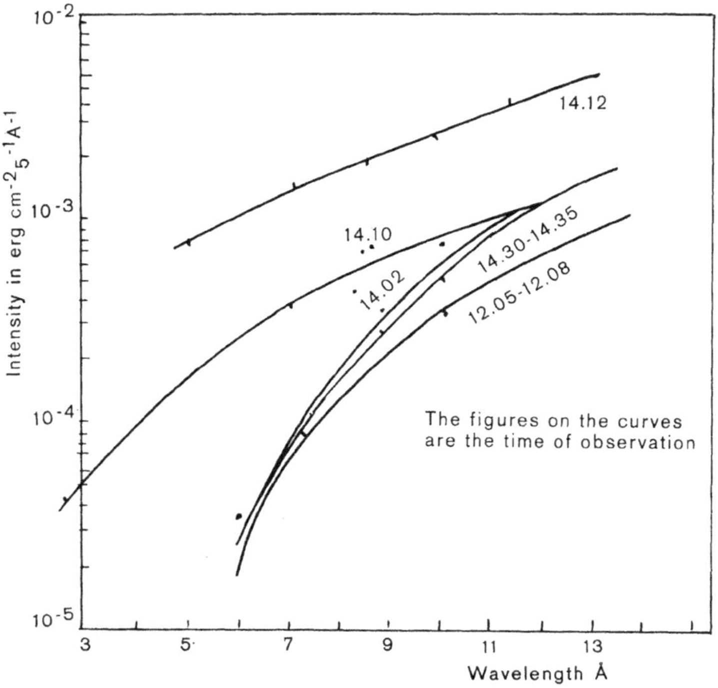

Fig. 30. Development of X-ray flux in a class 2 flare, 27 April 1962 (Ariel).

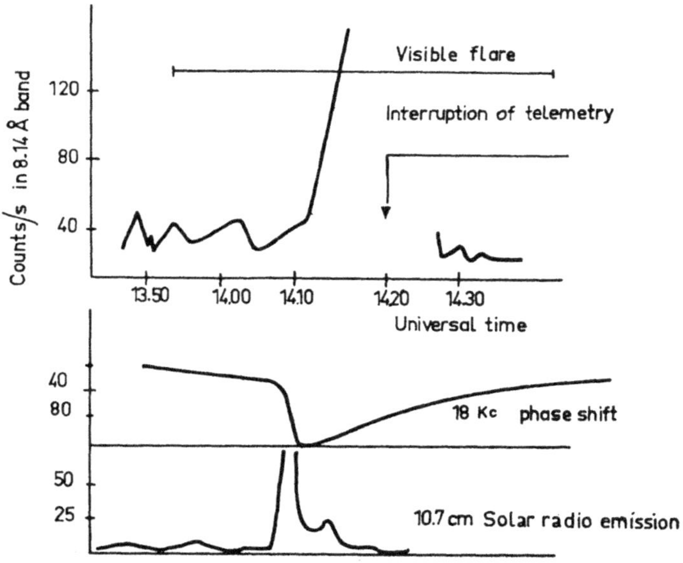

Fig. 31. Events associated with class 2 flare of 27 April 1962.

In Figure 31 is shown the correlation of the same X-ray flare (which was associated with an optical Class 2 flare, as well as a Class III radio burst S.I.D., S.P.A., etc.) with the time variation of the amplitude of the phase shift at 18 kHz. Also shown are the variations of the optical flare and of the solar 10.7 cm radio emission.

12.3 The OSO-1 satellite

This satellite carrying special instruments for solar radiation measurements was launched on 1606 UT, 7 March 1962, in a 550-km circular orbit. It allowed for the first time continuous measurements during 1 hour on each orbit with accurately pointed instruments.

a) *Instrumentation*

For the measurement of X-rays the following instruments were included on the payload:

A spectrometer for the 10–400 Å range [35] composed of a grazing-incidence concave grating and an open Bendix photomultiplier. More than 500 spectra have been recorded but they are not yet published.

An ion chamber measured the integrated X-ray flux in the range 0.1–10 Å. This ion chamber was filled with Xe and had a Be window [35]. Its efficiency as a function of wavelength is given in Figure 32. It has a maximum between 1 and 5 Å.

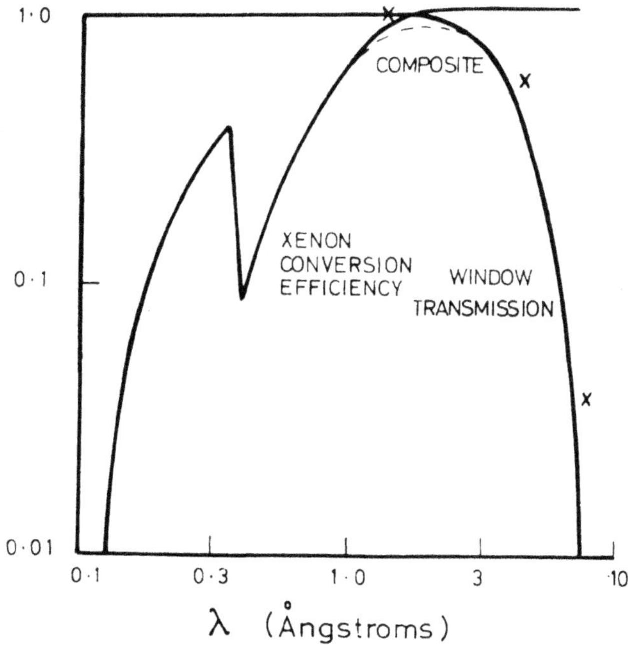

Fig. 32. Efficiency of ion chamber on OSO-1.

b) *Experimental results*

A slowly-varying component of the solar X-ray flux was observed which correlates well with the slowly-varying component of 2800 MHz solar radio flux and which can be accounted for by localised sources having the same extent as Ca plages. In Figure 33 is shown the slowly-varying components of 2800 MHz solar flux and of solar X-rays with wavelength smaller than 11 Å during the period 7 March to 29 May 1962. Moreover, it was concluded that the ratio of energy contained in lines to the energy contained in the continuum is at least 10, and probably 30, in the wavelength region smaller than 11 Å.

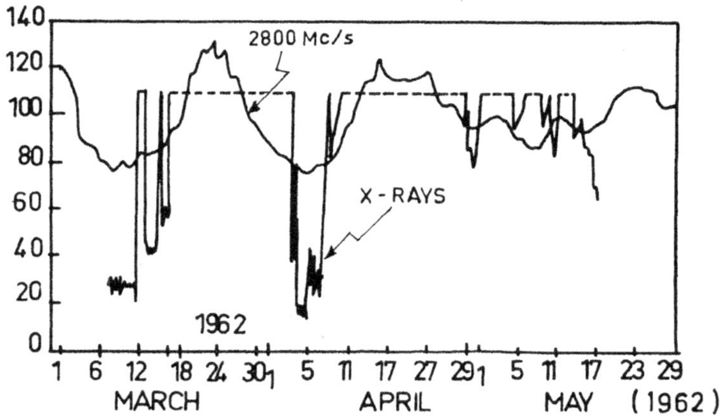

Fig. 33. Slowly varying components of 2800 MHz solar flux and $\lambda < 11$ Å solar X-rays (OSO-1).

Transient events (X-ray flares) lasting from 10 minutes to a few hours were frequently observed. Correlations with Hα flares, S.I.D., and 2800 MHz transients show that the OSO-1 X-ray experiment is far more sensitive than monitors of ionospheric effects or monitors of the solar radiation in the visible or radio wavelengths.

Bursts of hard X-rays in the 20–100 keV range were detected by a scintillator [36] (NaI-Tl crystal, 0.3 cm high and 2.24 cm diameter, in a copper shield of 1 cm thickness). The 3.8 cm^2 area of the front surface was kept normal to the solar direction. An aluminium disk 0.08 cm thickness covered the front surface of the crystal. Efficiency in the 20–100 keV range was approximately 0.90. Counting rate of the whole band was telemetered through a logarithmic rate-meter.

Five flares were observed, each lasting about 2 min. with flux increases 10 to 30 times over the background, in correlation with class 1^+ or class 2 optical flares, all of them associated with microwave bursts.

These X-ray flares and their association with microwave radio bursts have been analysed by Kundu [37]. No type III radio burst occurred during the X-ray events observed by OSO-1.

12.4 The NRL solar radiation monitoring satellite GREB-5 (1964 1D)

This X-ray monitoring satellite [38] was launched in January 1964. It is equipped with the standard ionisation chambers of the Naval Research Laboratory covering the ranges 1–8 Å, 8–12 Å, 44–60 Å. Data were transmitted in real time and several European laboratories have successfully recorded the telemetry. The magnetic broom for protection against trapped electrons is shown in Figure 34. In January and February 1964 the X-ray flux was as low as the minimum value measured by rockets during the last sunspot minimum period in 1953 and 1954. In the 1–8 Å band there was no detectable emission in excess of 10^{-4} erg cm^{-2} sec^{-1} in the absence of solar flares. In the 8–12 Å band the average flux was 4×10^{-4} erg cm^{-2} sec^{-1} and at 44–60 Å, 2.5×10^{-2} erg cm^{-2} sec^{-1}.

Fig. 34. The magnetic broom used with the 44–60 Å photometer.

Many bursts of X-rays were observed; an example of the variation of the X-ray flux during flares in different energy bands is given in Figure 35. This satellite brought about the discovery of the "Solar X-ray noise storm" which can last several days. This storm produces strong variations (several hundred percent) even at wavelengths in the vicinity of 50 Å. Typical variations of X-ray fluxes on the four wavelength bands during a class 1 optical flare are shown in Figure 36.

12.5 The Russian satellites "space-ships" II and III

a) *Instrumentation*

Mandel'shtam *et al.* [39] from the Lebedev Physical Institute have obtained data from end-window G.M. counters with mica and beryllium windows aboard Space-ship II on 19–20 August 1960, and on 1–2 December 1960 aboard Space-ship III. These counters were sensitive below 10 Å.

Transmitted data at the end of the 24-hour flight on 19–20 August 1960 indicated a flux of about 8×10^{-4} erg cm^{-2} sec^{-1} below 10 Å. During the flight of 1–2 December 1960 the mean flux value was 2.5×10^{-4} erg cm^{-2} sec^{-1}.

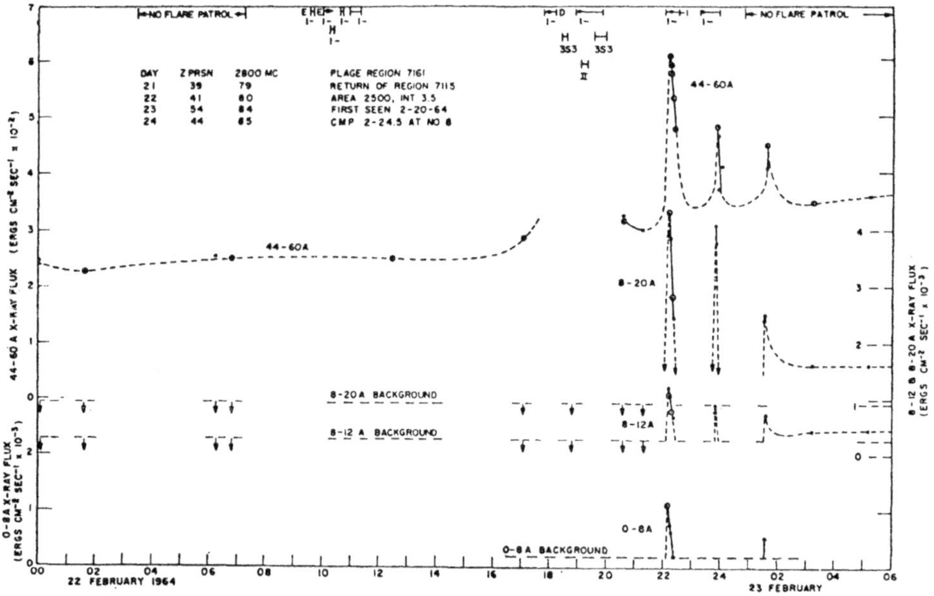

Fig. 35. Reduced solar X-ray flux data for February 22 and 23.

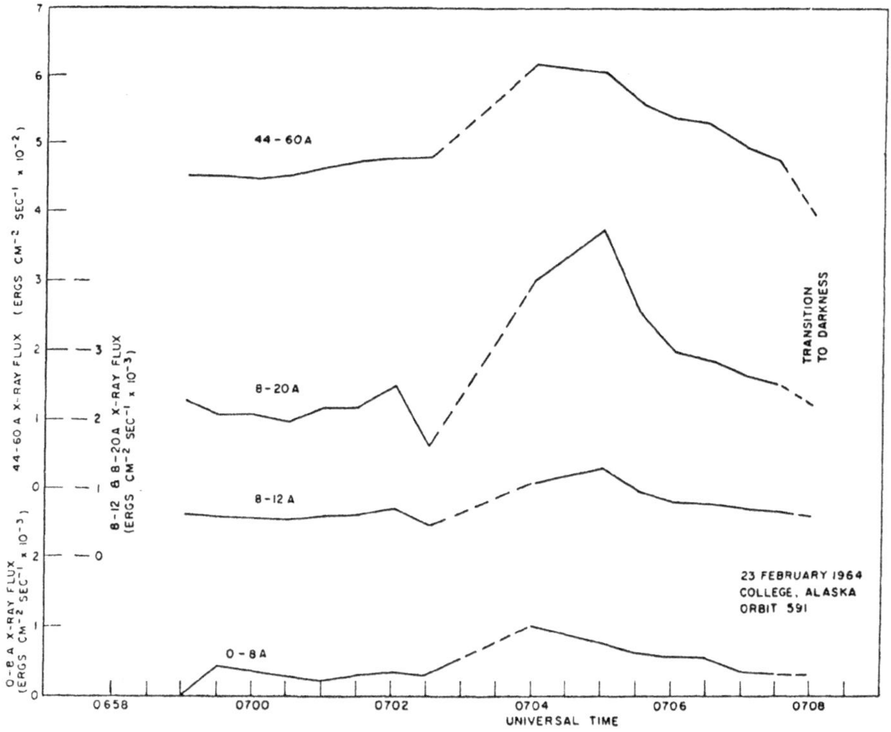

Fig. 36. Reduced telemetry record from College, Alaska made on 23 February 1964.

The flux in the region 2–10 Å was estimated at about 10^{-2} erg cm^{-2} sec^{-1} during a flare which occurred on 19 August 1960.

Russian physicists concluded that at an altitude of 200–300 km in the latitude range between approximately 35°N and 36°S, X-rays of solar origin are the main "hard" radiation. At higher latitudes, radiation caused by trapped particles is superimposed on the solar X-rays: at these higher latitudes special particle G.M. counters begin to count and an increase in X-rays as measured by the first counters is observed. A description of these counters is given by MANDEL'SHTAM *et al.* in [40].

Aboard Space-ship II another type of detector was also included, i.e. an open secondary-emission multiplier with BeO dynodes. By means of various filters (Al, Be, Cu and organic material) [41], a rough energy determination of the X-rays [42] was made.

b) *Experimental results*

The results may be summarised as follows:

i) Region 44–100 Å (CH_n filter): radiation always constant within ±8%, at 1.5×10^7 pulses sec^{-1} cm^{-2} = 0.15 erg cm^{-2} sec^{-1}.

ii) Region 8–21 Å (Al filter): constant most of the time at 6.2×10^4 pulses sec^{-1} cm^{-2}, but during flares one increase of 63% and one of 320% were observed.

iii) Region 5–10 Å (Be filter): radiation is very unsignificant during quiet sun but increased sharply (by factor 11) during the flare of 19 August at 1540 Moscow time.

iv) Region 1.4–3 Å (Cu filter): only background counts of non-solar origin.

In agreement with the results of Friedman *et al.*, Efremov *et al.* found the intensity of Lyman-α to be 5–6 erg cm^{-2} sec^{-1} which remained constant during the flare of 19 August.

For the quiet sun, the value $T = 0.9 \times 10^6$ °K gives the closest approximation of the equivalent gray body temperature to the results obtained with Al and CH_n filters.

Recent observations

As a first example of a recent achievement one can describe briefly a rocket flight on May 20th 1966 from White Sands [43]. The experiment package pointed at the sun by a control system stabilised about all three axes carried two Woltjer type I (conical approximation) glancing incidence telescopes to photograph the sun in wavelength regions (determined by band pass filter) between 3 and 75 Å, and two proportional counters to obtain flux data and rough spectral shapes in the region 2–11 Å and 8–11 Å. The spatial resolution obtained was about 20 arc seconds. The sun was moderately active on the 20th May and X-ray emission from both the quiet regions and from active centers is observed.

At short wavelengths (3–11 Å) only the most active regions show strong emission. These bright regions correspond with some of the calcium plages. These regions produced many small flares and strong radio emission during their passage across the disk.

In the 8–20 Å region, other regions appear in addition, always corresponding to Calcium plages. There is also a weak general emission from the disk, and faint limb brightening is visible. The 8–20 Å pictures have a more diffuse character than those taken in any other wavelength band.

In the 27–40 Å region which includes the strong line C VI Lyman-α at 33.7 Å (Tousey *et al.* [44]) all the numbered McMath plage areas show appreciable X-ray emission which corresponds closely in size and shape with the CaK areas. In addition, there are fainter patches of emission corresponding to weak unnumbered plage areas. Limb brightening is very pronounced, as is the darkening of the south polar cap above latitudes around 60°. The north pole above latitude 75° is occupied by a "tuft" of X-ray emission which has the appearance of a bunch of plumes, but between latitudes 60° and 75°N there is a dark gap which appears to be the remnant of the north polar darkening. On the NW limb in the region of the most northerly prominence an arch-like structure joins the north polar tuft with the lower latitude activity. This feature is also emitting at 9.1 cm. The X-ray feature extends about 2.1×10^5 km above the photosphere.

From the combined results of the X-ray photographs, the proportional counter data, and the data from other experiments performed on May 20th, the authors hope to obtain information on the electron density and temperature in different regions of the corona.

A second interesting example of recent work is given by the paper of Zhitnik *et al.* [45] who have flown successfully two rockets at 500 km on 20th September and 1st October 1965, with pinhole cameras and grazing incidence spectrograph. They were able to obtain simultaneously 18 photographs with wavelengths bands ranging from less than 10 Å to 200 Å and spectrograms from 9.5 Å to 72.26 Å.

For the first time, on a 1st October flight, the beginning of an X-ray flare has been photographed (and resolved in lines). This flare occurred in a region of the corona which had no apparent disturbances in the Hα and CaK lines. Many lines and groups of lines from the region where the flare occurred have been observed but some only are identified, because the resolution was only about 0.1 Å.

The flux in the flare for the group of lines 10.9 to 11.9 Å (Ne IX, Na X) is about 5×10^{-3} erg cm^{-2} sec^{-1}, and the flux of the line 14.9 Å (Fe XVII) is about 10^{-2} erg cm^{-2} sec^{-1}.

The fluxes in different wavelength bands are given for this flare on Table VI, together with flux values for two other regions of the sun and for the whole undisturbed corona. One of these two regions had given an optical Hα-flare of class 1 with a beginning, a maximum and an end respectively 18 min, 16 min and 2 min 20 sec before the beginning of the X-rays photographs. These two regions were also seen in Hα and CaK lines.

The appearance of these lines during a flare shows that it is not the synchrotron radiation which is responsible of the high energy X-ray emission, and that most of this energy comes from deeply ionized atoms, probably present in large amount at high altitudes in the sun's corona.

TABLE VI

The X-ray flux in the solar corona on 1st October 1965, in $erg\,cm^{-2}\,sec^{-1}$ [45]. Exposure time: from 03 h 59 m 20 s to 04 h 05 m 10 s UT.

Radiation region	$\lambda < 25$ Å	$\lambda < 20$ Å	$\lambda = 8$–15 Å	$\lambda < 10$ Å	Remarks
Bright region 1	2.6×10^{-1}	1.75×10^{-1}	5×10^{-2}	2×10^{-2}	X-ray flare begin: 04h 05 min
Bright region 2	$> 1 \times 10^{-1}$	$> 1 \times 10^{-1}$	$> 3 \times 10^{-2}$	1.5×10^{-2}	Optical flare class I End: 03h 57 min UT
Bright region 3	3×10^{-3}	3×10^{-3}	7×10^{-4}		
Undisturbed corona		1.5×10^{-2}	2×10^{-3}		

13. General Conclusion

During the last fifteen years many observations of solar X-rays have been carried out. Some important results were obtained, particularly the hardening of the spectrum during flares, the correlation of flare X-rays with radio emission and the great time variability of high-energy events.

Satellites are now engaged in a routine work in order to obtain more detailed data on the slow variations of the average flux of X-rays and on flares correlated with radio, optical and geophysical short-duration events.

However, much remains to be done; the most exciting problem in the years to come is probably the investigation of high-energy events. The exact location of the X-ray emission centres, the determination of their structure, the study of time- and energy variation will be achieved more and more accurately by means of increasingly refined spectral measurements and with the help of X-ray optics.

The investigation of the high-energy end of the sun's electromagnetic radiation will give more interesting results than the optical observations because the X-ray flux is much less constant, hence more correlated to other phenomena. Furthermore, the solar X-rays being closely connected with solar flares will yield important clues for the analysis of these still mysterious phenomena.

References

[1] De Jager, C.: Lecture given at this Summer School (see pp. 101-124 of this volume).
[2] Boyd, R. L. F.: 1965, 'Techniques for the measurement of extraterrestrial soft X-radiation', *Space Sci. Rev.* **4,** 35.
[3] Mandel'shtam, S. L., Tindo, I. P., Voron'ko, Yu. K., Vasil'ev, B. N., and Shurygin, A. I.: 1963, 'Investigations of the sun's X-radiation, II, Measurements with space satellites', *Planet. Space Sci.* **11,** 61.
[4] These papers are available on request from the Secrétariat du Service d'Electronique Physique, C.E.A., Boîte Postale No. 2, 91 Gif-sur-Yvette, France.
[5] Lyot, B.: 1930, *Compt. Rend. Acad. Sci.* **191,** 834.
[6] Hoyle, F. and Bates, D. R.: 1948, *Terr. Magnet. Atmos. Electr.* **53,** 51.
[7] Burnight, T. R.: 1949, *Phys. Rev.* **76,** 165.

[8] FRIEDMAN, H., LICHTMAN, S. W., and BYRAM, E. T.: 1951, *Phys. Rev.* **83,** 102.
[9] FRIEDMAN, H.: 1960, 'The sun's ionizing radiations', *Physics of the Upper Atmosphere* (J. RATCLIFFE, ed.), Academic Press, New York.
[10] CHUBB, T. A., FRIEDMAN, H., and KREPLIN, R. W.: 1960, 'Measurements made of high-energy X-rays accompanying three class 2+ solar flares', *J. Geophys. Res.* **65,** 1831.
[11] FRIEDMAN, H.: 1955, *Ann. Geoph.* **11,** 60.
[12] BYRAM, E. T., CHUBB, T. A., and FRIEDMAN, H.: 1956, *J. Geophys. Res.* **61,** 251.
[13] SMITH, H. J. and SMITH, E. V. P.: 1963, *Solar Flares*, McMillan, New York.
[14] CHUBB, T. A., FRIEDMAN, H., KREPLIN, R. W., BLAKE, R. L., and UNZICKER, A. E.: 1961, *Mem. Soc. Roy. Sci. Liège* **IV,** 228.
[15] MANDEL'SHTAM, S. L., TINDO, I. P., VORON'KO, JU. K., SHURYGIN, A. I., and VASIL'EV, B. N.: 1963, 'Measurements of solar X-ray radiation', *Space Res.* **3,** 822.
[16] BLAKE, R. L., CHUBB, T. A., FRIEDMAN, H., and KREPLIN, R. W.: 1963, 'Interpretation of X-ray photograph of the sun', *Astrophys. J.* **137,** 3.
[17] RUSSELL, P. C.: 1965, 'Soft X-ray image of the sun', *Nature* **205,** 684.
[18] RUSSELL, P. C.: 1965, 'Further soft X-ray images of the sun', *Nature* **206,** 281.
[19] GIACCONI, R., LINDSAY, J. S., MUNEY, W. S., REIDY, W. P., and ZEHNPFENNIG, T.: 1965, *Astrophys. J.* **142,** 1274.
[20] WOLTJER, H.: 1952, *Ann. Physik* **10,** 94.
[21] GIACCONI, R. and ROSSI, B.: 1960, *J. Geophys. Res.* **65,** 773.
[22] BOYD, R. L. F.: 1964, in ESRO Scientific Report No. 1.
[23] DELLINGER, J. H.: 1937, 'Sudden ionosphere disturbances', *Terr. Magnet. Atmos. Elect.* **42,** 49.
[24] CHUBB, T. A., FRIEDMAN, H., and KREPLIN, R. W.: 1964, 'Spectrum of solar X-ray emission from 2–20 keV during subflare activity', *Space Research* **4,** 759.
[25] CHUBB, T. A., FRIEDMAN, H., and KREPLIN, R. W.: 1961, 'X-ray emission accompanying solar flares and nonflare sunspot maximum conditions', *Space Research* **1,** 695.
[26] FRIEDMAN, H.: 1962, *Reports on Progress in Physics.* **XXV,** Institute of Physics and the Physical Society, London, p. 163.
[27] PETERSON, L. E. and WINCKLER, J. R.: 1959, *Phys. Rev. Letters* **1,** 205, and *J. Geophys. Res.* **64,** 697.
[28] VETTE, J. I. and CASAL, F. G.: 1961, *Phys. Rev. Letters* **6,** 334.
[29] ANDERSON, K. A. and WINCKLER, J. R.: 1962, 'Solar flare X-ray burst on September 28, 1961', *J. Geophys. Res.* **67,** 4103.
[30] DE JAGER, C. and KUNDU, M. R.: 1963, 'A note on bursts of radio emission and high energy (> 20 KeV) X-rays from solar flares', *Space Res.* **3,** 836.
[31] KREPLIN, R. W., CHUBB, T. A., and FRIEDMAN, H.: 1962, 'X-ray and Lyman-alpha emission from the sun as measured from the NRL SR-1 satellite', *J. Geoph. Phys.* **67,** 2231.
[32] BOWEN, P. J., NORMAN, K., SANFORD, P. W., POUNDS, K. A., and WILLMORE, A. P.: 1964, 'Measurement of the solar spectrum in the wavelength band 4–14 Å', *Proc. Roy. Soc.* **A 281,** 538.
[33] POUNDS, K. A. and WILLMORE, A. P.: 1963, 'Instrumentation of satellite UK-1 for obtaining low-resolution solar X-ray spectra', *Space Res.* **3,** 1195.
[34] BEHRING, W. E., NEUPERT, W. M., and LINDSAY, J. C.: 1963, 'Preliminary solar flare observations with a soft X-ray spectrometer on the orbiting solar observatory', *Space Res.* **3,** 814.
[35] WHITE, W. A.: 1964, 'Solar X-rays: slow variations and transient events', *Space Res.* **4,** 771.
[36] FROST, K. J.: 1965, 'Comments on high-energy X-ray bursts observed by OSO-1', *Space Res.* **5,** 513.
[37] KUNDU, M. R.: 1963, 'Centimeter-wave radio and X-ray emission from the sun', *Space Sci. Rev.* **2,** 438.
[38] KREPLIN, R. W.: 1965, 'NRL solar monitoring satellite – Description of instrumentation and preliminary results', *Space Res.* **5,** 951.
[39] MANDEL'SHTAM, S. L., VASIL'EV, B. N., VORONKO, J., TINDO, I., and SHURYGIN, A.: 1963, 'Measurements of solar X-ray radiation', *Space Res.* **3,** 822.
[40] MANDEL'SHTAM, S. L., TINDO, I. P., VORON'KO, YU. K., SHURYGIN, A. I., and VASIL'EV, B. N.: 1962, 'Investigation of the sun's X-radiation, I, Measurements with the aid of geophysical rockets', *Planet. Space Sci.* **9,** 977.
[41] EFREMOV, A. I., PODMOSHENSKII, A. L., EFIMOV, O. N., and LEBEDEV, A. A.: 1963, 'Investigations of solar X-rays and Lyman-alpha radiation on August 19–20, 1960', *Space Res.* **3,** 843.

[42] Efremov, A. I., Podmoshenskii, A. L., Ivanov, M. A., Nikiforov, V. N., and Efimov, O. N.: 1962, 'Filter apparatus for studying the sun's short-wave emission', *Planet. Space Sci.* **9**, 987.
[43] Underwood, J. H. and Muney, W. S.: 1967, 'A glancing incidence, solar telescope for the soft X ray region', *Solar Phys.* **1**, 129–144.
[44] Tousey, R., Austin, W. E., Purcell, J. D., and Widing, K. G.: 1965, 'The extreme ultraviolet emission from the sun between the Lyman-α lines of H I and C VI', *Ann. Astrophys.* **28**, 755.
[45] Zhitnik, I. A., Krutov, V. V., Maljavkin, L. P., Mandel'shtam, S. L., and Chereumkhin, G. S.: 1967, 'The X-ray photographs and the spectrum of the sun in the region 9.5–200Å, *Space Res.* **7**, 1263.

OBSERVATION OF STELLAR X-RAY SOURCES

B. ROSSI

*Based on lectures delivered by B. Rossi of the Massachusetts Institute of Technology, Cambridge, Mass., U.S.A.**

Abstract. The limitations of X-ray astronomy due to absorption and production of secondary photons in the atmosphere are discussed. All measurements have to be conducted from rockets, satellites, or balloons. The experiments which have been carried out so far are mentioned. In the second section the detectors employed (gas counters, scintillation crystals and photoelectric devices) are described as well as ancillary instrumentation. The analysis of data obtained from rocket flights is extensively discussed. In the last section the results of the observations performed until now are presented and compared in detail. The present knowledge on extra-terrestrial non-solar X-ray sources is critically reviewed.

1. Introduction

The discovery of strong X-ray sources outside the solar system was one of the first surprising results of space astronomy. X-rays coming from outer space cannot be observed with ground-based instruments, as is evident from Figure 1 which shows the absorption of X-rays in the earth's atmosphere. It is seen that observations of X-rays with energy smaller than 12 keV can only be made with rockets or satellites, while in the range 15–200 keV measurements with balloon-borne instruments are feasible. For still higher energies, the transmission continues to increase.

On the other hand, the luminosity of celestial X-ray sources appears to be a more rapidly decreasing function of energy than the intensity of X-rays produced by cosmic rays in the atmosphere. Hence, for $E > 1200$ keV the true celestial X rays are difficult to distinguish from the cosmic ray induced background.

Figure 2 shows that, outside the atmosphere, X-ray astronomy is limited by attenuation of the radiation in interstellar space: for $\lambda < 6$ Å the whole galaxy is essentially transparent, while X-rays with $\lambda > 100$ Å can reach the earth only from sources within a distance of the order of 100 light-years.

Non-solar X-rays were observed for the first time by Giacconi and his collaborators [1] in 1962 with a rocket which carried several Geiger-Müller counters with wide fields of view. The experiment revealed the existence of a strong source in the constellation of Scorpio near the galactic centre and a possible second source in Cygnus. Subsequent rocket flights confirmed these results and suggested still another source in the general region of the Crab Nebula [2]. In April 1963, a group of the Naval Research Laboratory under H. FRIEDMAN [3] also observed the Scorpio source. The improved spatial resolution in this experiment set an upper limit of 5° to its angular diameter and determined the location with an accuracy of 2°. This source is certainly not coincident with the galactic centre which was below the rocket's horizon

* Material for these lectures was prepared by B. Rossi in collaboration with R. Giacconi, H. Gursky and J.R. Waters of American Science and Engineering and with G. Clark, G. Garmire, M. Oda and M. Wada of the Massachusetts Institute of Technology.

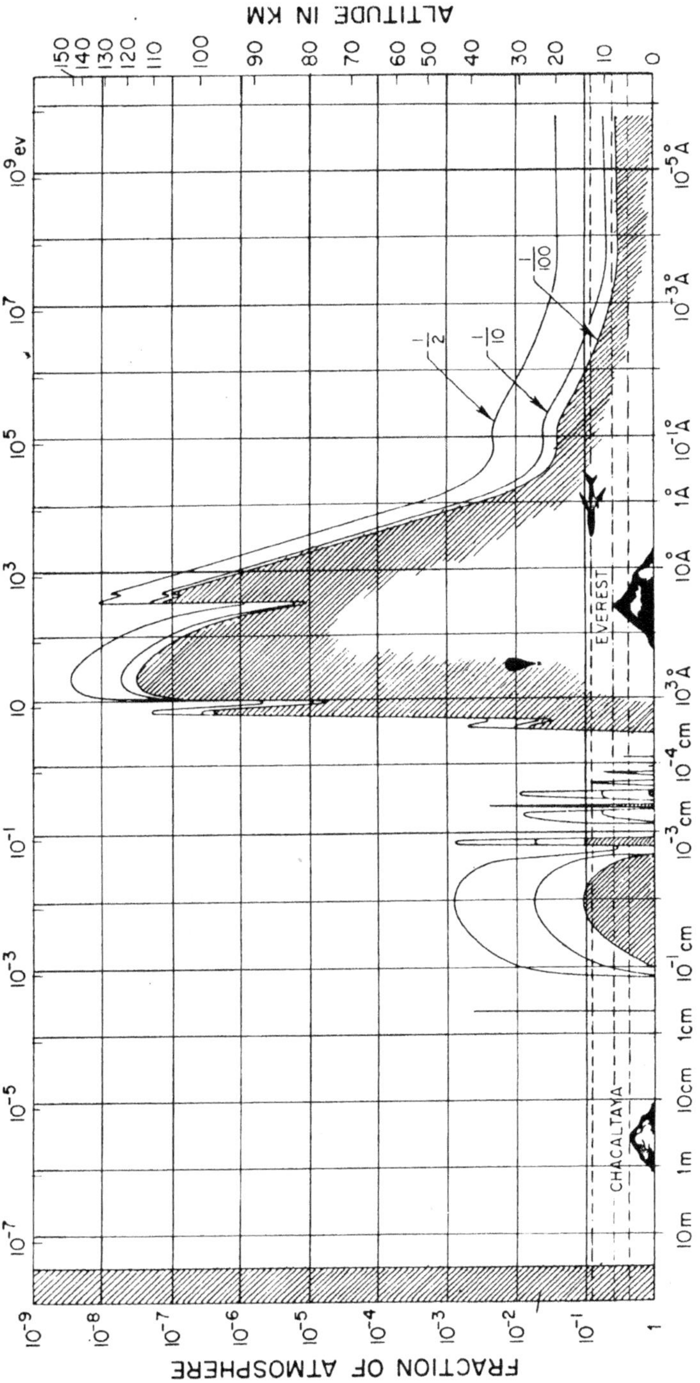

Fig. 1. Attenuation of radiation in the atmosphere.

during the whole flight. The same flight provided evidence for the existence of a weaker source in Taurus coincident with the Crab Nebula within 2°.

The successful observation of the lunar occultation of the Crab Nebula in July 1964 by the NRL group using a pointing rocket proved that the X-ray source in Taurus

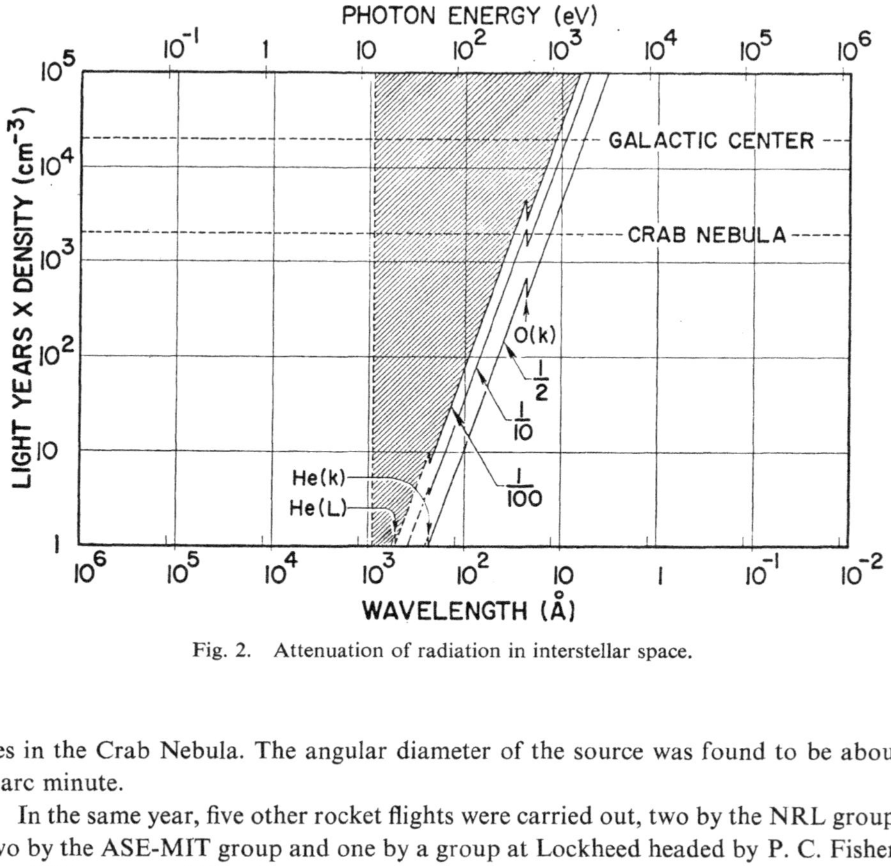

Fig. 2. Attenuation of radiation in interstellar space.

lies in the Crab Nebula. The angular diameter of the source was found to be about 1 arc minute.

In the same year, five other rocket flights were carried out, two by the NRL group, two by the ASE-MIT group and one by a group at Lockheed headed by P. C. Fisher. In addition to established results, a number of sources close to the galactic equator within 20° of the galactic centre and two additional sources in the constellations of Cygnus and Serpens were detected. Figure 3 shows the region of the sky which has been explored so far and the approximate locations of the X-ray sources. The only unexplored regions are those near the Magellanic clouds.

2. Instrumentation

2.1 X-RAY DETECTORS

Because of the weak fluxes of non-solar X-ray sources, detectors of high efficiency and large area (10–1000 cm^2) are required. Three different types have been used.

a) *Gas counters*

Figure 4 shows an example of a thin-window gas counter. The operation of a gas counter is based on the photoelectric effect of the X-photons in the gas and on the cascade multiplication of the ions in the electric field between the wire and the counter wall. The window of the counter must be sufficiently transparent to X-rays and also withstand the pressure of the counter gas. The efficiency of a counter is the product

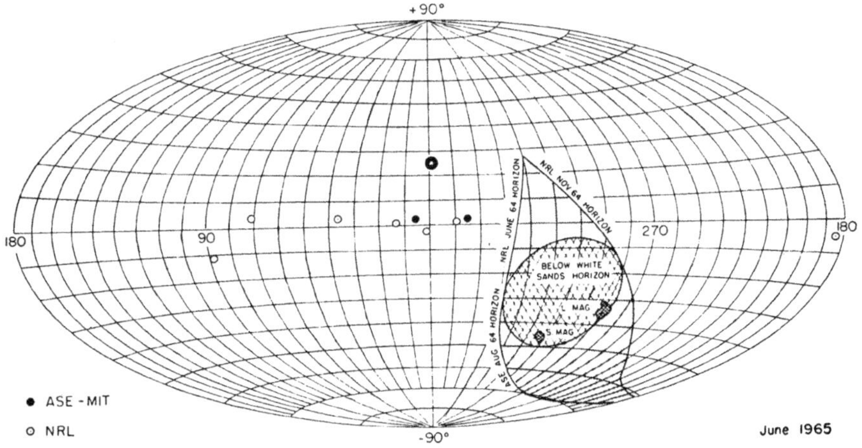

Fig. 3. X-ray source map in galactic coordinates.

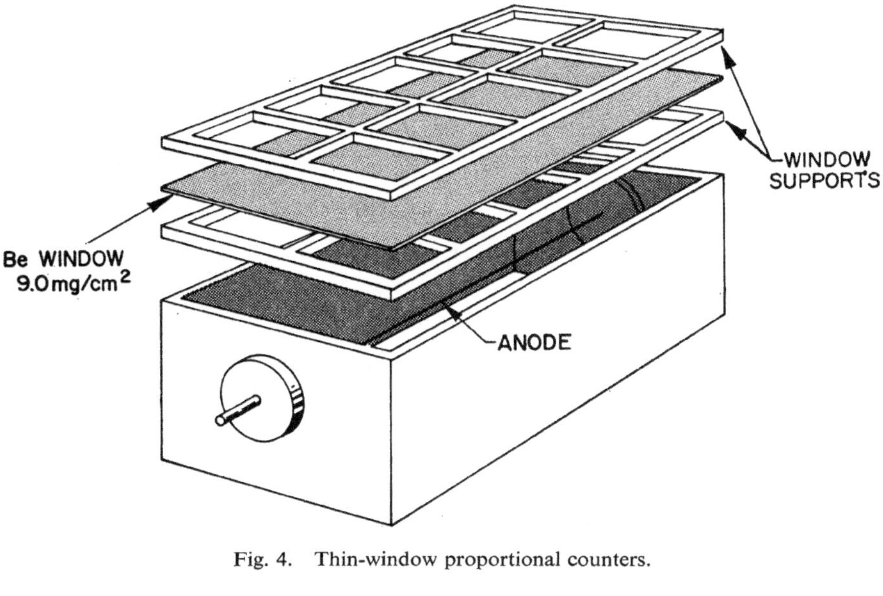

Fig. 4. Thin-window proportional counters.

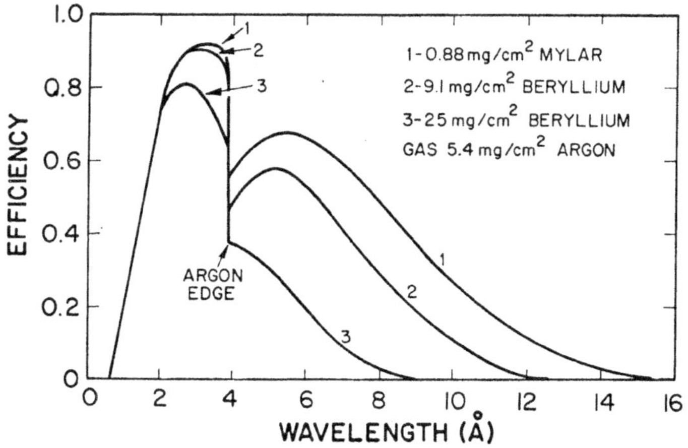

Fig. 5. Efficiency of gas counters.

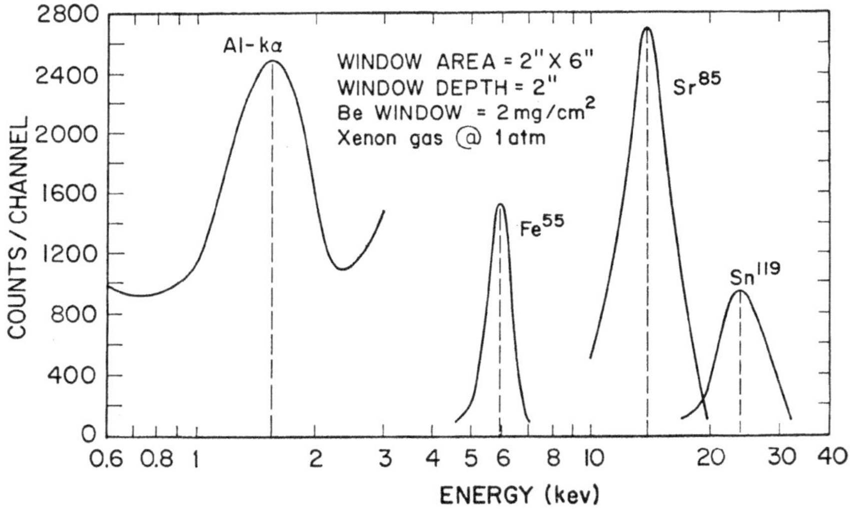

Fig. 6. Response of proportional counters to monoenergetic X-rays.

of the probability for an X-ray photon to traverse the window multiplied by the probability for absorption of a photon in the gas. Considering the efficiency as a function of wavelength, it is seen that the upper limit of the sensitive region is determined by the window becoming opaque and the lower limit by the gas becoming transparent. Figure 5 shows efficiency of gas counters with Argon filling and different windows.

The magnitude of the voltage between the wire and the wall of the counter determines whether the counter works as a proportional counter or as a Geiger-Müller counter (the first mode requires a lower tension than the second). In the Geiger range all pulses given by the counter are of equal size, while in the proportional range the size of the pulses is proportional to the energy of the incident photon. A typical response curve of a proportional counter is shown in Figure 6.

The average number of electrons, N_{av}, created in one pulse is given by

$$N_{av} = \frac{E}{E_0}$$

where $E = h\nu$ is the photon energy and E_0 is a characteristic energy which depends on the nature of the gas (for Argon $E_0 = 27$ eV). Thus a proportional counter can be used for pulse height spectrometry. However, the fluctuations in N_{av} become large for small photon energy. Therefore, the resolution decreases with decreasing photon energy.

In gas counters, noise is produced by UV photons, cosmic rays and particles of low energy. The counter can be protected against ultraviolet light by a suitable window. In order to reduce the cosmic ray background, the counter can be embedded in an anticoincidence shield which consists of a block of scintillating plastic material. However, low energy particles, especially electrons, cannot be eliminated and represent a source of background counts.

b) *Scintillation counters*

Figure 7 shows a scintillation counter. Incident X-ray quanta produce ionization in a thin slab of scintillating material. Photons emitted by recombination will reach the photocathode of the photomultiplier, thereby releasing N electrons. The photomultiplier output is closely proportional to N. The average energy per photoelectron is of the order of 1000 eV; it is thus much higher than the energy required for the creation

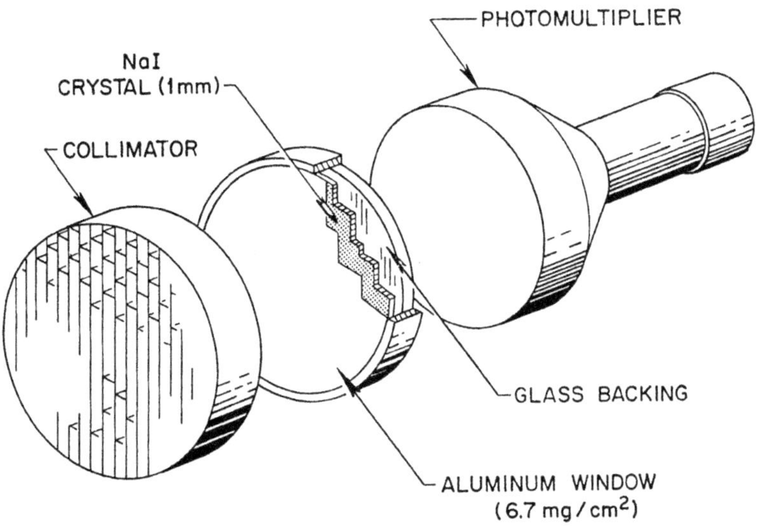

Fig. 7. Scintillation counter.

of an ion pair in a gas counter. Therefore, the smallest detectable photon energy is much higher for a scintillation counter than for a gas counter. Since the energy resolution is limited by the fluctuations of pulse height, a scintillation counter does not offer as good an energy resolution as may be obtained from a proportional counter. In X-ray astronomy the scintillation counters have been used only in the energy region above 10 keV. On the other hand, for sufficiently high photon energies, scintillation counters have both better efficiency and higher energy resolution than gas counters.

Scintillation counters must be covered with thin opaque films to protect them against ultraviolet and visible light. Cosmic rays can be distinguished from X-ray photons by pulse height analysis. Counts due to low-energy charged particles, in particular electrons, can be eliminated by comparison of the counting rates of two scintillators with scintillation materials of the same mass per cm^2 but of different Z (atomic number); these two counters have nearly the same response for electrons, but the response for photons varies with Z.

c) *Photoelectric detectors*

Gas counters cannot be operated in a vacuum without windows, and each window places a low-energy limit to the observable part of the X-ray spectrum. Photoelectric detectors do not need windows and can, therefore, be used to extend the measurements to lower photon energies. The discovery by a group of Russian scientists in 1960 that several alkali halides have an anomalously high photoelectric yield for X-rays enabled the ASE group to design the photoelectric detector for X-ray astronomy shown in Figure 8. The photocathode is covered with a layer of KCl, evaporated on a spherical surface. The electrons ejected from the photocathode are focussed on the first dynode

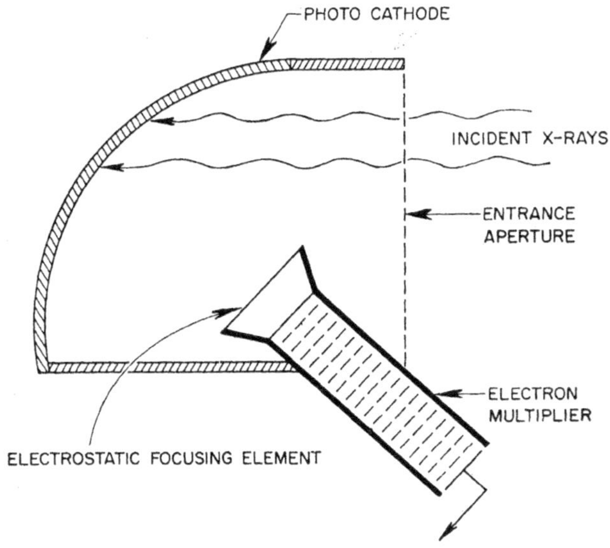

Fig. 8. Photoelectric detector.

of an electron multiplier whose output pulse is proportional to the number of these electrons. The sensitive area of this particular instrument is 40 cm^2. Its measured efficiency ranges from about 8% for 5 keV photons to about 25% for 0.8 keV photons. Visible light is no problem because of the high work function of KCl, while for protection against UV light a thin filter may be necessary.

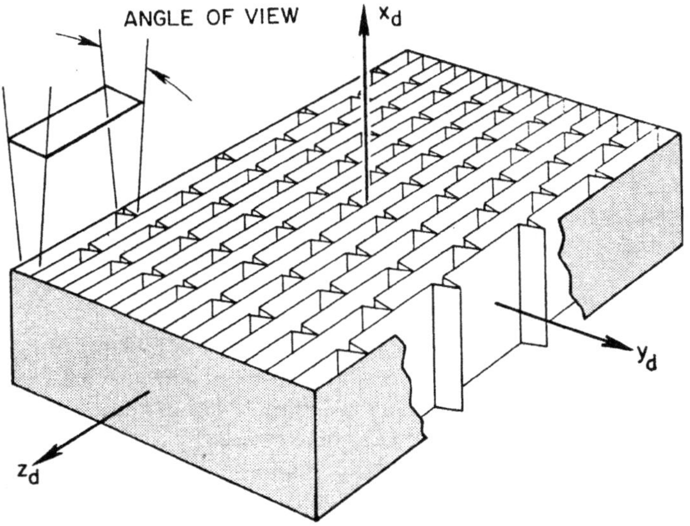

Fig. 9. Cellular collimator.

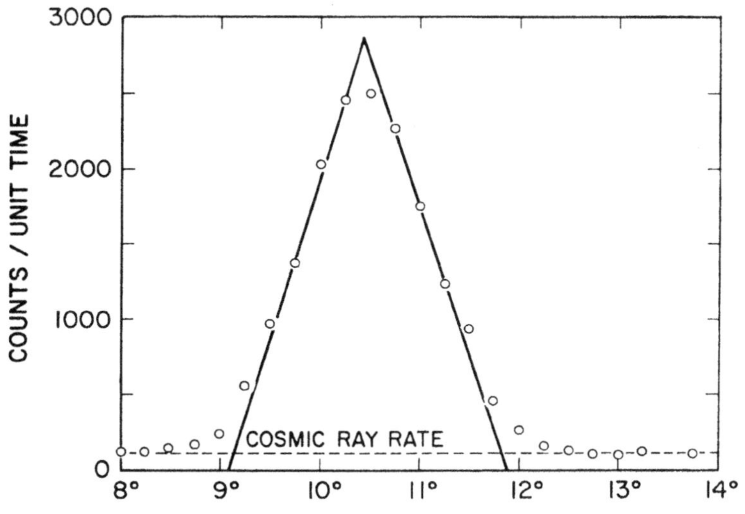

Fig. 10. Experimental response curve with X-rays of cellular collimator.

Detectors of this kind have already been flown. Although the results obtained so far are only of a preliminary character, they show that photoelectric detectors hold great promise for the future of X-ray astronomy.

2.2 COLLIMATORS

a) *Cellular collimators*

In order to obtain accurate information on the location of X-ray sources and on their angular size, it is necessary to limit the field of view of the detectors by means of suitable collimators. Because of the large area of the detectors and the limited space available, cellular collimators of the type shown in Figure 9 have been used. With these collimators one can easily obtain fields of view of different shapes and angular widths down to about 1°. As an example, Figure 10 shows the computed and the measured response curve of a collimator of the type shown in Figure 9. The deviation at the wings may be due to a glancing effect at the walls of the collimator.

b) *Modulation collimators*

In order to improve the angular resolution of a cellular collimator, it is necessary to make its field of view smaller. In this manner, however, one also reduces the number of counts, and, therefore, the statistical accuracy. There is thus a practical limit to the resolution that may be obtained with these devices.

A collimator of a different type, which combines high resolution with wide field of view, has been designed by M. Oda. Known as the "modulation collimator", it consists essentially of two plane grids of parallel wires placed one in front of the other at a suitable distance, D (see Figure 11). When the diameter, d, of the wires is nearly equal to the spacing, s, between adjacent wires, the transmission of the collimator

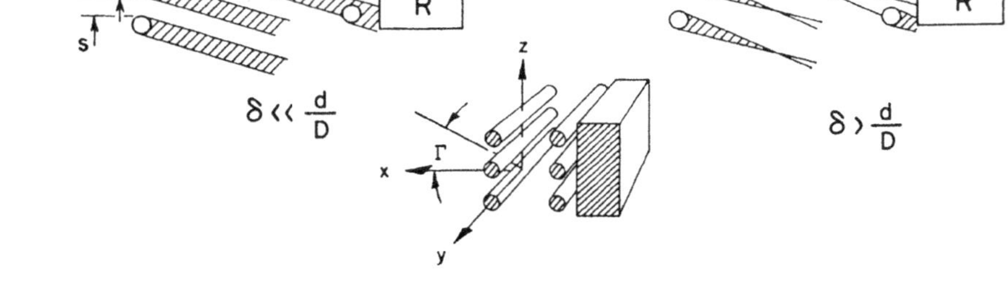

Fig. 11. Principle of modulation collimator.

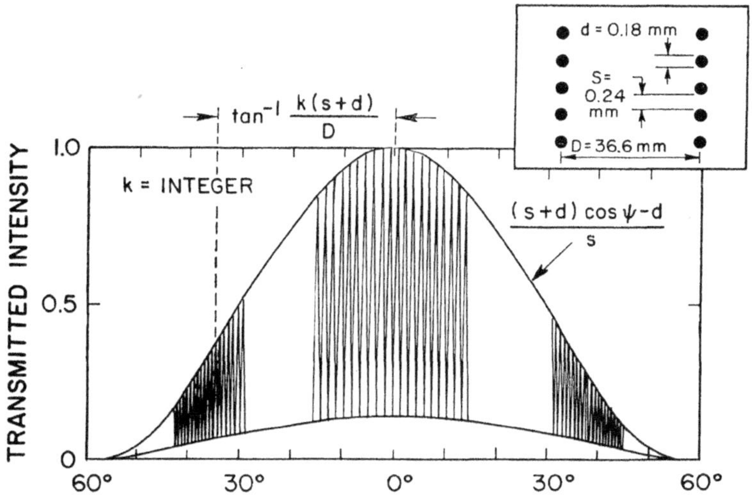

Fig. 12. Angular response of modulation collimator.

varies with the angle of incidence, Γ (Figure 11) of a parallel beam (see Figure 12). For small values of Γ the angular distance between two maxima is $\Delta\Gamma = (d+s)/D$. It is clear that the degree of modulation varies with the angular dimension, δ, of the source, it being zero when δ is large compared with $\Delta\Gamma$. The minimum angular size of a source that can be resolved depends on the mechanical precision achieved in the construction of the collimator, on the statistics of the data and on the noise level. Experience has shown that the smallest angle that can be resolved is about $\Delta\Gamma/4$.

Modulation collimators with resolving power as small as 8 arc minutes have been used (see Figures 13 and 14).

2.3 Image-forming telescopes

In order to improve the resolution further, and to investigate the actual structure of an extended source, it becomes necessary to use image-forming instruments. It is now possible to build instruments of this type using the phenomenon of total external

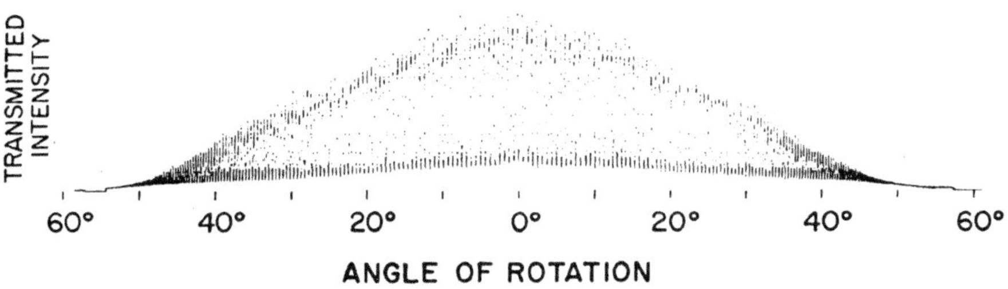

Fig. 13. Angular response of modulation collimator.

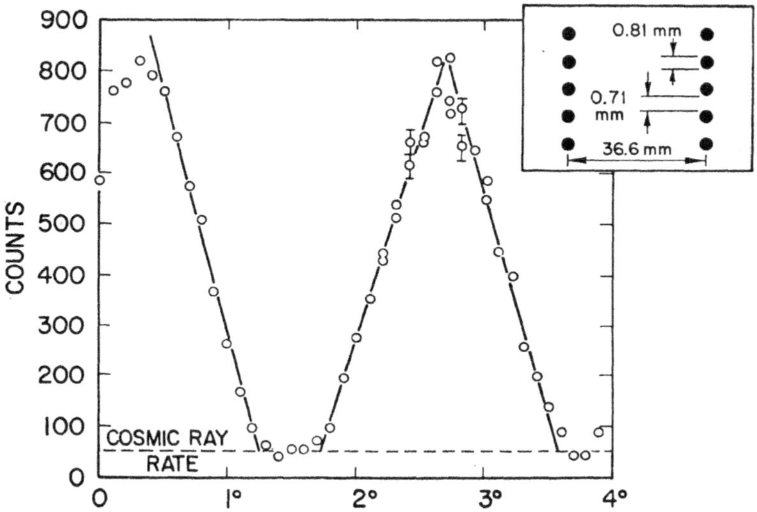

Fig. 14. Angular response of modulation collimator.

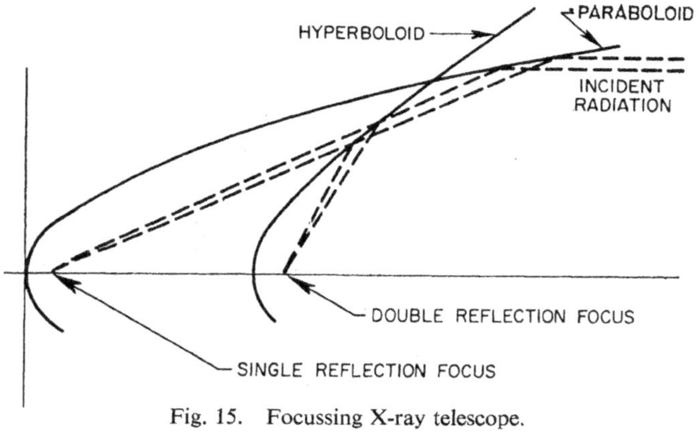

Fig. 15. Focussing X-ray telescope.

reflection of X-rays under grazing incidence. For an aplanatic image to be formed two reflections are necessary; a combination of a paraboloid and an hyperboloid with its external focus at the focus of the paraboloid (Figure 15) [4] has been used.

Image-forming telescopes have been used so far only for *solar* X-ray astronomy. Figure 16 shows such an instrument (sensitive area 1.6 cm^2; intensity after two reflections about 15% of the incident radiation).* Figure 17 shows a photograph of the sun in X-rays obtained with this telescope from a rocket and a picture of the sun in

* The resolution is about 1 minute of arc and is limited by the jitter of the rocket.

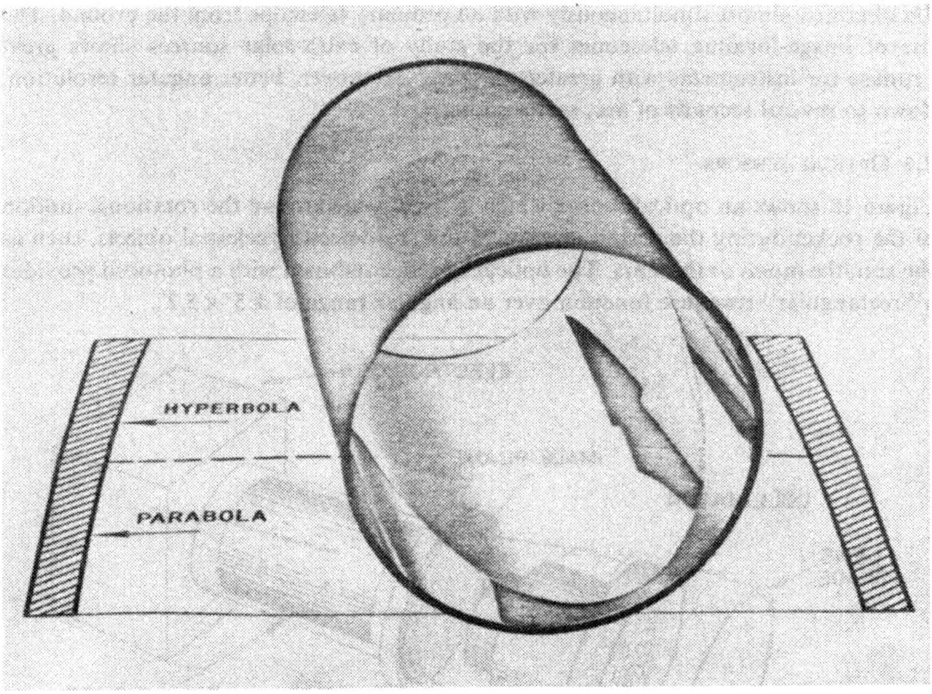

Fig. 16. X-ray telescope.

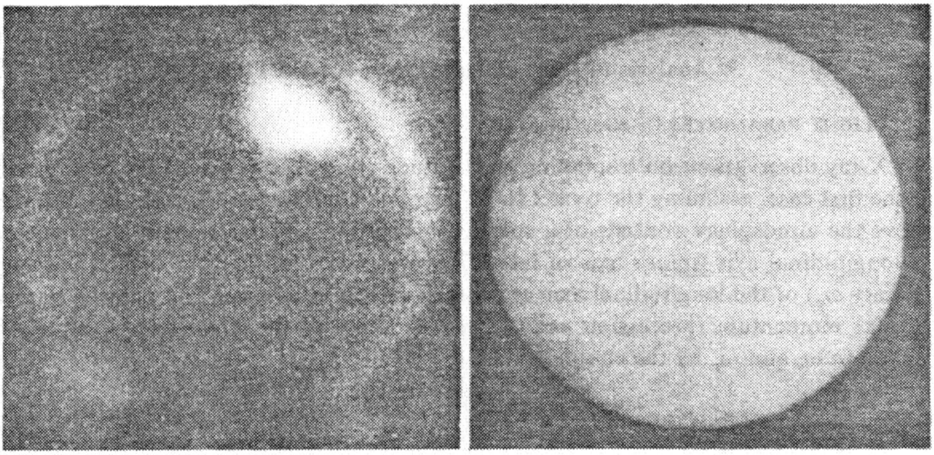

Fig. 17. Photographs of the sun in X-rays and in Hα light.

Hα obtained almost simultaneously with an ordinary telescope from the ground. The use of image-forming telescopes for the study of extra-solar sources shows great promise for instruments with greater aperture. Moreover, better angular resolution, down to several seconds of arc, seems to be possible.

2.4 Optical sensors

Figure 18 shows an optical sensor which is used to determine the rotational motion of the rocket during the period of observation, by detecting celestial objects, such as the sun, the moon or the stars. The optical system combined with a photocell provides a "rectangular" response function over an angular range of $1.5° \times 5.2°$.

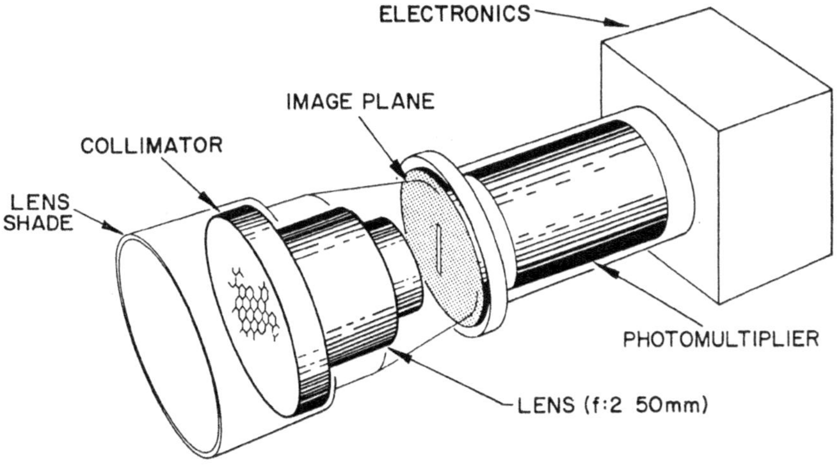

Fig. 18. Star aspect sensor.

3. Analysis of Data Obtained from Rocket Flights

3.1 Flight parameters of sounding rockets

For X-ray observations both spinning and attitude-controlled rockets have been used. In the first case, assuming the rocket to behave as a rigid body, its rotational motion above the atmosphere consists of a spin (with constant angular velocity ω_s) around its longitudinal axis (minor axis of inertia) and a precession (with constant angular velocity ω_p) of the longitudinal axis around an axis fixed in space and parallel to the angular momentum (precession axis). The half angle of the precession cone, α, is related to ω_s and ω_p by the equation:

$$\cos\alpha = \frac{\omega_s}{\omega_p}\frac{I_1}{I_2 - I_1}$$

where I_1 and I_2 are the moments of inertia relative to the longitudinal axis of the rocket and to an axis which is perpendicular to the longitudinal one. Actually, it is necessary to take into account the possibility that the motion of the rocket differs slightly from that given by the above equation, not because of the action of external torques but because the rocket is not really a rigid body. Thus, during flight, only the direction of the precession axis is fixed in space; other parameters, e.g. α, ω_s, ω_p, might deviate slightly from the rigid body values.

As was seen above, useful X-ray observations can be made only at heights greater than 80 km. The rockets used so far reach a maximum altitude of 200 km; the observing time is of the order of 5 minutes. At take-off these rockets are spun up to about two rotations per second for stabilisation. This spin is maintained during the entire flight since the atmospheric friction does not reduce it substantially. Owing to the high spin rate and because $I_2 \gg I_1$ and $\omega_s \gg \omega_p$, α must be small. It may sometimes be useful to reduce the spin rate by means of gas jets in order to scan a larger part of the sky.

The attitude-controlled rockets offer the possibility to study selected regions of the sky. Attitude-control systems which have been developed use gyroscopes as sensing devices and gas jets as torquers.

3.2 Frames of reference

To analyze the data obtained during a rocket flight, it is convenient to use various frames of reference. These are all Cartesian co-ordinate systems, specified by three mutually perpendicular unit vectors; to any one of them one may associate a polar system, specified by a direction and two angles, the elevation θ and the azimuth ϕ.

i) *Celestial frame* $\mathbf{x}_c\mathbf{y}_c\mathbf{z}_c$

The $(\mathbf{x}_c, \mathbf{y}_c)$ plane coincides with the celestial equator; $\mathbf{z}_c$ points towards the north and $\mathbf{x}_c$ towards the ascending node of the ecliptic.

ii) *Precession frame* $\mathbf{x}_p\mathbf{y}_p\mathbf{z}_p$

The $\mathbf{z}_p$ direction is parallel to the precession axis; $\mathbf{x}_p$ is parallel to the vector product $\mathbf{z}_c \times \mathbf{z}_p$, i.e., it lies in the $(\mathbf{x}_c, \mathbf{y}_c)$ plane and points in the direction of the so-called ascending node of the precession equator. If angular momentum is conserved (freely spinning rockets), the precession frame of reference is fixed with respect to the celestial frame.

iii) *Spin frame* $\mathbf{x}_s\mathbf{y}_s\mathbf{z}_s$

The $\mathbf{z}_s$ axis is parallel to the spin axis of the rocket; $\mathbf{x}_s$ lies parallel to the vector product $\mathbf{z}_p \times \mathbf{z}_s$. For freely spinning rockets (angular momentum and energy conserved), the angle $\theta_p^{(s)}$ between the $\mathbf{z}_s$ direction and the $(\mathbf{x}_p, \mathbf{y}_p)$ plane is constant and represents the supplementary angle of the apex angle α of the precession cone.

iv) *Rocket frame* $\mathbf{x}_R\mathbf{y}_R\mathbf{z}_R$

The $\mathbf{z}_R$ direction is parallel to $\mathbf{z}_s$; $\mathbf{x}_R$ is fixed with the rocket and, therefore, rotates with respect to $\mathbf{x}_s$ with angular velocity ω_s.

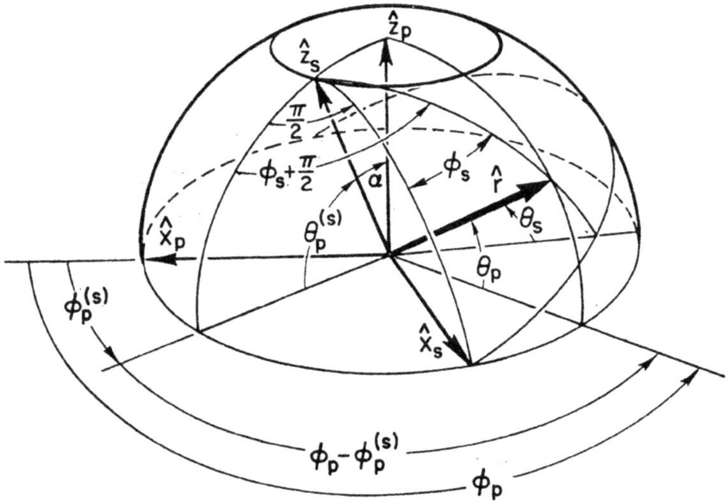

Fig. 19. Precession and spin coordinate systems.

The relation between the precession frame and the spin frame is illustrated in Figure 19, to which we shall refer.

3.3 Freely spinning rockets

When a freely spinning rocket carries a detector (X-ray counter or optical sensor) whose axis forms an angle $\theta_R^{(d)}$ with the equatorial plane of the rocket ($\mathbf{x}_R$, $\mathbf{y}_R$), the elevation of the detector's axis will in the precession frame oscillate between $(\theta_R^{(d)}-a)$ and $(\theta_R^{(d)}+a)$ during each spin. Hence, in a Mercator representation of the precession frame, the detector's axis will describe a nearly sinusoidal curve, with phase changing gradually because of the precession of the rocket (see Figure 20).

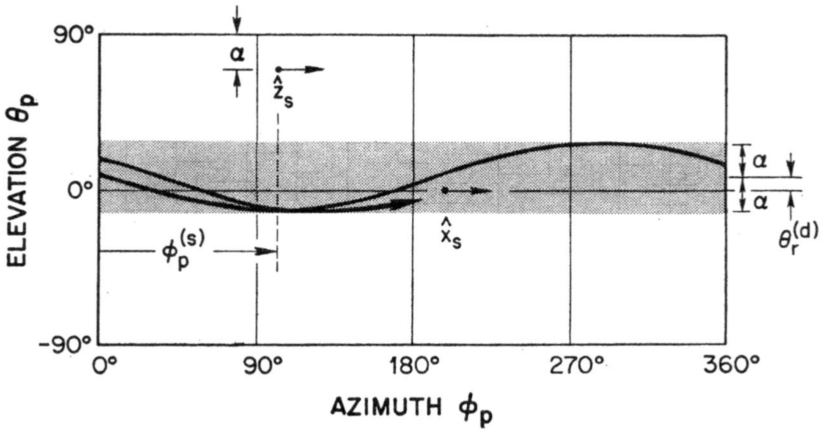

Fig. 20. Motion of detector axis in the precession frame of reference.

As a consequence of this, the region of the sky explored by the detector increases with angle α.

On the other hand, from the point of view of the detector, a celestial object moves in longitude at nearly constant angular velocity, slowly changing its elevation between $(\theta_p - \alpha)$ and $(\theta_p + \alpha)$ during a precession cycle. A point on the sky will then move, on a Mercator projection of the rocket frame, along an approximately horizontal line, oscillating as shown in Figure 21. The object will be observed if it crosses the detector field of view.

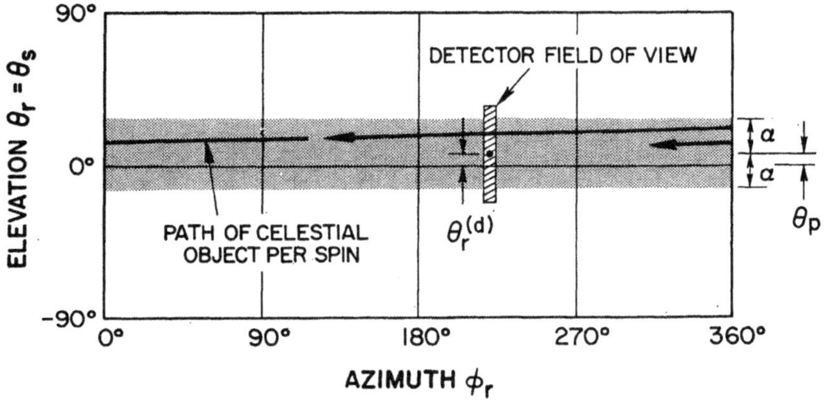

Fig. 21. Motion of celestial object in the rocket frame of reference.

The resulting data of a flight are signals (counts or photoelectric currents) recorded by the X-ray detectors or optical sensors. These data tell the position of the X-ray sources with respect to celestial objects seen by optical sensors and also give some information about the angular dimensions of the sources. The position of a source with respect to a celestial object must be obtained by means of optical sensors. Sufficiently bright stars give recognizable signals every time they pass through the field of view of an optical sensor. After these stars have been identified for all the phases of the flight, a computer programme can be set up to obtain all the parameters describing the motion of the rocket. Some of these parameters can also be obtained from the X-ray detectors.

Suppose, for example, that the rocket has a slow spin and, therefore, a large precession cone; further suppose that the field of view of the collimator has a circular shape (e.g., as was the case for the instrument of the NRL group). During one precession period the detector will scan a large portion of the sky above the local horizon (see Figure 22). If the angular diameter of the field of view is large, any source will be seen several times. If the motion of the detector's axis is known, the position of a source may be determined by making use of the fact that the number of counts per revolution due to this source decreases with decreasing minimum angular distance

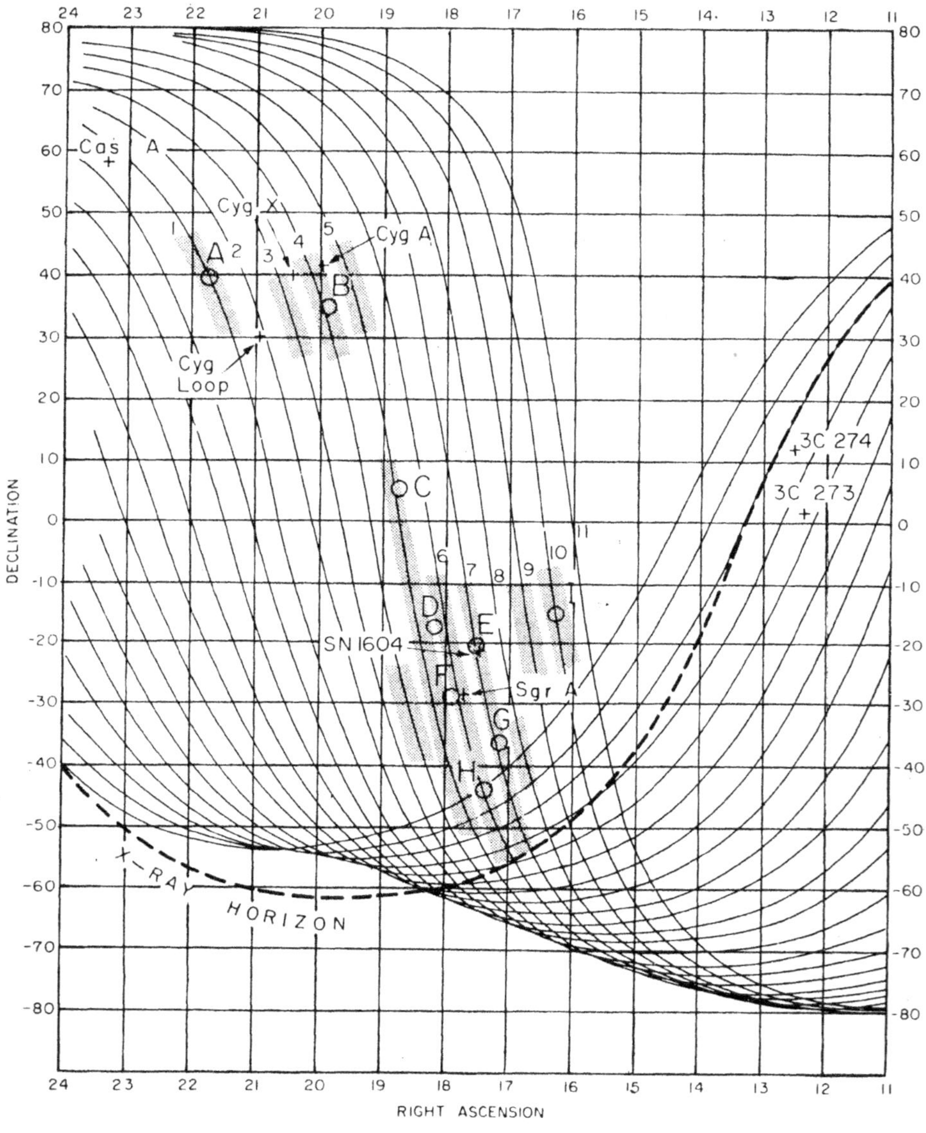

Fig. 22. X-ray source location as observed by an NRL rocket.

between the detector axis and the direction of the source. By a statistical analysis of the number of counts, it is also possible to place an upper limit to the angular diameter of the source if the angular aperture of the collimator is well known.

Let us now consider a fast-spinning rocket (used by the ASE-MIT group) which has a narrow precession cone. Suppose that the X-ray detector has a rectangular field of view with two sides perpendicular to the spin axis. Then, in the rocket frame, the transmission of the collimator along a circle of constant elevation goes through a maximum in the meridian plane containing the collimator axis. Therefore, if one can measure the counting rate as a function of the azimuth of the detector axis in the spin frame during each revolution (in view of the high speed of spin, the spin axis is practically fixed during a revolution), the maximum of the counting rate as function of azimuth determines an arc of a great circle on which the source must lie. The arcs corresponding to different positions of the spin axis along the precession cone intersect at a point which is the position of the source. For a complete precession cycle, these arcs are symmetrical with respect to the great circle representing the meridian plane of the source in the precession frame (see Figure 23). Naturally, for small values of α

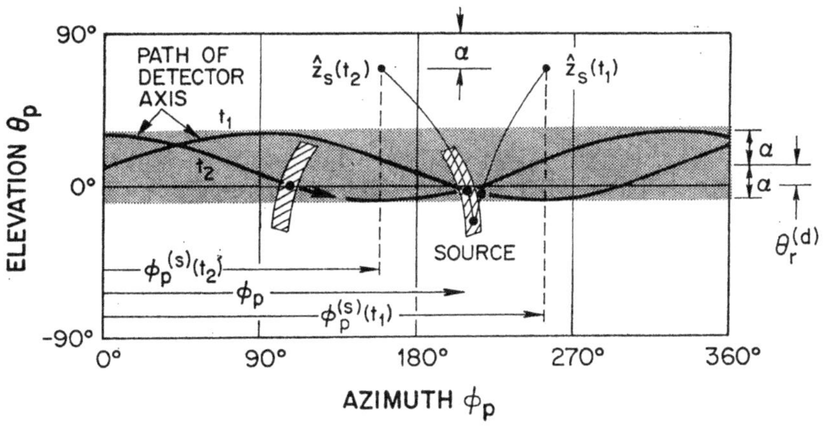

Fig. 23. Source location in the precession frame of reference.

the arcs intersect at small angles. Hence, even a small error in the measurement will seriously affect the result. Thus, while this method will yield a good value of the azimuth, only an approximate value of the elevation of the source in the precession frame is obtained. To improve the determination of the elevation, one may plot the maximum counting rate of each rocket rotation as a function of the elevation of the detector's axis in the precession frame. If the detector axis crosses the source, the curve will reach a maximum corresponding to the elevation of the source. If the detector axis does not cross the source, one can, nevertheless, estimate the elevation of the source by comparing this curve with the computed angular response of the collimator in the meridian plane of the detector.

The data which have been obtained by rapidly-spinning rockets, however, are not sufficient to provide statistically significant data for one spin period because of the small area of the detectors. It is thus necessary to superpose observations made during subsequent spins. This operation is not easy since the synchronization of the various passes is difficult. A powerful method of analysis is based on the use of the angle $\psi = \phi_s + \phi_p^{(s)}$ as an independent variable, where ϕ_s is the azimuth of a given vector in the spin frame, and $\phi_p^{(s)}$ is the azimuth of the spin axis in the precession frame. The angle ψ is called "bearing angle". If ω_s and ω_p are constant, the bearing angle of the detector $\psi^{(d)}$ is simply equal to

$$\psi^{(d)} = (\omega_s + \omega_p)\, t + \psi_0^{(d)}$$

where $\psi_0^{(d)}$ is the value of $\psi^{(d)}$ at $t = 0$. (It can be shown that if α is small, ψ is close to $\phi_p - \pi/2$, oscillating around this value).

Suppose now that we construct a "bearing angle distribution" of counts recorded during some precession cycles; we plot as abscissa the bearing angle $\psi^{(d)}$ of the detector axis and as ordinate the number of counts per unit interval of $\psi^{(d)}$. If we consider for one precession cycle the variation of azimuth $\phi_p^{(d)}$ and elevation $\theta_p^{(d)}$ of the detector axis for a given value of $\psi^{(d)}$, we see that the bearing angle distribution coincides with the azimuthal angle distribution (counting rate versus $\phi_p^{(d)} - \pi/2$) that would be observed if the rocket were spinning with the angular velocity $\omega_s + \omega_p$ around the fixed precession axis, while the X-ray source is oscillating slightly in azimuth around ϕ_p, and moving in elevation between $\theta_p + \alpha$ and $\theta_p - \alpha$. The abscissa at the maximum of the curve determines directly the azimuth of the source in the precession frame.

Let us now consider the case of a detector provided with a modulation collimator. It is necessary that the wires are placed at a large angle with respect to the spin axis; they might even be perpendicular to it. If this were not so, a large number of closely spaced maxima and minima at the counting rate would arise which could not be resolved. Apart from this, the system is very powerful in giving precise information on the angular dimensions of the source in elevation.

4. Results of Observations

4.1 The Crab Nebula

The successful observation of the lunar occultation of the Crab Nebula performed by the NRL group (which named the source Tau XR-1) has already been mentioned. Figure 24 shows how the edge of the moon swept across the Crab Nebula. Figure 25 shows the counting rates of the two detectors and the time derivative of the total counting rate, which represents the distribution of the X-ray source in the direction perpendicular to the advancing edge of the moon. These results show that the X-ray source is located near the centre of the Crab Nebula and that its angular diameter is about 1 arc minute.

4.2 Sources near the galactic centre

We shall now discuss the group of sources which lie within 25° of the galactic centre) including the first discovered X-ray source (called Sco X-1 by the ASE-MIT group,.

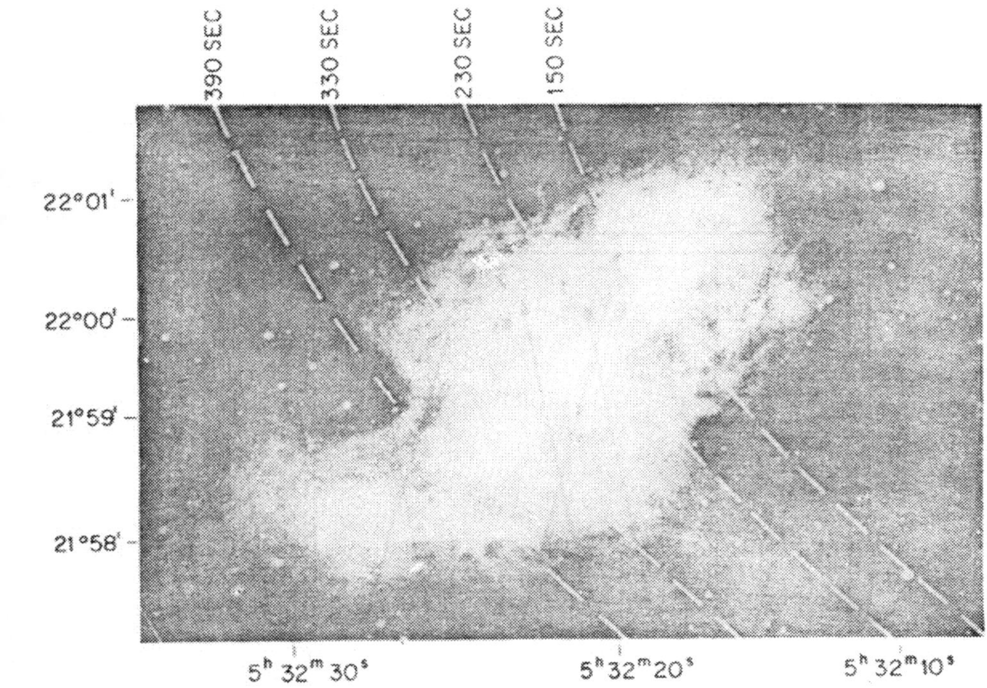

Fig. 24. Lunar occultation of Crab Nebula.

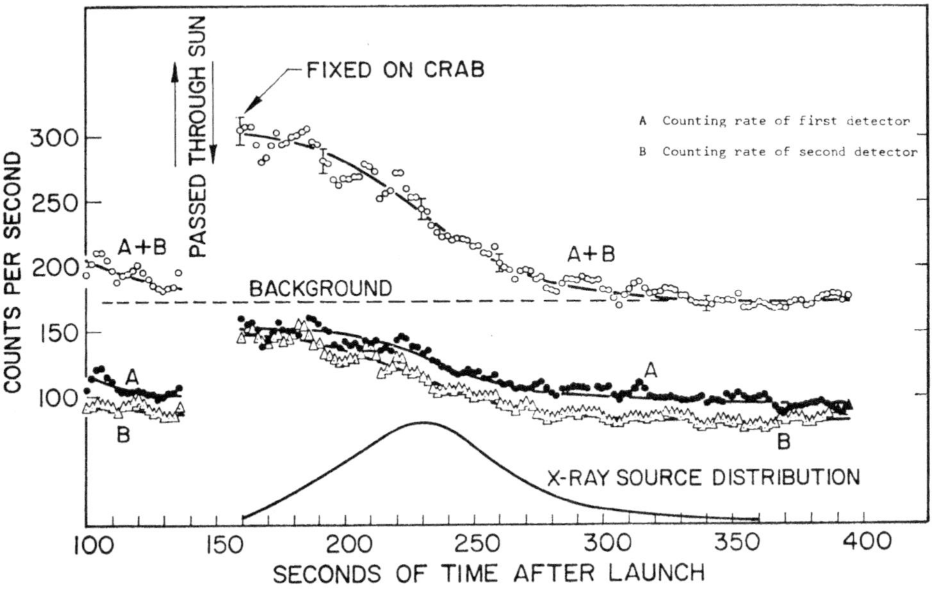

Fig. 25. Results of NRL lunar occultation experiment.

Figure 26 shows tracks of eight scans across the Scorpius region in an experiment performed in April 1963 by the NRL group [5]. The rocket was slowly rolling with a period of 4 sec, and the field of view was 10° (full width half maximum) defined by a honeycomb collimator. Numbers along tracks are counts per 0.09 sec. The circles are lines of equal counting rates, whose common centre represents the source position.

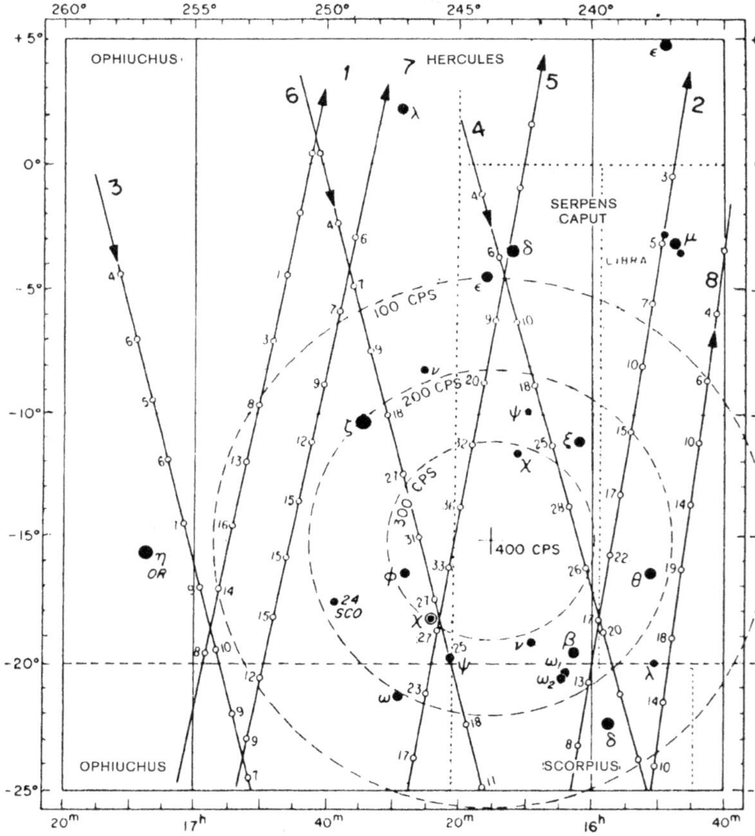

Fig. 26. Detection of X-ray source in Scorpius.

Figure 22 shows the results of another flight carried out by the same group with similar instrumentation in June 1964. In the shaded segments the X-ray counting rate was clearly above background but, of the sources near the galactic centre, only Sco X-1 (indicated as I) gave a clear peak; the others were unresolved. Two other clearly resolved sources were detected in Cygnus (Figure 22A and B) and one unresolved in Serpens which will be discussed later.

Accurate data for some of the X-ray sources appearing in the region of the sky near the galactic centre were obtained by the ASE MIT group with a rapidly spinning rocket flown in October 1964. This rocket carried three banks of Geiger-Müller counters with narrow slit collimators (GV10, GH20, GH0) and one bank provided with a modulation collimator (GMCO). The fields of view and the positions of the X-ray detectors and of the star sensors (SS30, SS0) are shown in Figure 27 in the frame of reference of the spinning rocket. Figure 28 shows the region of the sky scanned by the detectors with narrow-slit collimators, and the positions of three sources, Sco X-1, Sco X-2 and Sgr X-1, which were clearly identified by these observations.

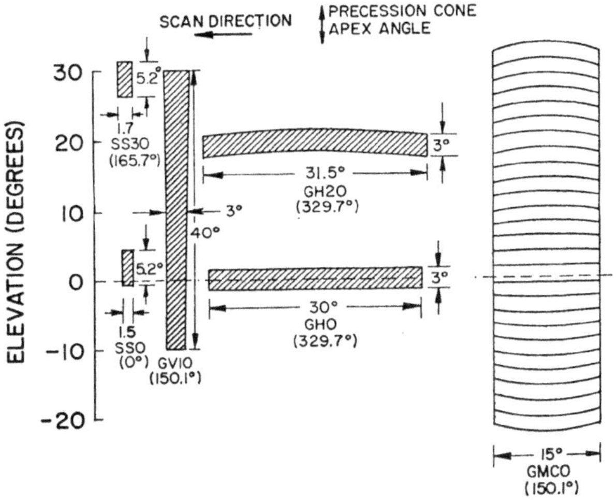

Fig. 27. Fields of view of collimators in Oct. 1964 flight.

The results shall now be discussed in detail. Consider, first, the results of counter GV10 which scanned the sky in the direction of the arrow (Figure 28), with the long side of the collimator perpendicular to it. Figure 29 shows the bearing angle distribution of the counts. The sharp maximum A (Figure 29) is due to Sco X-1; it affords an accurate determination of the azimuth of this source in the precession frame of reference placing it on the arc of great circle A (Figure 28). The possible range of elevation is restricted by the fact that the counting rate was maximum when the elevation in the precession frame was maximum (this places a lower bound to the elevation of Sco X-1; dotted line in Figure 28), and by the fact that the counter GH20 did not see Sco X-1. The smaller and broader maximum between B and C is due to a number of unresolved sources, including Sgr X-1, Sco X-2 and probably other sources near the galactic plane between the arcs of great circles B and C (Figure 28). Since the long dimensions of the slits limiting the field of view of the counters GH0 and GH20 are in the direction of the scan, the maxima are rather flat and the azimuth determinations provided by these counters are not as precise as those provided by counter GV10. The maxima determine the azimuth of the sources Sgr X-1 and Sco X-2 placing them along the arcs of the great circles which are marked E and

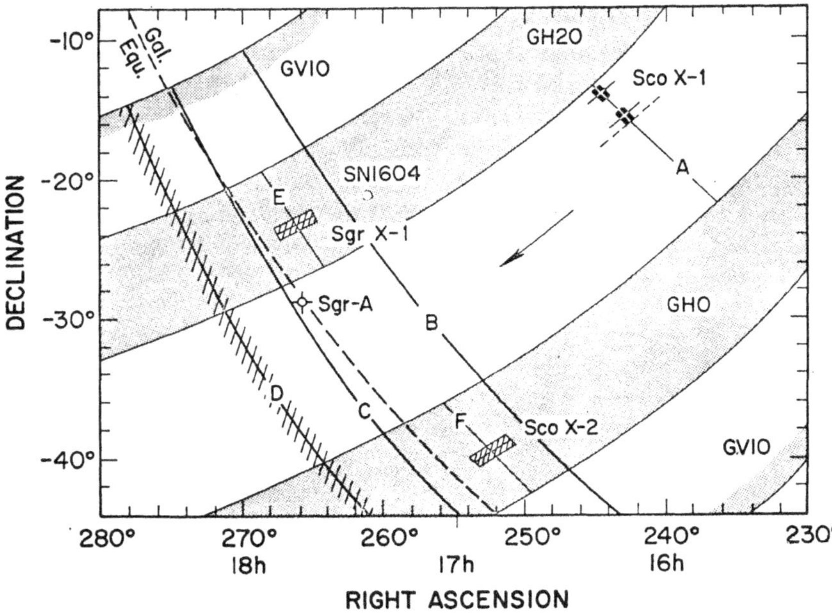

Fig. 28. ASE-MIT source locations.

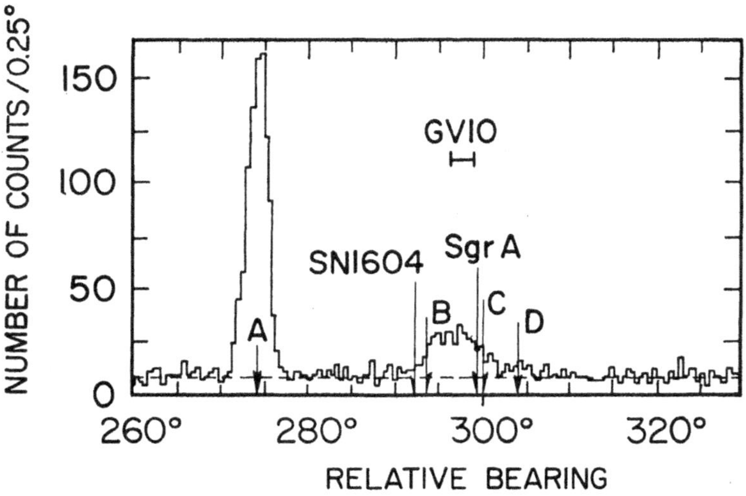

Fig. 29. Bearing distribution, Oct. 1964.

F in Figure 28. The counting rate of the counters GH0 and GH20 plotted against the elevation of the detector's axis in Figure 30 shows that the counting is maximum near the middle of the precession cycle. Thus, Sgr X-1 lies near the centre of the band scanned by GH20, and Sco X-2 lies near the centre of the band scanned by GH0.

Probably another source is situated near the arc of great circle D (Figure 28) where a small but statistically significant maximum appears.

Figure 31 shows the variation of the counting rate of GMCO during the precession cycle; the resolution was about 30 arc minutes. A previous experiment with a modulation collimator of higher resolution had shown that the angular dimensions of Sco X-1 are less than 7 arc minutes. The strong modulation observed with the present detector is entirely consistent with this result. Moreover, the position of the maxima makes it possible to place Sco X-1 on one of the circles of constant elevation corre-

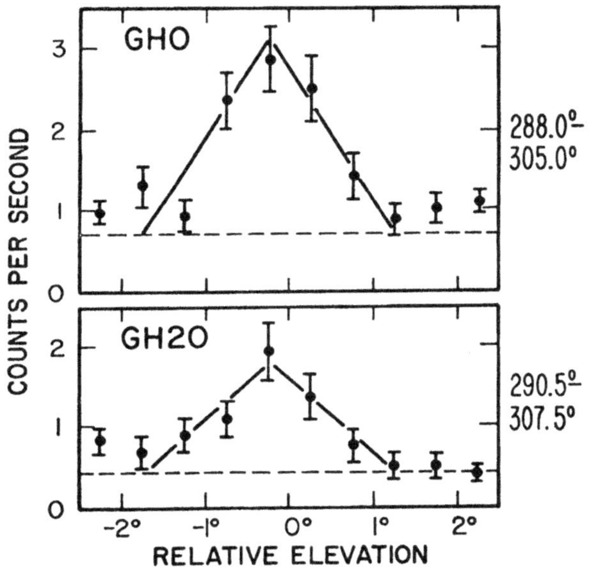

Fig. 30. Elevation distribution of counting rate.

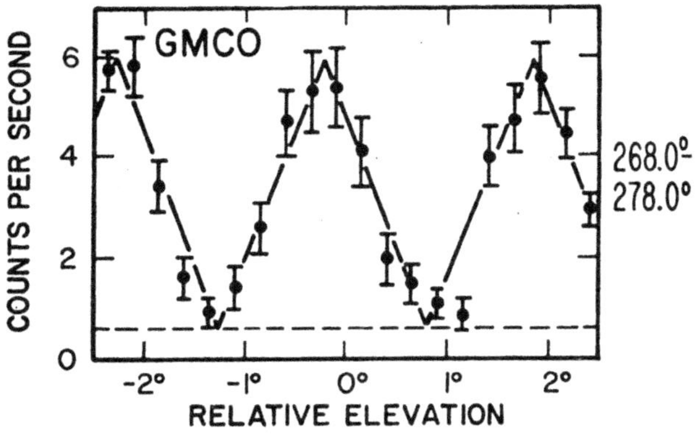

Fig. 31. Elevation distribution of counting rate.

sponding to the transmission maxima of the modulation collimator at the time when a maximum of the counting rate is observed. Two of these circles fall between the limits of elevation previously determined, thus giving the Sco X-1 the two possible locations shown in Figure 28.

Table I lists the most likely positions of Sco X-1, Sco X-2 and Sgr X-1, as obtained from the observations just described. The experimental errors arise from the uncertainty with which the orientations of the detectors and sensors in the rocket frame of reference are known.

TABLE I

X-Ray Source Locations – ASE/MIT Results

Source	Right ascension	Declination
Sco X-1	$16^h12^m \pm 2^m$ or	$-15.6° \pm 0.5°$
	$16^h19^m \pm 2^m$	$-14.0° \pm 0.5°$
Sco X-2	$16^h50^m \pm 7^m$	$-39.6° \pm 0.9°$
Sgr X-1	$17^h44^m \pm 7^m$	$-23.2° \pm 0.9°$

The same region of sky was scanned by the Lockheed group, by means of an attitude-controlled rocket flown in October 1964. Figure 32 shows how the slits of the collimators of the counters were arranged. Information on the elevation of the source in the rocket's frame of reference was derived from the delay between the signals of the various detectors. The rocket was programmed to execute four different scans during its flight. Sco X-1 was detected during scans Nos. 2, 3 and 4. All three counters, T, C and B, gave useful data during the slower scans (Nos. 3 and 4). The

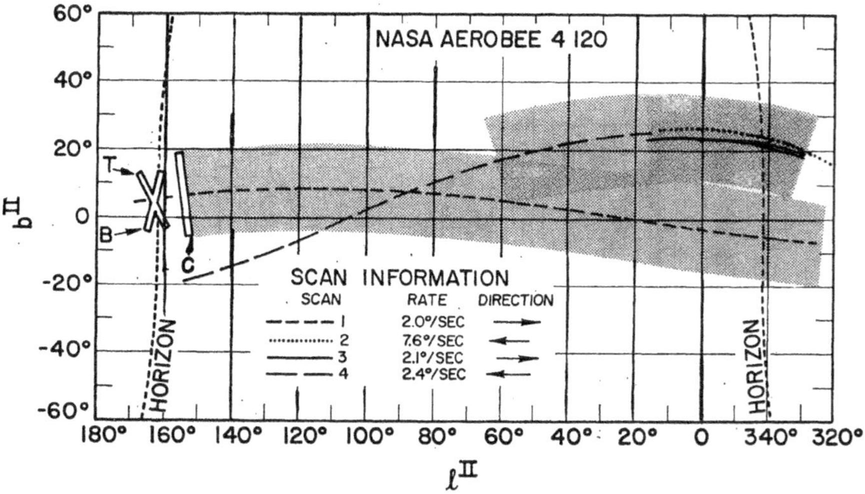

Fig. 32. Rocket flight plan, Lockheed experiment.

following data on the position of Sco X-1 was obtained:

Right ascension $16^h 14^m \pm 1^m$
Declination $-15°36' \pm 15'$.

Only counter C provided useful data during scan No. 1 which ran along the galactic equator. The counting rate showed several maxima where the corresponding azimuthal angles could be determined. The arcs of great circle corresponding to the sources near the galactic centre are shown in Figure 33.

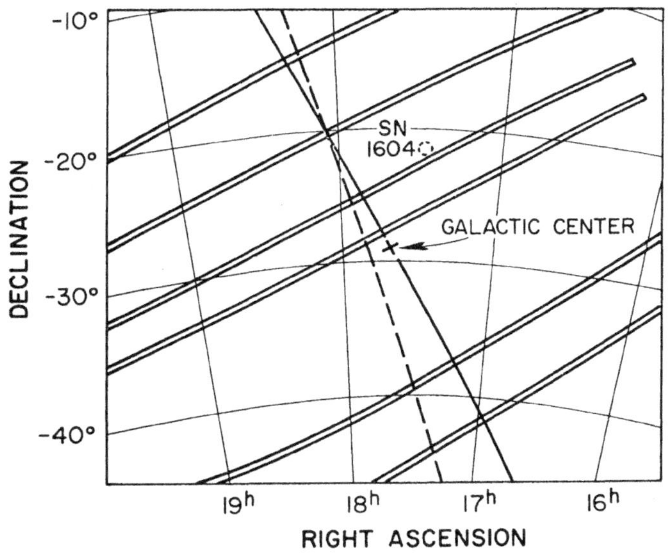

Fig. 33. Lockheed source locations.

In Figure 34, the results of the NRL group, the ASE-MIT group and the Lockheed group are summarised. The following sources near the galactic centre seem to be well established:

a) *Sco X-1*: The two possible positions found by the ASE-MIT group fall within the circle of uncertainty of the measurements carried out by the NRL group. One of them is consistent with the position given by the Lockheed group. The angular size is less than 7 arc minutes.

b) *Sgr X-1*: The position as determined by the ASE-MIT group is about 1° from the arc of great circle corresponding to one of the sources found by the Lockheed group. This discrepancy may be due to experimental errors. However, Sgr X-1 does not lie within the circle of uncertainty of any of the sources found by the NRL group.

c) *SN 1604*: No X-ray source found by ASE-MIT coincides with the Kepler supernova remnant, and the tentative identification by the NRL group of their Oph X-1 source with SN 1604 does not seem to be correct.

d) *Sco X-2*: The positions found by the ASE-MIT group and by the Lockheed group are coincident as in (b). The nearest source reported by the NRL group, Sco XR-2, is situated about 5° away.

e) In addition to (a) and (d), the Lockheed results indicate the existence of at least two other sources. One of them may coincide with NRL source Sgr XR-1.

f) Sgr A, the radio centre of the galaxy, is at most a weak X-ray source compared with the other sources nearby.

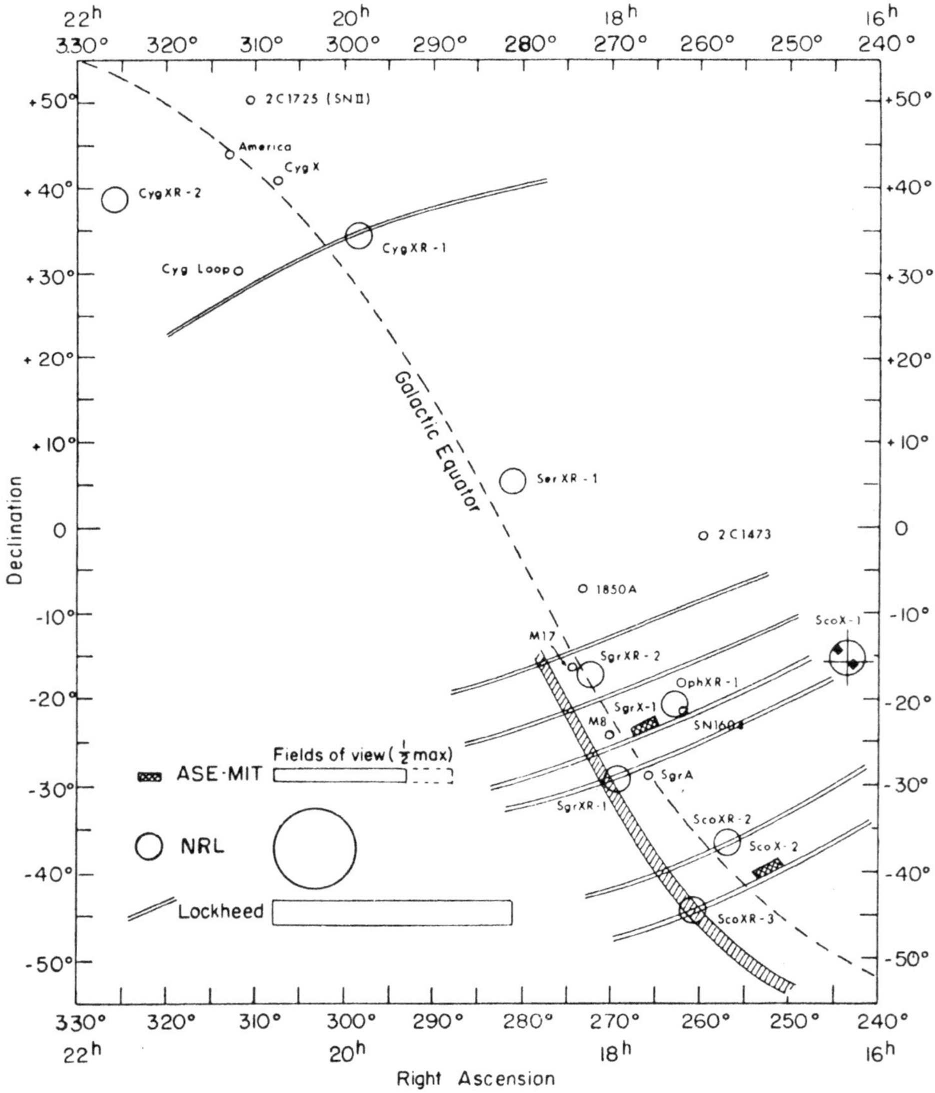

Fig. 34. X-ray source locations.

4.3 Other sources

In addition to the sources found near the centre of the galaxy, the ASE-MIT group suggested that there are X-ray sources in the region of Cygnus. Actually, the NRL group detected two sources, Cyg XR-2 and Cyg XR-1 (see Figure 34), in that region. Cyg XR-1 lies on one arc of great circle where the Lockheed group also located an X-ray source. The NRL observations indicate the existence of a source in Serpens; a small maximum in the counting rate observed by the Lockheed group in scan No. 1 may be due to the same source.

4.4 Spectral information and flux estimates

Information on the spectra of the observed X-ray sources was obtained from pulse height measurements with scintillation and proportional counters and from absorption measurements of Geiger-Müller counters of different spectral response. But the available data does not allow a direct determination of the spectral distribution $j(E)\mathrm{d}E$ of the sources (giving the energy flux of X-rays with photon energy between E and $E+\mathrm{d}E$). After certain assumptions had been made on the form of the spectral function $j(E)$, the experimental results could be used to compute the free parameters entering in the expressions for $j(E)$ and, in some cases, rule out certain types of spectral functions.

The spectral functions usually considered are:

i) Power law $j(E)\mathrm{d}E=(\text{constant}/E^{\alpha})\mathrm{d}E$ referring to synchrotron radiation and inverse Compton effect.

ii) Exponential law $j(E)\mathrm{d}E=\text{constant}\times\exp(-E/kT)\mathrm{d}E$ characteristic for Bremsstrahlung emitted by fully ionized, optically thin plasma clouds where the electrons have a Maxwellian velocity distribution corresponding to a temperature, T.

iii) Planck's law $j(E)\mathrm{d}E=\text{constant}\times E^3\mathrm{d}E/[\exp(E/kT)-1]$ giving the radiation of optically thick objects at a surface temperature, T.

a) *The Crab Nebula*

The NRL group, in their November 1964 flight, observed the Crab Nebula with two X-ray counters fitted with different Mylar windows. The counting rates are consistent with a spectrum of type (i) with $\alpha=1.1$ in the energy range 1.2 to 4.0 keV.

The balloon measurements made by G. Clark in July 1964 are consistent with a spectrum of the same type with $\alpha=2$ for the energy range 15–62 keV. These measurements combined with flux measurements in the visible and radio regions are summarised in Figure 35. It appears that the spectrum may be approximated by a form of type (i) with $\alpha\simeq1$ from about 1 eV to a few keV. For higher energies, however, the spectrum becomes steeper. Moreover, interstellar absorption must produce a "gap" in the spectrum whose detailed experimental investigation will yield important information on the composition of the interstellar gases.

b) *The sources near the galactic centre*

In addition to earlier measurements, results were obtained in 1964–1965 by the

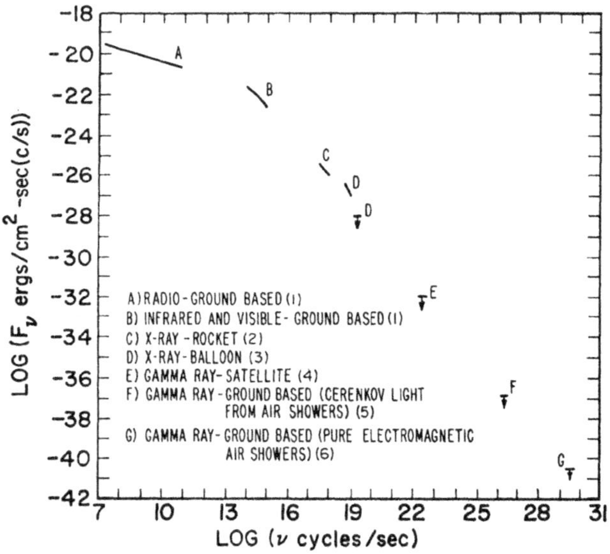

Fig. 35. Summary of data on the electromagnetic spectrum of the Crab Nebula.

(1) O'DELL, C. R.: 1962, *Astrophys. J.* **136,** 809.
(2) FRIEDMAN, H. (unpublished).
(3) CLARK, G.: 1965, Phys. Rev. Letters **14,** 91.
(4) KRAUSHAAR, W., CLARK, G., GARMIRE, G., HELMKEN, H., HIGBIE, P., and AGOGINO, M.: 1965, *Astrophys. J.* **141,** 845.
(5) CHUDAKOV, A. E., ZATESPIN, V. I., NESTEROVA, N. M., and DADIKEN, D. L.: 1962, *Proc. of Fifth Interamerican Seminar on Cosmic Rays*, La Paz, Vol. V, p. XLIV–1.
(6) SUGA, K., ESCOBAR, I., MURAKAMI, K., DOMINGO, V., TOYODA, Y., CLARK, G., and LAPOINTE, M.: 1963, *Proceedings of the International Conference on Cosmic Rays*, Jaipur.

Lockheed group, the ASE-MIT group and the NRL group. The results of the Lockheed group were obtained with three proportional counters (see Figure 32). The observed pulse height distributions are shown in Figure 36; it has been reported privately that the interpretation of the measurements in the 2–4 keV channel seems somewhat uncertain.

Spectral data were obtained by the ASE-MIT group with rocket flights in August and October 1964. The instrumentation included a bank of Geiger-Müller counters with Be windows of 9.0 mg/cm^2 thickness, filled with Argon to a gas thickness of 5.4 mg/cm^2. The computed efficiency curve was checked in the laboratory with an X-ray tube operated at different voltages. In front of the counters were two filters F-1 (7.0 mg/cm^2 Be) and F-2 (1.72 mg/cm^2 Mylar) which were removed one after the other during the flight. The results are summarised in Table II. The spectral information which is obtained from these data refers approximately to the energy region 1 to 10 keV. Note that there is no statistically significant difference between the attenuations observed from Sco X-1 and for the sources near the galactic plane. It turns out that the experimental results do not allow the distinction between the three types of spectra mentioned above. However, they provide a determination of

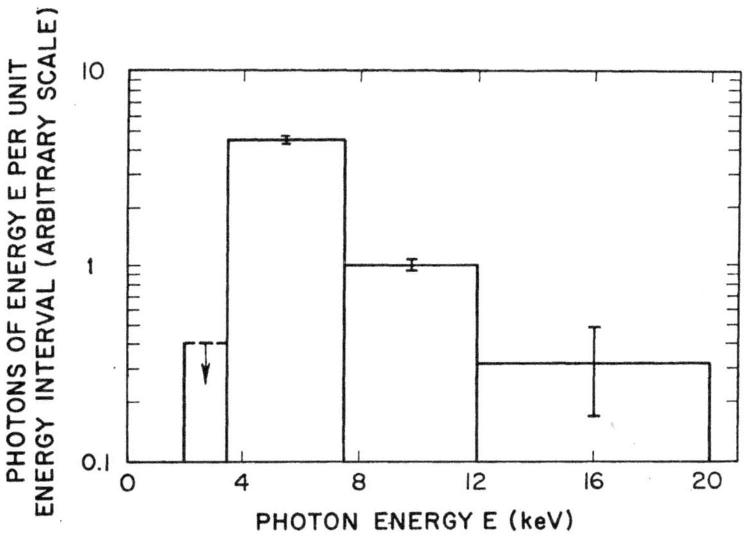

Fig. 36. Pulse height distribution obtained in the Lockheed experiment.

TABLE II

Filter Absorption Data on X-Ray Sources
(from GIACCONI and GURSKY: 1965, *Space Sci. Rev.* **4**, 151)

Filter Condition	Source Region	Counting Rate Measured cts/sec	Attenuation Relative to No Filter
No filter	Sco X-1	620 ± 20	1
F1	Sco X-1	440 ± 18	0.71 ± 0.04
F1 + F2	Sco X-1	350 ± 17	0.56 ± 0.03
No filter	Gal Cen	350 ± 18	1
F1	Gal Cen	240 ± 18	0.70 ± 0.06
F1 + F2	Gal Cen	210 ± 15	0.62 ± 0.05
Typical background rate		30 cts/sec	

the characteristic parameter, in the sense that, assuming a spectrum of type (i), one obtains $\alpha = 1.1 \pm 0.3$; of type (ii) one obtains $T = (3.8 \pm 1.8) \times 10^7$ °K; and of type (iii), one obtains $T = (9.1 \pm 0.9) \times 10^6$ °K. The data are not consistent with the sharp decrease of the spectrum below 4 keV which was reported by the Lockheed group.

Figure 37 shows the pulse height distributions, given by a NaI scintillation counter, of the two source regions observed and of the background. The cut-off below 10 keV results from the electronic threshold. Above 10 keV, the three pulse height distributions are significantly different: the sources near the galactic centre indicate a possible peak near 20 keV, while the Sco X-1 spectrum falls below the detection level at about 18 keV.

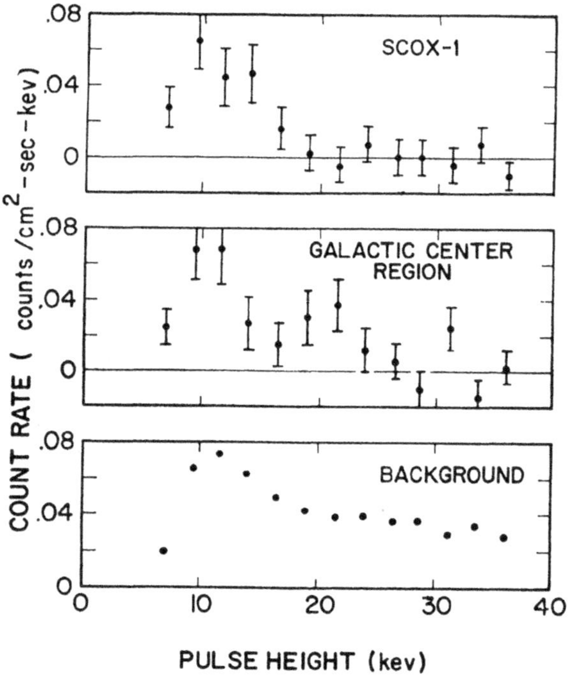

Fig. 37. Pulse height distribution of NaI for source regions.

TABLE III

Integrated Energy Flux for Various X-Ray Sources and two Different Ranges of Photon Energy

Source	Spectral Range (keV)	Energy Flux (ergs)/cm^2 sec	Remarks
Crab	1 to 10	2×10^{-8}	NRL Data
	20 to 30	$(6 \pm 2) \times 10^{-9}$ *	G. Clark Balloon Data
Sco X-1	1 to 10	$(1.6 \pm 0.4) \times 10^{-7}$	ASE Geiger Counter Data
	20 to 30	$< 6 \times 10^{-9}$	ASE Scintillation Counter Data
Galactic equator (including Sgr X-1, Sco X-2 and probably several additional sources)	1 to 10	$(0.4 \pm 0.1) \times 10^{-7}$	ASE Geiger Counter Data
	20 to 30	$(16 \pm 5) \times 10^{-9}$	ASE Scintillation Counter Data

* Represents 2σ deviation above background; no significant excess was observed.

A black body spectrum cannot fit both sets of data for Sco X-1 and the sources near the galactic equator, since such a spectrum falls off with increasing energy much more rapidly than the experimental results indicate. On the other hand, both a power law and an exponential law, with the approximate values of the parameters mentioned previously, can account for the data obtained with the scintillation counter as well as with the Geiger-Müller counters. There is evidence, however, that the combined spectra of the sources near the galactic equator are somewhat harder than that of Sco X-1, or perhaps that the former contain a narrow line near 20 keV.

In the April 1965 flight, the NRL group used four counters with the following characteristics: (a) Argon filling, 1 mil Mylar window; (b) Argon filling, 0.25 mil Mylar window; (c) Neon filling, 0.25 mil Mylar window; (d) Helium filling, 0.25 mil Mylar window. The relative counting rates of these detectors, which had very different spectral response curves, were found to be significantly different for different X-ray sources. These differences indicate differences in spectra, Sco X-1 being one of the sources which appears to have the softest spectrum. The results concerning Sco X-1 could not be accounted for by a black body spectrum.

In Table III some estimates are summarised for the X-ray fluxes in two different spectral regions from the Crab Nebula, from Sco X-1 and from the group of sources along the galactic equator, within 20° of the galactic centre. Note that these estimates depend, among other parameters, on the assumed shapes of the spectra which are still quite uncertain.

Notes compiled by Vittoria Borla and Attilio Ferrari

References

[1] Giacconi, R., Gursky, H., Paolini, F. R., and Rossi, B. B.: 1962, 'Evidence for X-rays from Sources Outside the Solar System', *Phys. Rev. Letters* **9**, 439.

[2] Gursky, H., Giacconi, R., Paolini, F. R., and Rossi, B. B.: 1963, 'Further Evidence for the Existence of Galactic X-rays', *Phys. Rev. Letters* **11**, 530.

[3] Friedman, H., Bowyer, S., Byram, E. T., and Chubb, T. A.: 1964, 'X-ray Sources in the Galaxy', Nature **201**, 1307.

[4] Giacconi, R. and Gursky, H.: 1965, 'Observation of X-ray Sources Outside the Solar System', *Space Sci. Rev.* **4**, 151.

[5] Friedman, H.: 1964, 'X-ray Astronomy', *Scientific American* June 1964, p. 36.

FURTHER RECOMMENDED LITERATURE

Bowyer, S., Byram, E. T., Chubb, T. A., and Friedman, H.: 1964, 'The Lunar Occultation of X-ray Emission from the Crab Nebula', Science **146**, 912.

Bowyer, S., Byram, E. T., Chubb, T. A., and Friedman, H.: 1965, 'Cosmic X-ray Sources', *Science* **147**, 394.

Clark, G.: 1965, 'Balloon Observation of the X-ray Spectrum of the Crab Nebula above 15 keV', Phys. Rev. Letters **14**, 91.

Fisher, P. C., Johnson, H. M., Jordan, W. C., Meyerott, A. J., and Waters, J. R.: 'Positions of Three X-ray Sources in Scorpio and Sagittarius' (preprint).

Giacconi, R., Gursky, H., Waters, J. R., Clark, G., and Rossi, B. B.: 1964, 'Observation of Two Sources of Cosmic X-rays in Scorpio and Sagittarius', *Nature* **204**, 981.

Oda, M., Clark, G., Garmire, G., Wada, M., Giacconi, R., Gursky, H., and Waters, J. R.: 1965, 'The Angular Sizes of the X-Ray Sources in Scorpio and Sagittarius', *Nature* **205**, 554.

Rossi, B. B.: 1964, 'X-ray and Gamma-ray Astronomy', Proc. 5th Intl. Space Sci. Symp. (COSPAR), Florence, May 1964.

PART III

COSMIC RADIATION

RAYONS COSMIQUES ET RAYONNEMENT SYNCHROTRON*

J. LEQUEUX

Observatoire de Paris, Meudon, France

Résumé. Le rayonnement synchrotron est une source de renseignements particulièrement intéressants sur les électrons relativistes et les zones où la concentration en rayons cosmiques et en champs magnétiques est forte, et qui sont probablement des sources de rayons cosmiques. Après avoir revu rapidement les propriétés du rayonnement synchrotron, l'auteur passe en revue les radiosources discrètes, lesquelles émettent toutes par rayonnement synchrotron et sont donc très probablement des sources de rayons cosmiques. La deuxième leçon étudie l'accélération, l'injection et la propagation des rayonnements cosmiques. Deux processus d'accélération ont une grande importance: le processus bétatron et le processus de Fermi. Quant aux pertes et aux fuites, elles déterminent en grande partie le spectre d'énergie. La propagation est ensuite étudiée grâce à deux modèles possibles: celui des lignes de force et celui des inhomogénéités. Dans la troisième leçon, l'auteur étudie les pertes d'énergie subies au cours de la propagation. La fragmentation par collisions et la création d'électrons secondaires sont examinées. Les résultats expérimentaux mettent en évidence la surabondance des éléments légers et renseignent sur la composition du rayonnement électronique. Dans sa conclusion l'auteur estime que pour connaître l'origine des électrons cosmiques, il faudrait: a) perfectionner le modèle galactique; b) mesurer le rapport e^+/e^-; c) mesurer le spectre des électrons dans le domaine de 0,25 GeV à 10 GeV, aux fins de comparaison avec le spectre du rayonnement synchrotron.

PREMIÈRE LEÇON

RAYONNEMENT SYNCHROTRON ET SOURCES DE RAYONS COSMIQUES

1. Introduction

L'étude qui va suivre doit être comprise plutôt d'un point de vue phénoménologique. On possède deux sortes de données sur les rayons cosmiques. Ce sont:

1) *des données directes*, relatives au voisinage de la terre, et qui se rapportent au spectre énergétique, à la composition chimique, au caractère isotrope des rayons cosmiques. L'isotropie observée s'entache peut-être d'une légère anisotropie vers la direction du bras galactique local.

2) *des données indirectes*, concernant la distribution dans l'univers de différents rayonnements corpusculaires ou électromagnétiques, obtenues par observation:

a) *du rayonnement X*. Les divers processus qui lui donnent naissance peuvent être:

le bremsstrahlung (rayonnement de freinage) des électrons d'énergie $\simeq 100$ keV, ou bien celui des protons.

le rayonnement synchrotron des électrons de très grande énergie.

l'effet Compton inverse (collision particule-photon thermique).

b) *du rayonnement γ*. En plus du bremsstrahlung, ou de l'effet Compton inverse, d'autres processus importants peuvent engendrer ce rayonnement, par exemple:

la désintégration électromagnétique du méson π^0, $\pi^0 \to 2\gamma$, et des interactions nucléaires où interviennent π^+ et π^- et la composante nucléonique du rayonnement cosmique.

* Leçons données à l'Ecole d'Eté de l'ESRO/CERS à Alpbach (Autriche) en Juillet 1965.

les collisions, aux hautes énergies, entre protons et photons, avec production de pions.
la désexcitation radiative des noyaux résultant de la fragmentation (par collisions) de noyaux plus lourds.

c) *des neutrinos*. Quoiqu'on possède peu de renseignements jusqu'ici, on peut prévoir qu'ils résultent de la désintégration, via des interactions faibles, des divers pions et muons dans le rayonnement cosmique.

d) *du rayonnement synchrotron*, c'est à dire le rayonnement électromagnétique, allant du domaine des fréquences radio au domaine optique, voire X (Nébuleuse du Crabe) et qui constitue le spectre continu émis par les électrons positifs ou négatifs, se déplaçant avec des énergies relativistes dans un champ magnétique.

L'intérêt qui s'attache à ce rayonnement est qu'il donne des renseignements sur les électrons relativistes, intimement liés aux rayons cosmiques lourds, soit parce qu'ils subissent simultanément les processus d'accélération, soit parce que ce sont des électrons secondaires.

En outre, il met en évidence les zones où la concentration en rayons cosmiques *et* en champs magnétiques est forte. Ces régions sont probablement des sources de rayons cosmiques.

Enfin, dans le cas de la Galaxie, ce rayonnement permet d'avoir une idée de la répartition spatiale des électrons, donc donne des indications sur celle des rayons cosmiques. En effet, les conditions de propagation sont peu différentes.

L'étude du spectre électromagnétique permet de remonter au spectre d'énergie des électrons, ce qui renseigne sur les conditions de leur accélération ou de leur propagation.

2. Propriétés du rayonnement synchrotron – (voir Leroux [5])

On a déjà indiqué qu'il s'agit du rayonnement émis par des électrons très relativistes, se déplaçant dans un champ magnétique. L'énergie des électrons qui rayonnent est supérieure à $\simeq 100\ m_0 c^2$, c'est à dire supérieure à 50 MeV.

2.1 Caractéristiques du rayonnement pour un électron isolé

a) *La directivité*

Le rayonnement est émis dans un cône étroit, entourant la vitesse $\boldsymbol{v}$, de demi-angle au sommet $\alpha \simeq m_0 c^2/E$, où E est l'énergie de l'électron (Figure 1).

b) *Le spectre dans une direction donnée*

Il s'obtient par transformation de Fourier des impulsions brusques de rayonnement reçues par l'observateur chaque fois que la vitesse de l'électron pointe vers lui. Ces pics très aigus donnent par transformation de Fourier un spectre très étalé en fréquences (Figure 2).

La fréquence caractéristique s'obtient par la formule:

$$\nu_c = 1{,}61 \times 10^{13} H_\perp E^2$$

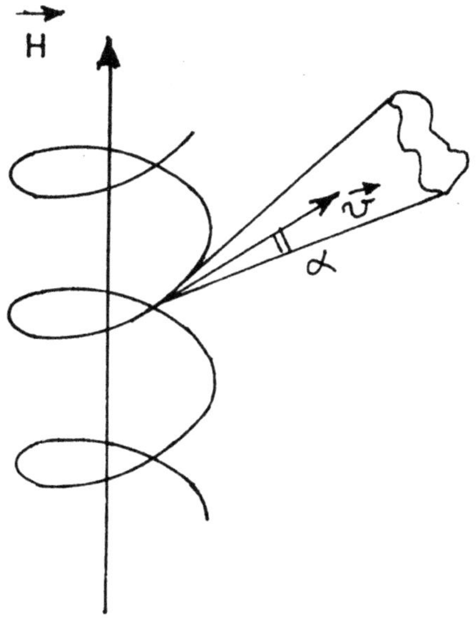

Fig. 1. Direction d'émission du rayonnement synchrotron.

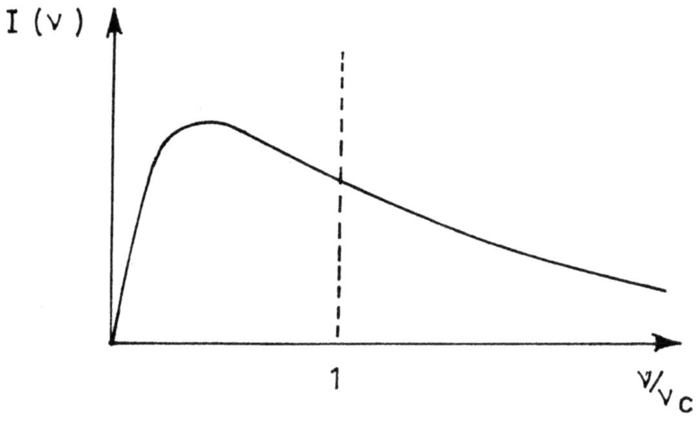

Fig. 2. Spectre du rayonnement synchrotron. ν est la fréquence, $I(\nu)$ l'intensité, ν_c une fréquence caractéristique.

où ν_c est exprimé en Hertz, $H_\perp$ en oersteds ($H_\perp$ représente la composante du champ orthogonale à la vitesse de la particule d'énergie E) et E en GeV. On peut donner des expressions asymptotiques pour $I(\nu)$, qui sont:

$$\text{pour } \nu \ll \nu_c, \qquad I(\nu) \sim \left(\frac{\nu}{\nu_c}\right)^{1/3}$$

$$\text{pour } \nu \gg \nu_c, \qquad I(\nu) \sim \sqrt{\frac{\nu}{\nu_c}}\, e^{-(\nu/\nu_c)}.$$

Enfin, l'énergie totale rayonnée, en eV/s, a pour expression

$$\frac{\mathrm{d}E}{\mathrm{d}t} = -\,3{,}79 \times 10^3 H_\perp^2 E^2$$

où $H_\perp$ et E sont en oersted et en GeV respectivement.

La valeur de la durée de vie d'un électron donné, dans un champ magnétique fixé, se trouve alors être, en années,

$$T_{1/2} = \frac{7{,}7 \times 10^{-3}}{H_\perp^2 E}.$$

Pour des électrons d'énergie $\simeq 1$ GeV, la durée de vie est de l'ordre de 10^8 ans, qui est aussi l'ordre de grandeur de la durée de vie des rayons cosmiques vis-à-vis de la fragmentation.

c) *La polarisation*

La polarisation a beaucoup d'importance en pratique. La Figure 3 montre la répartition des polarisations dans le cône entourant la vitesse $\boldsymbol{v}$. La polarisation, linéaire sur l'axe portant $\boldsymbol{v}$, est elliptique en dehors; dans la direction de $\boldsymbol{v}$, la polarisation est orthogonale à $\boldsymbol{H}$.

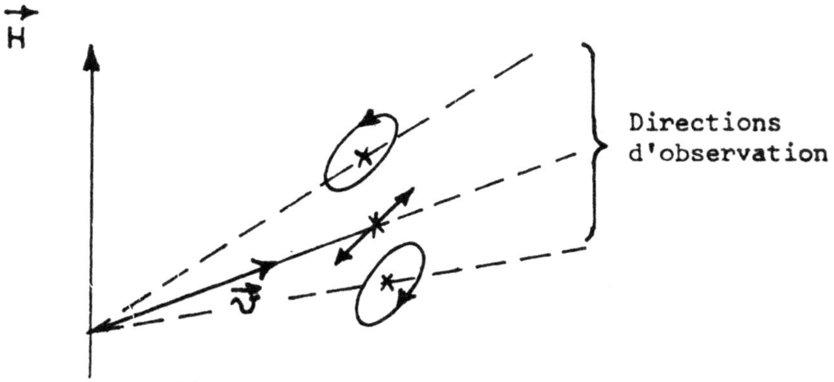

Fig. 3. Polarisation du rayonnement synchrotron.

2.2 Rayonnement d'un ensemble d'électrons isotropes

On supposera que le spectre d'énergie est donné par une loi de puissance de la forme

$$n(E)\,\mathrm{d}E = kE^{-\gamma}\,\mathrm{d}E \quad (E \text{ en GeV})$$

rapporté à l'unité de volume.

a) *L'énergie rayonnée*

L'énergie rayonnée par unité de volume et de bande passante, dans un stéradian, s'exprime par

$$J(\nu)\,\mathrm{d}\nu \sim k\nu^{-(\gamma-1)/2} H_\perp^{(\gamma+1)/2} \mathrm{d}\nu.$$

Cette fois, $H_\perp$ représente la projection de H perpendiculairement à la direction de visée.

En particulier, si le spectre des électrons est de la forme donnée par la Figure 4(a), avec un indice γ, celui du rayonnement émis a la forme montrée à la Figure 4(b), avec un indice α.

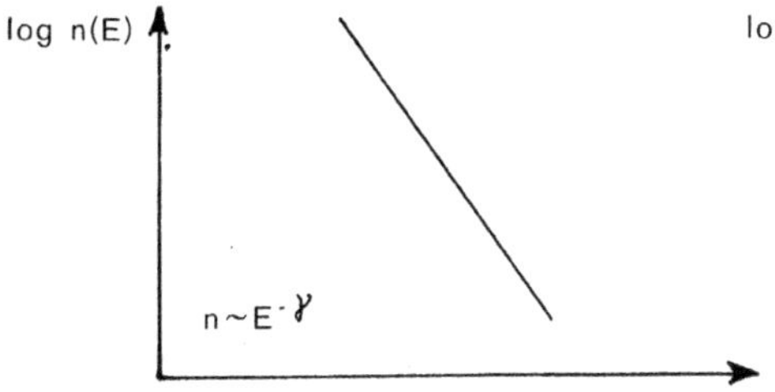

Fig. 4a. Spectre des électrons.

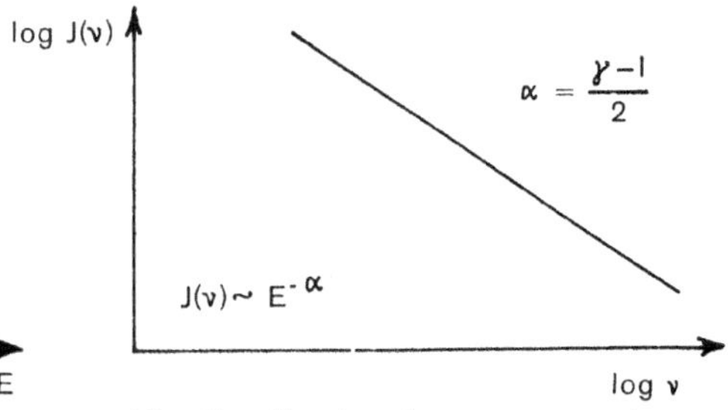

Fig. 4b. Spectre du rayonnement synchrotron émis par un groupe d'électrons dont le spectre d'énergie est celui de la Figure 4a.

Pour la plupart des sources de rayonnement synchrotron dans la galaxie α est de l'ordre de 0,7 et γ de 2,4.

b) *Taux de polarisation*

On doit sommer les contributions, pour des $\mathbf{v}$ de directions diverses, de polarisations elliptiques de vibrations sans relation de phase; ceci conduit à une polarisation *linéaire*, en moyenne, perpendiculaire au champ magnétique. Même si la distribution des vitesses est anisotrope, elle est suffisamment isotrope dans le cône étroit entourant $\mathbf{v}$ pour que la polarisation soit linéaire.

Pour un champ uniforme, le taux de polarisation du rayonnement d'électrons dont le spectre énergétique a l'indice γ est:

$$p = \frac{J_{\text{pol}}}{J_{\text{pol}} + J_{\text{nonpol}}} = \frac{3\gamma + 3}{3\gamma + 7}.$$

Il est indépendent de la fréquence, et sa valeur, pour les valeurs habituelles de γ, varie de 50 à 80%. Mais le champ n'est la plupart du temps pas uniforme; pour la Nébuleuse du Crabe, on a par exemple $p = 9\%$. Ceci amène à considérer divers facteurs de dépolarisation; ainsi, la *rotation Faraday* différentielle, qui apparaît entre les différents trajets optiques durant la propagation. Au cours de la traversée d'une région S où la densité électronique est N_e, le plan de polarisation d'une vibration rectiligne tourne d'un angle θ, proportionnel à l'intégrale:

$$\theta \sim \int_S N_e H_{//} \, ds$$

prise le long du trajet dans S, et variant avec lui (Figure 5).

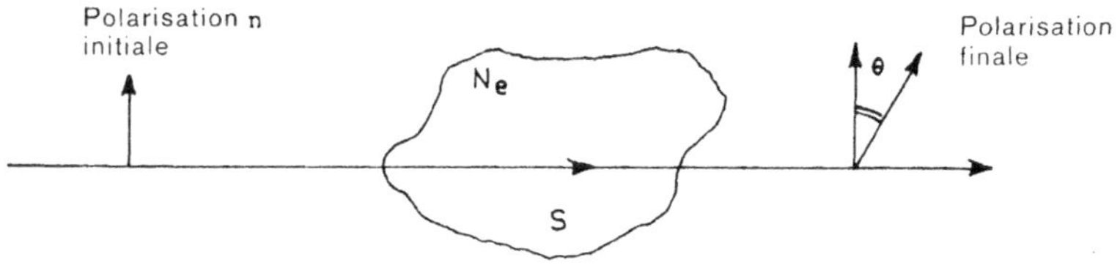

Fig. 5. Rotation du plan de polarisation lors de la traversée d'une région de densité électronique N_e.

Si la rotation Faraday est différente pour les différentes parties de la source, le rayonnement global de celle-ci est dépolarisé.

3. Radiosources discrètes

Les sources de rayonnement synchrotron et celles de rayons cosmiques sont probablement les mêmes; en tout cas, à l'exception d'une partie du rayonnement radio solaire, de celui des "flare-stars", et aussi des régions H_{II} galactiques, toutes les radiosources émettent par rayonnement synchrotron. Il est par conséquent utile de les passer en revue, ainsi que les énergies correspondantes (voir Tableau I).

On peut noter en outre que les supernovae avaient été considérées comme des sources possibles de rayons cosmiques *avant* la découverte de leur rayonnement radio.

TABLEAU I

Sources de rayonnement synchrotron

Type	Indice spectral γ	Puissance rayonnée (erg/s)	$\varepsilon_{\text{équip}}$ $\alpha = 100$ (erg)	$H_{\text{équip}}$ (gauss)
Sources galactiques				
Soleil (sursauts IV)	2 à 5	–	–	–
Flare-Star	6 (si synchrotron)	–	–	–
Supernova (restes)	1,7	10^{31} à 10^{36}	10^{47} à 10^{50}	10^{-5} à 10^{-3}
Galaxie normale	$\simeq$ 2,5	10^{38} à 10^{39}	10^{56} à 10^{57}	3×10^{-6} à 10^{-5}
Radio-Galaxie	2,5	10^{40} à 10^{45}	10^{58} à 5×10^{60}	5×10^{-6} à 10^{-4}
Radio sources Quasi-Stellaires (QSS)	2,5	10^{44} à 10^{45}	$> 10^{61}$ (?)	10^{-5} à 10^{-3} (?)

A fin de comparaison

Energie d'explosion d'une supernova (optique + radio + cinétique)	10^{49} à 10^{51} ergs
Energie thermonucléaire de 1 $M_{\odot}$ d'hydrogène	$1{,}2 \times 10^{52}$ ergs
Energie de masse de 1 $M_{\odot}$	$1{,}8 \times 10^{54}$ ergs
Energie thermonucléaire de la Galaxie	3×10^{63} ergs
Energie de masse de la Galaxie	4×10^{65} ergs

3.1 Généralités sur l'énergie d'une radiosource

Il est possible d'estimer l'énergie contenue dans une source de rayonnement synchrotron dont on connaît le volume V, ainsi que la puissance totale émise sous forme radio, P (obtenue par intégration de $J(\nu)$ et par la connaissance de la distance). L'énergie y existe sous deux formes au moins, qui sont:

l'énergie magnétique

$$\varepsilon_m = V\frac{H^2}{8\pi}$$

l'énergie des particules cosmiques ε_{CR}

Si α est le rapport des énergies des rayons cosmiques lourds et des électrons, on déduit l'énergie ε_{CR} à partir de l'énergie des électrons ε_{el}:

$$\varepsilon_{CR} = (1 + \alpha)\,\varepsilon_{el}\,.$$

Or ε_{el} est liée à la puissance totale rayonnée dans le domaine radio, par une relation déduite de la théorie du rayonnement synchrotron:

$$\varepsilon_{el} = CPH^{-3/2}$$

C est une constante connue, dépendant de l'indice spectral γ du rayonnement; H est le champ régnant dans la source, qui est en général inconnu. On a finalement:

$$\varepsilon_{totale} = (1 + \alpha)\,CPH^{-3/2} + V\frac{H^2}{8\pi}\,.$$

La fraction α est mal connue, mais généralement prise égale à 100 par analogie avec la galaxie. Quant à H, s'il est pris pour paramètre, il donne une énergie totale qui passe par un minimum pour une certaine valeur; pour ce minimum, on constate approximativement l'égalité $\varepsilon_m \simeq \varepsilon_{CR}$. Cette équipartition approchée de l'énergie entre les deux formes, électromagnétique et corpusculaire, peut faire penser à un équilibre entre les rayons cosmiques et le champ électromagnétique. La valeur du champ et de l'énergie dans ces conditions d'équipartition sont alors:

$$H_{\text{équip}} = \left(6\pi(1 + \alpha)\frac{CP}{V}\right)^{2/7}$$

$$\varepsilon_{\text{équip}} = VH^2_{\text{équip}}/4\pi\,.$$

On a effectué le calcul de $H_{\text{équip}}$ et $\varepsilon_{\text{équip}}$ pour de nombreuses radiosources (Tableau I). On obtient des valeurs du champ raisonnables, et des énergies déjà très élevées. Il semble difficile de s'écarter beaucoup de ces conditions, qui sont d'ailleurs celles vers lesquelles le système tend de lui-même; mais ce dernier point ne serait vrai que pour un système stationnaire ou clos, ce qui n'est évidemment pas le cas de beaucoup de radiosources.

On peut obtenir un ordre de grandeur de la durée de vie d'une source en divisant $\varepsilon_{\text{équip}}$ par P; les durées de vie obtenues sont généralement courtes (10^6 à 10^8 ans).

3.2 Cas du soleil

Ce n'est certainement pas une source importante de rayons cosmiques; mais on y voit à l'oeuvre pendant les éruptions un processus d'accélération qui fournit des particules dont l'énergie va de 100 MeV à parfois quelques GeV, responsables de phénomènes ionosphériques (PCA), et dont le spectre présente un indice γ allant de 2 à 5. Il fournit également les électrons qui sont à l'origine des sursauts de type IV, dont l'énergie est de quelques MeV. L'accélération est *très rapide*, de quelques minutes à quelques dizaines de minutes.

3.3 Les "flare-stars"

Un certain nombre d'étoiles ont des éruptions optiques très intenses, qui s'accompagnent d'un rayonnement radio intense en $\nu^{-2,5}$ environ. La puissance de ces émissions est de l'ordre de 10^6 fois plus grande que celle d'une grosse éruption solaire. Si on suppose que le rayonnement est synchrotron, l'indice du spectre d'énergie des électrons est $\gamma = 6$, mais l'origine synchrotron du rayonnement n'est pas sûre; en tout cas, on doit avoir ici encore une accélération très rapide des particules, en quelques minutes.

Aux faibles énergies, la contribution de telles sources aux rayons cosmiques galactiques peut ne pas être négligeable.

3.4 Les restes de supernovae

On en connaît une trentaine de sûrs ou possibles. En général ils se présentent sous forme d'enveloppe sphérique. Dans le cas le mieux connu (Cassiopée A) le flux *décroît* avec l'âge, de l'ordre de 1 à 1,5% par an. Dans cette source, qui n'émet pas dans le domaine optique, les électrons relativistes ne sont pas renouvelés au cours de l'expansion de l'enveloppe. Par contre, dans la Nébuleuse du Crabe, dont l'âge est environ 900 ans, existent des électrons de plus de 100 GeV rayonnant dans le domaine optique, et dont la durée de vie vis-à-vis de l'émission synchrotron est inférieure à la centaine d'années pour $H = 10^{-3}$ gauss. Ceci laisse supposer que des processus d'accélération sont encore présents.

Si on envisage que de tels restes sont des sources de rayons cosmiques pour la Galaxie, ils doivent être capables de fournir une puissance $\simeq 3 \times 10^{40}$ erg/s sous forme de particules de haute énergie. Or s'il y avait dans la Galaxie une explosion de supernova par 50 ans, elle donnerait 10^{40} erg/s en moyenne temporelle; mais ce taux d'explosion semble exagéré car on n'observe en moyenne qu'une supernova par 300 ans dans les galaxies extérieures. La théorie qui voit dans les supernovae la source unique des rayons cosmiques – sans être complètement impossible – rencontre des difficultés supplémentaires puisqu'il n'existe aucune concentration en supernovae au centre des galaxies, alors qu'on y observe un maximum d'émission synchrotron.

3.5 Galaxies normales

La plupart des galaxies spirales émettent un rayonnement radio comparable à celui de la Galaxie. Mais en raison de leur faiblesse ces sources sont mal connues; le fait le plus saillant est la petitesse de leur diamètre radio, la plupart se réduisant à un disque; la présence d'une couronne n'est possible que dans quelques cas.

3.6 Radio-galaxies et sources quasi-stellaires (QSS)

Les radio-galaxies sont optiquement des galaxies elliptiques géantes. Le phénomène qui donne les électrons relativistes est visiblement une explosion au centre de la Galaxie, probablement récurrente.

Les sources quasi-stellaires (QSS) sont en général des sources plus petites mais plus intenses. Leur aspect optique étant très différent de celui des radio-galaxies, la relation entre radio-galaxies et QSS n'est pas évidente, bien qu'on puisse penser que la source d'énergie est de même nature dans les deux cas. L'énergie est probablement en réalité beaucoup plus grande que l'énergie d'équipartition, si on tient compte du rendement de l'accélération, de l'existence d'une expansion adiabatique, et des valeurs trop élevées que l'on trouve de $H_{\text{équip}}$ pour les sources larges.

4. Distribution du rayonnement radio dans la Galaxie

4.1 Modèle de la galaxie

Les observations du fond continu sont rendues difficiles en raison de la structure chaotique de l'émission et de l'importance des structures locales d'une part, et à cause de l'existence d'une composante extragalactique d'autre part.

Les Figures 6 et 7 montrent deux modèles possibles, qui comportent ou non une couronne étendue. Le modèle sans couronne est rendu plausible par les observations récentes.

Si on calcule l'énergie totale de la Galaxie dans un modèle avec couronne, on trouve les données suivantes, auxquelles on a adjoint quelques autres données utiles:

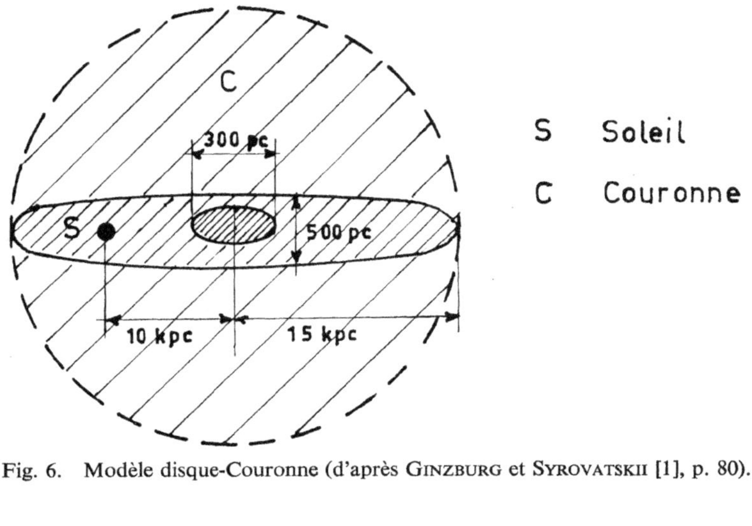

Fig. 6. Modèle disque-Couronne (d'après Ginzburg et Syrovatskii [1], p. 80).

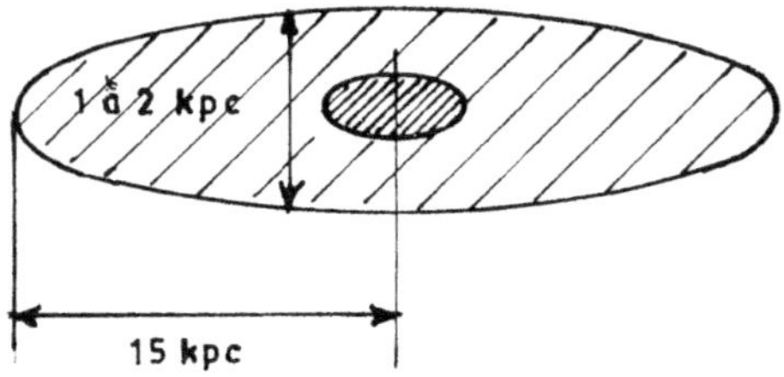

Fig. 7. Modèle disque seul.

TABLEAU II

Densité des sources extragalactiques

Type de la source	Densité Mpc^{-3}	Densité relative
QSS	2×10^{-9}	1/15000000
Galaxies quasi-stellaires (non radio)	10^{-7}	1/300000
Radiogalaxies	2×10^{-5}	1/1500
Galaxies elliptiques géantes	7×10^{-5}	1/400
Galaxies brillantes	3×10^{-2}	1

TABLEAU III

La Galaxie

Volume (avec couronne)	2×10^{68} cm^3	Masse	$2 \times 10^{11} M_{\odot} = 6 \times 10^{44}$g
Densité d'énergie des rayons cosmiques	0,3 à 1 eV cm^{-3} $\simeq 10^{-12}$ erg cm^{-3}	Energie des électrons	2×10^{54} ergs
Energie totale ($\alpha \simeq 100$)	$W = 2 \times 10^{56}$ ergs	Durée de vie vis-à-vis des fuites	$T \simeq 3 \times 10^8$ ans $= 10^{16}$ s
Puissance des sources	$W/T = 3 \times 10^{40}$ erg s^{-1}	Puissance radio (synchrotron)	3×10^{38} erg s^{-1}
$H_{\text{équip}}$	5×10^{-6} gauss	$H_{\text{près du soleil}}$	3×10^{-5} gauss

Les valeurs figurant dans le tableau III ont été calculées en partant de l'hypothèse que le spectre énergétique du rayonnement synchrotron en provenance de la Galaxie correspond à la courbe illustrée dans la Figure 8 et est limité entre 4 et 1000 MHz.

4.2 Inhomogénéités de la matière et du champ magnétique dans la galaxie

Dans le disque, on a une masse de 2 à 3×10^{42}g de gaz (réparti dans des nuages), dont très peu au centre, et qui correspond à une densité moyenne $\bar{\rho} \simeq 1$ atome cm^{-3}; le champ magnétique H est en général assez parallèle au plan de symétrie; mais pour le

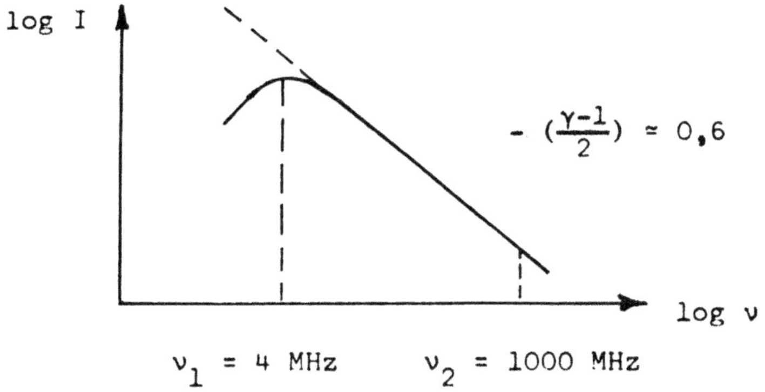

Fig. 8. Spectre du rayonnement synchrotron en provenance de la Galaxie.

gaz comme pour le champ on doit s'attendre à une microstructure compliquée dans les nuages mêmes, à l'échelle du parsec.

Caractéristiques moyennes des nuages:
- Diamètre: $\simeq 6$ pc
- Densité des atomes: 20 cm^{-3}
- Densité: $\simeq 10$ par kpc sur la ligne de visée

La microstructure dans ces nuages concernant à la fois ρ et H entraîne une structure compliquée pour le rayonnement des régions proches, ainsi que des variations d'un point à l'autre de la polarisation de ce dernier; on observe aussi une dépolarisation par effet Faraday différentiel même pour des trajets optiques séparés angulairement par environ 1 minute d'arc, ce qui correspond à des inhomogénéités de l'ordre de 1 à 0,1 pc.

DEUXIÈME LEÇON

ACCÉLÉRATION, INJECTION ET PROPAGATION DES RAYONS COSMIQUES

1. Mécanismes d'accélération

1.1 Généralités

C'est un problème très difficile, surtout quantitativement; on doit en effet expliquer des énergies de 10^{20} eV, soit plus de 1 joule, pour certaines particules. Si l'on refuse l'expédient de considérer les rayons cosmiques comme des "fossiles" de la création de l'Univers, en cherchant des processus susceptibles de s'appliquer également aux rayons cosmiques du soleil ou des supernovae, on est conduit à deux types d'hypothèses:

soit *une accélération systématique* (ou séculaire) due à un processus d'échange d'énergie *réversible* dont une modification de structure du milieu dans la source est responsable. Une compression adiabatique de la source est un exemple; il faut supposer que les particules quittent le milieu assez tôt pour échapper à la réversibilité.

SCHÉMA DES RELATIONS ÉNERGÉTIQUES DANS LA SOURCE DE RAYONS COSMIQUES

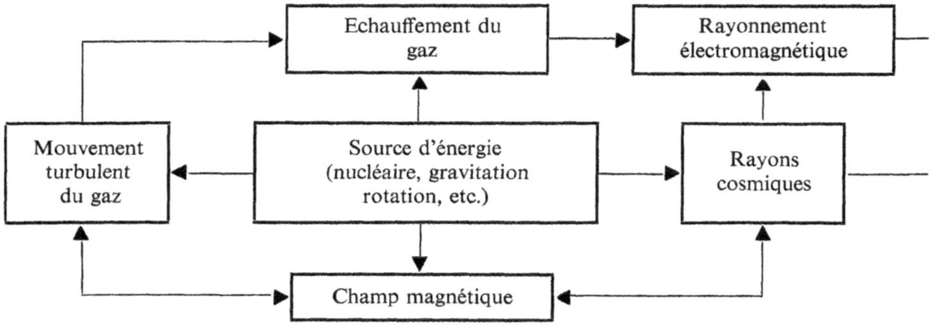

SCHÉMA DE "L'HISTOIRE" DES RAYONS COSMIQUES

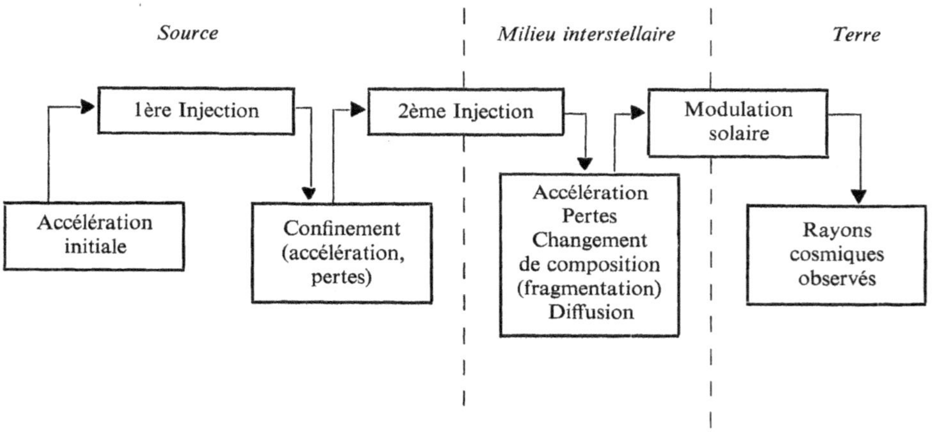

soit *une accélération statistique.* Le transfert d'énergie macroscopique du milieu aux particules s'effectue irréversiblement à travers des gains et des pertes successives, les gains l'emportant en moyenne. Par exemple, l'équipartition, par collisions multiples, entre des particules et des nuages de gaz magnétique peut conduire à l'accélération des particules.

En général, l'accélération est possible par transfert d'énergie du milieu aux particules lorsque les collisions avec les inhomogénéités du milieu sont plus fréquentes que les collisions entre particules. Celles dont l'énergie est assez grande initialement s'écartent de plus en plus de la distribution maxwellienne; il y a tendance vers l'équipartition entre l'énergie cinétique macroscopique du milieu, et l'énergie des particules supra-thermiques. Mais il reste à supposer un "chauffage" préliminaire assez puissant compensant les mécanismes de pertes d'énergie importants aux faibles énergies, si on veut expliquer la présence de nombreuses particules de haute énergie.

Tous ces processus sont électromagnétiques; il ne semble pas, en effet, que dans le milieu généralement ionisé, donc conducteur, de la source, les champs électrostatiques puissent être élevés; l'accélération est donc liée à la variation spatiale ou temporelle du champ magnétique.

L'accélération est toujours *lente* comparée à la gyrofréquence; en tous cas le gain d'énergie ne peut se faire par à-coups pour les noyaux lourds, car ils se désintégreraient.

Rappel des relations entre moment et énergie

	Moment p	Energie cinétique E
	$p = \frac{E^{1/2}(E + 2m_0c^2)^{1/2}}{c}$	$E = \sqrt{(pc)^2 + (m_0c^2)^2} - m_0c^2$
cas classique	$p \simeq \sqrt{2m_0E}$	$E \simeq \frac{p^2}{2m_0}$
cas très relativiste	$p \simeq \frac{E}{c} \simeq \frac{W}{c}$	$E \simeq pc$

Fréquence cyclotron et rayon des trajectoires

$$\omega_H = \frac{ZeH}{mc} \qquad r_H = \frac{v \sin\theta}{\omega_H} = \frac{cp_\perp}{ZeH}$$

$$r_H \simeq \frac{c \sin\theta \sqrt{2m_0E}}{ZeH} \qquad \text{cas classique}$$

$$r_H \simeq \frac{E \sin\theta}{300\, ZH} \qquad \text{cas très relativiste}$$

r_H est exprimé en cm, H en oersteds, E en eV, Z étant le nombre de charge.

Deux processus ont une grande importance pratique: ce sont le processus bêtatron et le processus de Fermi.

1.2 Processus bêtatron

Ce processus a été initialement proposé en 1933 par Swann; il peut être systématique ou statistique.

a) *Augmentation lente du champ*

Ecrivons l'invariabilité du moment magnétique de la particule décrivant une orbite spiralée:

$$\frac{p_\perp^2}{H} = \frac{p^2 \sin^2\theta}{H} = \mathrm{C^{te}} \qquad p_{//} = \mathrm{C^{te}}$$

ce qui est valable au 2e ordre lorsque $H(\mathrm{d}t/\mathrm{d}H) \gg T_{\mathrm{gyr}}$. La quantité $p_\perp$ est la projection du moment sur un plan orthogonal au champ. On a donc:

$$p_\perp^2 \sim H$$

$$r_H = \frac{cp_\perp}{eH} \quad \text{et} \quad H \sim \frac{1}{r_H^2}.$$

Si H augmente, $p_\perp$ aussi et $r_\perp$ diminue. C'est ce qui se produit dans une compression transversale du milieu, qui augmente $p_\perp$ mais non $p_{//}$; le volume varie comme r_H^2 et $p_\perp$ comme $1/V$. Mais il se peut que les chocs avec le milieu rétablissent l'isotropie entre les trois degrés de liberté et alors V varie comme r_H^3. En général, si E est l'énergie cinétique, on a:

$$E = \frac{1}{V^{\gamma-1}} \quad \text{avec } \gamma = \begin{cases} \textit{2 dimensions} \\ 2 \text{ cas non relativiste} \\ 3/2 \text{ cas relativiste} \\ \textit{3 dimensions} \\ 5/3 \text{ cas non relativiste} \\ 4/3 \text{ cas relativiste} \end{cases}$$

Ceci montre que le processus bêtatron équivaut à une compression adiabatique.

b) *Mécanisme bêtatron statistique*

La particule peut rencontrer des variations périodiques de H, qui sont par exemple des sursauts rapides devant la fréquence de collision; il y a à chaque cycle de variation du champ magnétique un gain >1 évalué à

$$K = \frac{1}{3}\sqrt{5 + 2k + \frac{2}{k}} \quad \text{où} \quad k = \frac{H_{\max}}{H_{\min}}.$$

Mais le taux d'accélération ainsi fourni est en général faible et ce processus paraît peu important.

1.3 Processus de fermi

Cette accélération se produit lors de collisions de la particule avec un nuage en mouvement contenant un champ magnétique ("miroir" magnétique). Le miroir se déplace dans le système du laboratoire où l'on mesure l'énergie de la particule. Nous étudierons ce processus dans des cas schématiques.

a) *Collision transversale*

Il y a transfert d'impulsion aux particules à partir du miroir, grâce au champ électrique

$$\mathbf{E} = \frac{-\mathbf{u} \times \mathbf{H}}{c}$$

rencontré à droite (Figure 9). D'où un gain d'énergie:

$$\Delta W = Ze\int \mathbf{E}\cdot d\mathbf{s} = 2Ze\frac{uH}{c} r_H \sin\theta$$

$$\Delta W = -\frac{2W}{c^2}\,\mathbf{u}\cdot\mathbf{v}$$

W étant l'énergie totale de la particule mc^2.

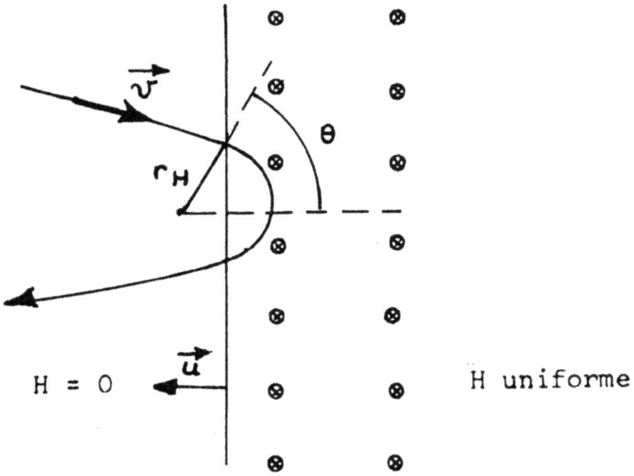

Fig. 9. Réflexion d'une particule par un champ.

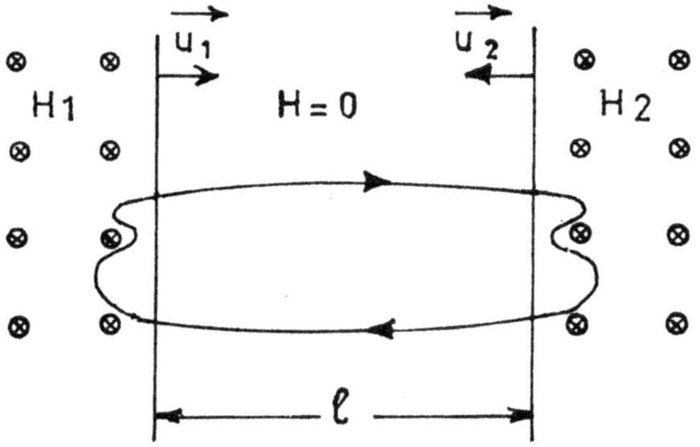

Fig. 10. Collision entre deux "miroirs magnétiques" qui se rapprochent.

b) *Collisions entre deux miroirs qui se rapprochent*

C'est le cas de la Figure 10. On a alors

$$\frac{\mathrm{d}W}{\mathrm{d}t} = \frac{\Delta W_1 + \Delta W_2}{\Delta t}$$

$$\frac{\mathrm{d}W}{\mathrm{d}t} = - W \frac{v^2}{c^2} \frac{1}{l} \frac{\mathrm{d}l}{\mathrm{d}t}.$$

On peut montrer l'équivalence avec une compression à une dimension.

$$E \sim \frac{1}{V^{\gamma - 1}} \quad \text{où} \quad \begin{cases} \gamma = 3 & \text{non relativiste} \\ \gamma = 2 & \text{cas relativiste} \end{cases}$$

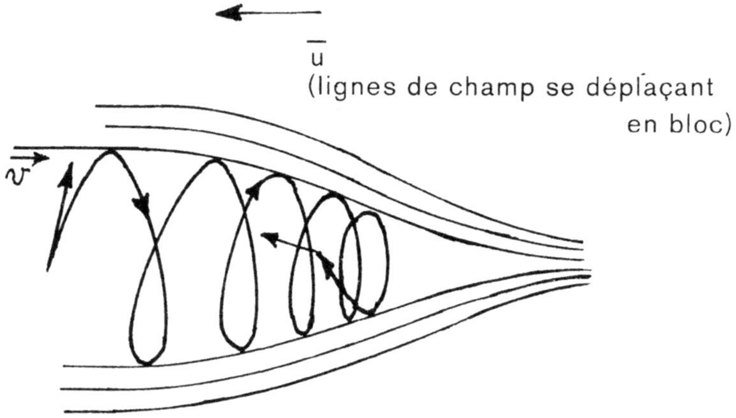

Fig 11. Mouvement d'une particule dans le "col" d'une "bouteille magnétique".

c) *Collision longitudinale*

On peut imaginer une "bouteille" magnétique se déplaçant et qui réfléchit les particules; le calcul donne le même résultat que pour une collision transversale. Le phénomène est illustré par la Figure 11.

d) *Processus de Fermi statistique*

On considère un ensemble de nuages réfléchissants se mouvant de façon aléatoire. Si on se limite pour simplifier aux collisions de face pour lesquelles

$$\Delta W = \frac{2W}{c^2} uv$$

et que l'on remarque que la probabilité d'un collision avec rapprochement est $p_1 = (u + v/2v)$; mais que la probabilité avec éloignement est seulement $p_2 = (v + u)/2v$, on constate en moyenne un gain par collision

$$\overline{\Delta W} = \frac{2u^2 W}{c^2}.$$

Si on tient compte des chocs obliques, on a

$$\overline{\Delta W} = \zeta \frac{u^2}{c^2} W$$

avec $\zeta \simeq 1$; par exemple $\zeta = \frac{4}{3}$ pour des collisions diffusantes élastiques. Grâce à l'expression du libre parcours moyen $l = v\tau$, où v est la vitesse entre deux collisions et τ le temps qui les sépare, on peut écrire

$$\frac{dW}{dt} = \zeta \frac{u^2}{c^2} \frac{W}{\tau} = \alpha \frac{v}{c} W$$

où $\alpha=\zeta(u^2/lc)$; si l_0 est le diamètre des nuages, dont la densité est N_c, on a $l=(\pi/3\, l_0^2\, N_c)^{-1}$.

1.4 Cas réels d'accélération

Ils sont tous en réalité plus complexes; on a étudié par exemple la compression d'une "bouteille" magnétique, ou un processus de Fermi par réflexion sur une onde de choc et des nuages diffusants fixes.

2. Le rôle des pertes et des fuites

2.1 Accélération et pertes d'énergie

Les pertes déterminent en grande partie le spectre d'énergie. En effet, pour qu'une particule soit accélérée, le taux d'accélération doit évidemment surpasser le taux des pertes. Supposons schématiquement l'existence des deux processus concurrents:

$$\text{Accélération} \quad \frac{\mathrm{d}E}{\mathrm{d}t} = f(E)$$

$$\text{Pertes} \quad \frac{\mathrm{d}E}{\mathrm{d}t} = -\,g(E).$$

La condition nécessaire pour qu'on ait une accélération s'écrit

$$f(E) > g(E).$$

En général, f et g sont de la forme:

$$f(E) \sim aE^{\alpha} \qquad \alpha \geqq 0$$

$$g(E) \sim bE^{-\beta} \qquad \beta \geqq 0 \quad \text{aux basses énergies.}$$

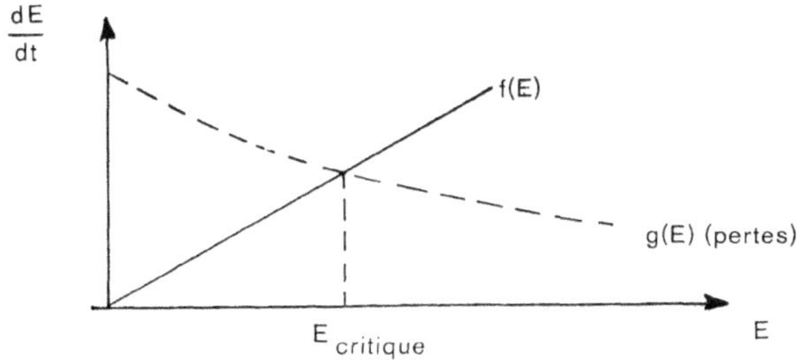

Fig. 12. Accélération et pertes d'énergie en fonction de l'énergie de la particule.

On voit sur la Figure 12 que seules les particules dont l'énergie est supérieure à l'énergie critique pour laquelle f et g sont égaux seront accélérées. Ceci oblige à envisager l'existence préliminaire d'un "préchauffage" afin qu'il y ait dès le début des particules de vitesse suffisante en nombre important; ce peut être les particules

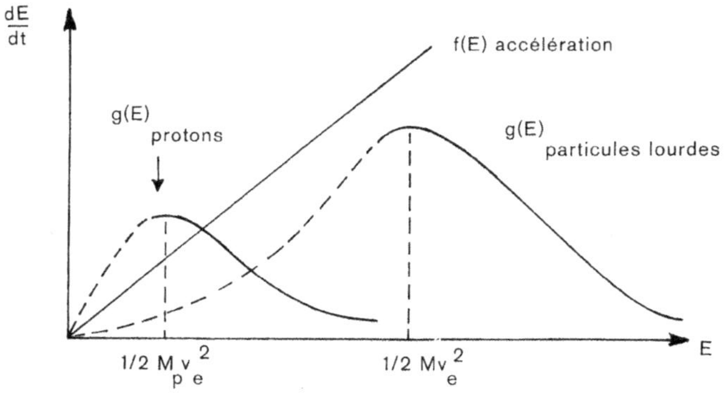

Fig. 13. Accélération $f(E)$ et pertes d'énergie $g(E)$ des protons et des particules plus lourdes que des protons. v_e est de l'ordre de la vitesse de l'électron sur la première orbite de Bohr de l'hydrogène, pour un milieu neutre. Dans un milieu ionisé,

$$v_e = \sqrt{\frac{3kT}{m_e}}.$$

provenant de réactions nucléaires. En outre le processus initial doit être rapide, les processus statistiques n'intervenant qu'en 2ème étape.

L'énergie caractéristique dépend du type de la particule, (Z et M_0); les pertes principales sont dues à l'ionisation et ont en réalité l'aspect de la Figure 13. La valeur de v_e est de l'ordre de la vitesse de l'électron sur la première orbite de Bohr de l'hydrogène, pour un milieu neutre; dans un milieu ionisé, $v_e = \sqrt{3kT/m_e}$. On constate qu'il n'y a pas de seuil pour les particules très lourdes, qui peuvent être accélérées même si leur énergie initiale est faible: ceci pourrait expliquer la surabondance de particules lourdes dans le rayonnement cosmique. Pour les électrons et les protons les pertes par ionisation sont semblables.

2.2 Spectre d'énergie après l'injection initiale

Un calcul ancien dû à Fermi est fondé sur un équilibre fuites-accélération: on suppose que l'accélération est due au processus de Fermi statistique et que les particules ont initialement toutes la même énergie totale $W_0 \simeq m_0 c^2$. On a alors $\mathrm{d}W/\mathrm{d}t = \alpha W$ et $W = m_0 c^2\, e^{\alpha t}$.

On exprime les pertes en supposant que la probabilité $\omega(t)\mathrm{d}t$ pour qu'une particule disparaisse à l'âge t est de la forme:

$$\omega(t)\,\mathrm{d}t = \frac{1}{T}\, e^{-t/T}\,\mathrm{d}t$$

et est indépendante de W. Ceci correspond à des annihilations par collisions. On trouve alors que le nombre $N(W)\mathrm{d}W$ des particules dont l'énergie est comprise entre W et $W + \mathrm{d}W$ est:

$$N(W)\,\mathrm{d}W = \frac{W_0^{1/\alpha T}}{\alpha T}\, W^{-(1+1/\alpha T)}.$$

On a donc un spectre en loi de puissance d'indice spectral

$$\gamma = 1 + \frac{1}{\alpha T}.$$

L'application à la Galaxie échoue si on prend pour la durée de vie une valeur $T \ll 3 \times 10^8$ ans (indépendante de la particule et de la source); en effet pour un libre parcours moyen $l > 1$ parsec et $u = 10$ km sec^{-1} et en se souvenant que $\alpha = u^2/lc$, on obtient $1/\alpha T \gg 10$, c'est à dire un profil trop abrupt ($\gamma = 2{,}5$ est la valeur fournie par l'observation).

Néanmoins le principe du calcul reste valable: d'une façon générale, le spectre *dans la source*, est partiellement défini par le temps que les particules passent dans celle-ci; le problème du spectre est lié à celui de l'injection.

3. L'injection (étude thermodynamique)

Supposons que dans la source il y ait constamment équipartition entre l'énergie des rayons cosmiques, l'énergie de turbulence et l'énergie magnétique. Soit N le nombre total de particules cosmiques, $\bar{E}$ leur énergie moyenne. On a:

$$W_{\mathrm{Rc}} = \bar{E}N \quad W_{\mathrm{tot}} = 3N\bar{E}.$$

La perte d'énergie est due aux fuites; elle vaut:

$$\mathrm{d}W = \bar{E}\,\mathrm{d}N = \mathrm{d}(3N\bar{E})$$

ce qui donne

$$N = a(\bar{E})^{-1,5}.$$

Ceci implique en certains cas un spectre en loi de puissance qui est calculable à partier du spectre d'énergie dans la source. Par exemple, si celui-ci s'écrit

$$N(E, t)\,\mathrm{d}E = N(t) f(E, t)\,\mathrm{d}E \quad \text{avec} \quad \int_0^\infty f(E, t)\,\mathrm{d}E = 1$$

où $N(t)$ est la densité totale, on obtient

$$\bar{E}(t) = \int_0^\infty f(E, t) E\,\mathrm{d}E$$

qu'on insérera dans l'expression $N(t) \sim (\bar{E}(t))^{-(\gamma-1)}$ où $\gamma = 2{,}5$.

Le spectre des particules qui ont quitté la source est:

$$N(E) = \int_{t=0}^\infty f(E, t) \frac{\mathrm{d}N(t)}{\mathrm{d}t}\,\mathrm{d}t \sim \int_{\bar{E}(0)}^\infty f(E, t)(\bar{E}(t))^{-\gamma}\,\mathrm{d}\bar{E}(t).$$

Dans le cas simple où $f(E, t) = \delta(E - \bar{E}(t))$ on trouve bien $N(E) \sim E^{-\gamma}$.

Mais ce n'est pas la seule forme de $f(E, t)$ qui fournisse une telle loi de puissance pour les particules injectées; on peut aussi obtenir d'autres valeurs de γ si on tient compte d'une expansion éventuelle. En tout cas, on a une latitude suffisante pour supposer que le spectre des particules injectées a peu de rapport avec le spectre interne; ceci pouvait expliquer l'universalité des spectres des rayons cosmiques et des électrons relativistes issus de sources de nature très différente.

4. Propagation des rayons cosmiques dans la Galaxie. Rôle du champ magnétique

Au cours de la propagation des rayons cosmiques, à partir de la source, deux modèles extrêmes de mouvement à travers la Galaxie sont possibles:

Soit qu'ils "suivent" les lignes de forces (mouvement "régulier" ou "adiabatique").

Soit qu'ils sont diffusés par les inhomogénéités et ne suivent pas les lignes de forces (mouvement avec diffusion).

4.1 Mouvement "régulier" ou "adiabatique"

Le champ n'a alors que des variations à grande échelle; le fait que les rayons cosmiques "suivent" les lignes de force a des répercussions sur leur isotropie. Par exemple, au cours du passage d'une région de champ faible H_0 à une région de champ fort H_1, certaines particules sont réfléchies. En effet, le moment magnétique transversal est constant:

$$\frac{p_\perp^2}{H} = \frac{p^2 \sin^2 \theta}{H} = \mathrm{C^{te}}.$$

Comme le montre la Figure 14, on a réflexion des particules si $\theta_1 = \pi/2$, donc de celles pour qui $\theta_0 < \arcsin\sqrt{H_0/H_1}$. On en déduit que l'isotropie en A entraîne celle en B; mais si les particules sont isotropes en B, elles ne le seront pas en A. En conséquence, si les particules proviennent, comme il est probable, des régions de champ fort, elles ne devraient pas être isotropes.

En outre, il est difficile de rendre compte d'un rayonnement synchrotron dans les régions de champ faible. En effet, la densité des électrons relativistes y est plus faible

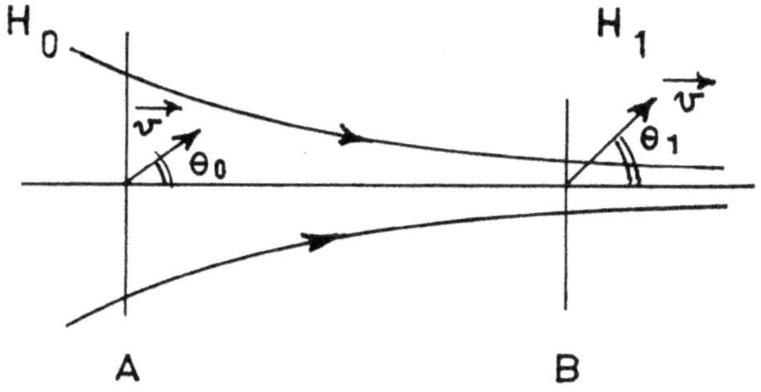

Fig. 14. Réflexion magnétique des particules.

(d'après le théorème de Liouville); d'autre part l'intensité du rayonnement est faible si θ est faible puisque

$$I \sim H^{(\gamma+1)/2} (\sin\theta)^{(\gamma+1)/2}$$

$\sin\theta$ est proportionnel à $H^{1/2}$; si on fait intervenir la densité, on trouve une variation avec H de l'intensité de la forme:

$$I \sim H^{(3\gamma+7)/4}$$

l'exposant vaut 3,6 pour $\gamma = 2{,}5$.

Le modèle régulier est finalement peu plausible dans la Galaxie.

4.2 Mouvement avec diffusion

Dans ce modèle, la particule est diffusée par des inhomogénéités magnétiques, ce qui suppose la non-conservation de l'invariant adiabatique $P_{\perp}^2 H$; ceci nécessite que le champ varie sensiblement sur des distances de l'ordre du rayon de giration. Plusieurs mécanismes physiques peuvent être responsables de la diffusion: ainsi la diffusion par un front d'ondes de choc, ou bien celle résultant d'un gradient de champ. On a surtout envisagé le cas de nuages diffusants à champ fort, séparés par des régions de champ faible; ce processus est efficace si on a la condition $r_{H_0} \lessgtr l_0$, où r_{H_0} est le rayon de giration et l_0 le diamètre moyen des nuages, dont le nombre par unité de volume est N_c. Alors, le libre parcours moyen l, et le coefficient de diffusion D sont donnés par les expressions:

$$l = \left(\frac{\pi}{3} l_0^2 N_c\right)^{-1} \qquad D = \frac{1}{3} l v \simeq \frac{C}{3 l_0^2 N_c}$$

d'où le flux F_0 à travers une surface normale au gradient de la densité ∇N:

$$\mathbf{F}_0 = -D \nabla N.$$

Dans le cas où la source est ponctuelle, on a, à la distance r:

$$\mathbf{F}_0 = -D \frac{\mathrm{d}N}{\mathrm{d}r}.$$

Lorsque $r_{H_0} \gg l_0$, on a au contraire une déflexion faible $\sim l_0/r_{H_0}$, et une longueur caractéristique de diffusion l_d remplace alors le "libre parcours moyen". Cette longueur a pour valeur $l_d = (r_{H_0}^2 / l_0^2) l$; le coefficient de diffusion est alors:

$$D = \tfrac{1}{3} l_d v$$

(l'expression de r_H est donnée plus haut, au paragraphe 1.1 de la deuxième leçon).

Les valeurs du coefficient D sont assez faibles pour que l'approximation de diffusion soit valable, on peut en déduire le degré d'anisotropie des rayons cosmiques.

$$\delta = \frac{I_{\max} - I_{\min}}{I_{\max} + I_{\min}} \qquad I(\theta) = I + \delta I \cos\theta$$

δ est lié au flux F_0 et à l'intensité I par

$$F_0 = \frac{4\pi}{3}\,\delta I = 2\pi \int_0^{\pi} I(\theta) \cos\theta \sin\theta \, d\theta .$$

Dans le cas d'une source ponctuelle, et puisque $N = 4\pi I/c$ on a:

$$\delta = \frac{3D}{cN}\left|\frac{dN}{dr}\right| .$$

4.3 Cas réels

Ils sont probablement intermédiaires entre ces modèles, la diffusion pouvant être préférentielle le long du champ. Dans un modèle correct, D devrait donc dépendre de la direction.

Ce qu'on connaît de la structure du champ dans la Galaxie montre que la diffusion est tout à fait possible dans la Galaxie jusqu'aux énergies élevées, peut-être 10^{15} eV.

TROISIÈME LEÇON

PROPAGATION: AUTRES EFFETS

Ayant considéré l'effet du champ magnétique, sur la propagation des rayons cosmiques, il reste à étudier l'effet des collisions avec les noyaux interstellaires, et les pertes d'énergie au cours de la propagation.

1. Fragmentation par collisions

Ce sont surtout des collisions avec les protons; 10% du nombre total de collisions ont lieu également avec les noyaux de l'hélium interstellaire.

1.1 Paramètres liés aux collisions

La collision de type i est caractérisée par sa section efficace σ_i; pour les collisions $p-p$ et $p-\alpha$, les sections efficaces ont été mesurées; pour les autres on peut en gros prendre $\sigma = \pi r^2$, avec

$$r = 1{,}26 \times 10^{-13} A^{1/3} \text{ cm}$$

pour les collisions contre des protons, A étant le numéro atomique.

Le libre parcours moyen de collision est $l_i = 1/n\sigma_i$. On exprime plutôt la quantité en masse par cm^2 de milieu traversé, soit un nombre Λ_i obtenu en multipliant l_i par la densité ρ du gaz:

$$\Lambda_i = \frac{\rho}{n\sigma_i} = \frac{M}{\sigma_i} \quad (\text{en g cm}^{-2})$$

où M est la masse des noyaux. On trouvera certaines valeurs au Tableau IV.

1.2 Fragmentation

Les collisions peuvent conduire à une fragmentation en noyaux plus petits, au cours de laquelle on a conservation de l'énergie par nucléon. Ces processus ne sont en général connus que grâce à la fragmentation dans les émulsions nucléaires. Les probabilités P_{ik} de former un noyau i à partir d'une collision entre une particule k et un noyau d'hydrogène sont données au tableau 4, dans lequel on a l'inégalité $i \leq k$.

TABLEAU IV

Probabilités de fragmentation P_{ik} (voir signification des symboles d'espèce au Tableau V)

Espèces i \ k	H	M	L	α
H	0,31	0,36	0,12	1,35
M		0,11	0,28	1,22
L			0,15	1,09
α				0,41

TABLEAU V

(Avec $n = 0{,}01\ \text{cm}^{-3}$; $\rho = 2{,}2 \times 10^{-26}\ \text{gcm}^{-3}$)

Espèce composante	A_{moyen}	$\Lambda\,(\text{g cm}^{-2})$	$\lambda_{\text{absorption}}$	$T_{\text{absorption}}$ (ans)	Composition sur Terre	Composition dans la source
p	1	72	72	$3{,}6 \times 10^9$	680	63–360
α	4	20	34	$1{,}7 \times 10^9$	46	4,5–21
L	10	8,7	10	$5{,}0 \times 10^8$	1	0
M	14	6,9	7,8	$3{,}9 \times 10^8$	3,0	1,6–2,6
H	31	4,2	6,1	$3{,}0 \times 10^8$	1	1
VH	51	3,0	4,4	$2{,}2 \times 10^8$	0,3	0,34–0,48

1.3 Libre parcours d'absorption

C'est le libre parcours moyen au bout duquel le noyau se transforme en des fragments d'autres espèces. On a:

$$\lambda_i = \frac{\Lambda_i}{1 - P_{ii}}$$

La durée de vie vis-à-vis de l'absorption est

$$T_i = \frac{\lambda_i}{\rho c} \quad \text{si} \quad v \approx c\,.$$

2. Création des électrons secondaires

Ces électrons proviennent de la désintégration des mésons; le processus le plus efficace est la collision $p-p$:

$$p+p \begin{cases} \nearrow p+p+\pi^0 & \text{suivi de } \pi^0 \to 2\gamma \\ \searrow p+n+\pi^+ & \text{suivi de } \pi^+ \to \mu^+ + \nu \\ & \text{suivi de } \mu^+ \to e^+ + \nu + \bar{\nu} \\ & \searrow \text{désintégration } \beta \text{ donnant des électrons de faible énergie.} \end{cases}$$

Dans le cas des noyaux lourds, on a aussi:

$$p+n \begin{cases} p+p+\pi^- \\ p+n+\pi^0 \\ n+n+\pi^+ \end{cases} \quad \text{et} \quad \begin{cases} n^- \to \mu^- + \nu, \text{ suivi de} \\ \mu^- \to e^- + \nu + \bar{\nu} \\ \pi^0 \to 2\gamma \\ \pi^+ \text{ désintégration déjà vue} \end{cases}$$

La fraction de l'énergie d'un proton transferée aux électrons crées pendant une collision $p-p$, κ_e, est environ 0,05; cette valeur est presque independante de l'énergie du proton incident. La multiplicité (nombre de particules secondaires issues d'une collision primaire) est donnée par n_μ, qui varie avec l'énergie selon la loi:

$$n_\mu = 2E^\delta \quad \begin{array}{l} E \text{ étant exprimé en GeV} \\ \delta = 0{,}20 \pm 0{,}05 \end{array}$$

On peut calculer[1] le spectre d'énergie des électrons secondaires

$$N_e(E)\,dE = K_e E^{-\gamma_e}\,dE$$

à partir du spectre des protons incidents

$$N(E)\,dE = KE^{-\gamma}\,dE.$$

En effet, les protons d'énergie E produisent des électrons dont chacun a l'énergie $E' = \kappa_e E/2E^\delta$, avec une densité $N_e(E')$ telle que

$$N_e(E')\,dE' = 2E^\delta N(E)\,dE$$

d'où

$$N_e(E') = 2E^\delta N(E)\frac{dE}{dE'} = 2E^\delta KE^{-\gamma}\left(\frac{2E^\delta}{\kappa_e(1-\delta)}\right).$$

En exprimant E en fonction de E' dans cette expression, on obtient bien pour les électrons secondaires un spectre de la forme

$$N_e(E) = K_e E^{-\gamma_e}$$

avec:

$$\gamma_e = \frac{\gamma - 2\delta}{1 - \delta}$$

$$K_e = \frac{2K}{1 - \delta}\left(\frac{\kappa_e}{2}\right)^{\frac{\gamma - 1 - \delta}{1 - \delta}}.$$

Dans l'hypothèse où $\gamma = 2{,}5$ et avec $\kappa_e = 0{,}05$ et $\delta = 0{,}20$, on a:

$$\gamma_e = 2{,}64 \quad \text{et} \quad \frac{K_e}{K} = 6 \times 10^{-3}.$$

Le spectre des électrons est donc très semblable à celui des protons; mais il y a toujours moins de $\frac{1}{100}$ d'électrons secondaires. En dessous de 1 GeV, les approximations ci-dessus ne sont plus valables, le spectre a l'allure de la Figure 15.

Le rapport positrons/électrons (e^+/e^-) est supérieur à 50%, compte tenu des collisions avec les noyaux lourds; lorsque E augmente, ce rapport tend vers 1. Pollack et Fazio[2] ont trouvé, en tenant compte des collisions $p-p$ et $p-\alpha$, les valeurs suivantes pour les électrons secondaires:

TABLEAU VI

E (GeV)	e^+/e^-	% de e^+
0,5	1,88	65
1,0	1,58	61
2,0	1,41	58
4,0	1,29	56

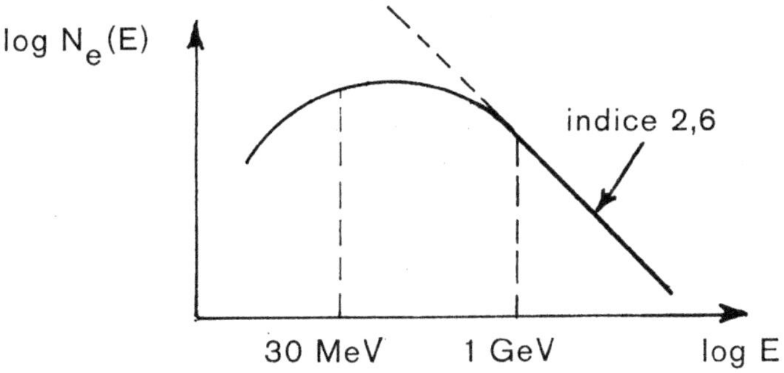

Fig. 15. Spectre des électrons.

3. Quelques conséquences immédiates des résultats expérimentaux

3.1 Surabondance des éléments légers (L)

Les éléments L ne peuvent être que secondaires, car ils sont détruits par les réactions nucléaires et sont très rares dans les sources; ceci indique que les rayons cosmiques ont traversé quelques $\mathrm{g\,cm^{-2}}$ de matière. L'abondance élevée de l'hélium 3 a la même origine; si $He^3/(He^4+He^3)=0{,}1$, l'épaisseur moyenne traversée est 4 $\mathrm{g\,cm^{-2}}$ environ. Ces nombres sont quelque peu fonction du modèle choisi pour la propagation dans la Galaxie.

3.2 Composition du rayonnement électronique

Nous reviendrons plus loin sur ce problème, mais on peut déjà faire les remarques suivantes: il ne semble pas y avoir plus de 20% de positrons, en dessous de 1 GeV et plus de 30% au dessus de 4,5 GeV. Si ces résultats expérimentaux sont exacts, la plupart des électrons doivent donc être primaires. Ceci parait confirmé d'autre part:

a) par le fait que les rayons cosmiques contiennent 1% d'électrons, alors que la théorie donne une valeur moindre si les électrons sont secondaires; également, par le fait que les électrons perdent leur énergie plus vite que les rayons cosmiques. Cependant, si le trajet des protons est de 5 à 10 $\mathrm{g\,cm^{-2}}$, ils produisent 10 à 20 fois plus d'électrons secondaires qu'il n'en est observé. Ces derniers doivent donc fuir ou perdre rapidement de l'énergie.

b) par le fait que les électrons aux basses énergies ne montrent pas, vers 30 MeV, le maximum spectral caractéristique des électrons secondaires (Cline *et al.* [6]). Mais ceci n'est pas absolument sûr.

c) enfin, si la Galaxie contient $10^9\,M_\odot$ de gaz et si la densité des rayons cosmiques est uniforme, la création d'électrons secondaires d'énergie >1 GeV ne correspond pas à un apport d'énergie supérieur à 10^{37} $\mathrm{erg\,sec^{-1}}$. Or, l'émission synchrotron de la Galaxie donne probablement 3 à 5×10^{38} $\mathrm{erg\,sec^{-1}}$. Cependant ce dernier chiffre est peut-être surestimé si la contribution extragalactique, qu'il est difficile de séparer de l'émission galactique, est importante.

4. Les pertes d'énergie

4.1 Noyaux lourds

Les pertes se produisent par 4 mécanismes essentiels:

a) Les interactions nucléaires (étudiées précédemment)

b) l'ionisation: les pertes par ionisation sont données par l'expression

$$-\frac{\mathrm{d}E}{\mathrm{d}t}\sim\frac{Z^2}{A}v$$

aux très basses énergies; aux très hautes énergies, le processus devient inefficace. La Figure 16 donne l'allure des pertes par ionisation.

c) la photofission subie au cours d'un choc sur un photon "thermique" (c'est à dire photon lumineux ou radio). Ce mécanisme intervient pour des énergies supérieures à 10^{14} eV/nucléon et se distingue de l'effet Compton inverse, qui ne donne pas de fission. Hayakawa et Matsuoka [3] y voient la cause des rayons γ interstellaires.

d) expansion adiabatique.

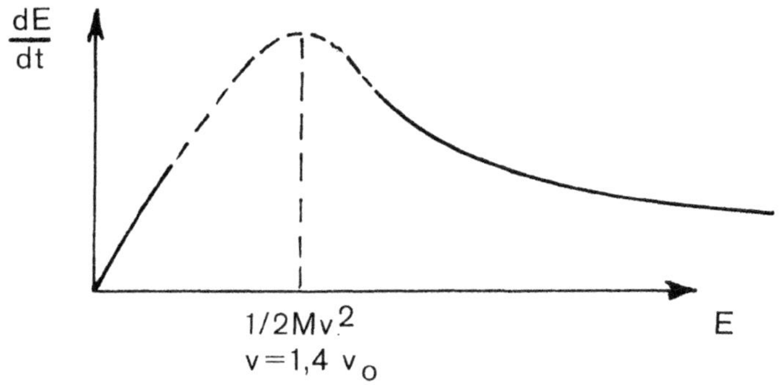

Fig. 16. Pertes d'énergie des noyaux lourds par ionisation. v est de l'ordre de la vitesse de l'électron sur la première orbite de Bohr de l'hydrogène.

4.2 Electrons

Les pertes d'énergie sont plus importantes que pour les noyaux. On peut donner quelques expressions des pertes essentielles:

a) *Ionisation* (pour des électrons relativistes)

$$-\frac{\mathrm{d}E}{\mathrm{d}t} = 7.62 \times 10^{-9} n \left\{18.8 + 3 \log \frac{E}{m_0 c^2}\right\}$$

où $\mathrm{d}E/\mathrm{d}t$ est exprimé en eV sec^{-1} et n en nombre d'atomes cm^{-3}. Dans un milieu ionisé:

$$-\frac{\mathrm{d}E}{\mathrm{d}t} = 7.62 \times 10^{-9} n \left\{73{,}4 + \log \frac{E}{m_0 c^2} - \log n\right\}.$$

b) *Bremsstrahlung* (rayonnement de freinage)

$$-\frac{\mathrm{d}E}{\mathrm{d}t} = 7.26 \times 10^{-16} nE \quad \text{où } t \text{ est exprimé en secondes.}$$

c) *Rayonnement synchrotron* (champ magnétique)

$$-\frac{\mathrm{d}E}{\mathrm{d}t} = 0.98 \times 10^{-3} H_\perp^2 \left(\frac{E}{m_0 c^2}\right)^2$$

où $\mathrm{d}E/\mathrm{d}t$ est exprimé en eV sec^{-1} et $H_\perp$ en gauss.

d) *Diffusion Compton inverse*

$$-\frac{\mathrm{d}E}{\mathrm{d}t} = 2 \times 10^{-14} W_{\mathrm{ph}} \left(\frac{E}{m_0 c^2}\right)^2 \quad \text{(en eV sec}^{-1}\text{)}$$

W_{ph} étant la densité d'énergie du champ électromagnétique, en eV cm^{-3}. Pour la Galaxie, on a $H \approx 10^{-5}$ gauss et $W_{\mathrm{ph}} \simeq 1$ eV cm^{-3}. Cette perte est légèrement moins importante que la précédente.

5. Equations de transfert (d'après Ginzburg et Syrovatskii [1])

5.1 Forme générale

L'inconnue est, pour chaque espèce i de particule, le nombre de particules i par unité de volume N_i $(E, \boldsymbol{r}, t)$, fonction du temps t, de la position $\boldsymbol{r}$ et de l'énergie E. Le système d'équations différentielles donnant N_i est très complexe, si on veut tenir compte de tous les facteurs qui sont susceptibles d'intervenir:

$$\frac{\partial N_i}{\partial t} - \nabla(D_i \nabla N_i) + \frac{\partial}{\partial E}\left(\left(\frac{\mathrm{d}E}{\mathrm{d}t}\right)_i N_i\right) - \frac{1}{2}\frac{\partial^2}{\partial E^2}\left(\frac{\mathrm{d}}{\mathrm{d}t}\overline{(\Delta E^2)}\,N_i\right)$$

(termes) (1) (2) (3) (4)

$$= Q_i(E, \boldsymbol{r}, t) - p_i N_i + \sum_{k>i} \int p_{ik}(E', E)\, N_k(E', \boldsymbol{r}, t)\, \mathrm{d}E' .$$

(5) (6) (7)

Les termes représentent

(2) la diffusion; si $D_i = 0$, le mouvement est régulier (pas de diffusion).

(3) les pertes et gains systématiques, supposés continus.

(4) les fluctuations dans ces pertes et ces gains (généralement négligées).

(5) la fonction source (nombre de particules i injectées par unité de volume et de temps).

(6) les pertes d'énergie "catastrophiques" avec disparition de la particule d'espèce i:

$$p_i = \frac{1}{T_i} = nv\sigma_i$$

(7) le flux de particules d'espèce i et d'énergie E, provenant de la dégradation de particules d'espèce k et d'énergie E'. P_{ik} est la probabilité qu'une particule k d'énergie E' donne une particule i d'énergie E.

5.2 Simplifications

a) *L'espèce i est un noyau*

On prend comme variable l'énergie par nucléon ε, qui se conserve dans la fragmentation, on a alors:

$$p_{ik}(\varepsilon', \varepsilon) = p_i^k \delta(\varepsilon' - \varepsilon)$$

$$p_i^k = \frac{P_{ik}}{T_k}$$

où T_k est la durée de vie d'absorption pour l'espèce k. Le terme (7) devient $\sum_k p_i^k N_k$ avec p_i^k indépendant de ε. Les quantities P_{ik} sont données dans le Tableau IV.

On suppose que la fonction source est de la forme

$$Q_i = q_i x(E, \boldsymbol{r}, t)$$

la fonction $x(E, \boldsymbol{r}, t)$ étant universelle; q_i est un nombre.

Les termes (3) et (4) sont négligés, et on suppose que D_i est indépendant de i. Il reste un système d'équations

$$\frac{\partial N_i}{\partial t} - D_i \nabla^2 N_i = Q_i - p_i N_i + \sum_k p_i^k N_k$$

que l'on peut résoudre analytiquement ou numériquement dans des modèles simples.

b) *L'espèce i est l'électron*

On néglige les termes (4) et (7) et il reste:

$$\frac{\partial N_e}{\partial t} - D\nabla^2 N_e + \frac{\partial}{\partial E}\left(\frac{\mathrm{d}E}{\mathrm{d}t} N_e\right) + \frac{N_e}{T} = Q(E, \boldsymbol{r}, t)$$

5.3 Quelques solutions dans le cas de la Galaxie; application à la composition chimique du rayonnement cosmique

Pour résoudre le système d'équations donnant N_i, on se fixe un modèle; Ginzburg et Syrovatskii [1] en proposent sept; nous en retiendrons six.

	Modèle	Source	Particularités
(I)	Diffusion	Source Centrale	Départ libre ($N_i = 0$ pour $r = R$) $R \simeq 10$ kpc
(II)	Diffusion	Source Centrale	Diffusion dans un espace infini
(III)	Régulier	Source à x gcm^{-2}	Mouvement le long des lignes de force
(IV)	Régulier	Sources *extérieures*	N_i donné pour $r = R$
(V)	Régulier	Source Centrale	Particules piégées
(VI)	Homogène	Sources partout	Particules piégées

La quantité de matière traversée est:

$$x = \rho c t_D = \frac{\rho c r^2}{2D} \quad (r^2 = 2Dt_D)$$

On se fixe pour étudier le modèle:

(i) le modèle

(ii) la densité et la répartition du gaz dans la Galaxie

(iii) l'absence initiale d'éléments L ($q_L = 0$)

(iv) la composition observée des rayons cosmiques sur terre

On en déduit:

(1) x(gcm^{-2}) et D à partir de (i) et (iii)

(2) la composition q_i dans la source

On peut éliminer d'emblée les modèles (V) et (VI) qui donnent $q_L < 0$ (on devrait observer dans ces modèles encore bien davantage d'éléments L que dans les rayons cosmiques).

Les autres modèles restent possibles, à cause des imprécisions sur les probabilités de fragmentation, etc. On trouve approximativement:

(I) $x = 13$ gcm^{-2}
(II) $x = 9$ gcm^{-2}
(III) $x = 6$ gcm^{-2}
(IV) $x = 30$ gcm^{-2}

L'étude de ces modèles conduit aux remarques générales suivantes:

a) Il faut qu'il y ait des fuites hors de la Galaxie, ce qui est d'ailleurs naturel dans un modèle à diffusion.

b) Dans tous les cas, il y a dès l'origine des éléments lourds en excès.

c) On ne peut éliminer par des arguments de composition chimique ni le modèle régulier (III) ni le modèle extragalactique (IV) pour lequel on doit pourtant supposer l'existence d'éléments L avant l'arrivée dans la Galaxie.

Dans le cas des modèles (I) et (II) à source centrale, on peut évaluer des valeurs limites pour le coefficient de diffusion.

D'une part

$$D = \frac{\rho c r^2}{2x} = 2 \times 10^{31} \frac{n}{x}$$

(en C.G.S.; $r = 10$ kpc, distance du soleil au centre galactique). Avec $x = 10$ gcm^{-2} et $n \approx 10^{-2}$ cm^{-3} on obtient $D \approx 2 \times 10^{28}$ cm^{2}sec^{-1}.

D'autre part l'absence d'anisotropie implique $D < 10^{29}$ cm^{2}sec^{-1}. Si l'on suppose enfin $v = 10^{10}$ cmsec^{-1} et $D = \frac{1}{3} l \bar{v}$, et un ordre de grandeur raisonnable pour l (1 à 10 pc) on obtient des valeurs de D de même ordre. On a donc:

$$10^{28} \text{ cm}^2 \text{ sec}^{-1} < D \leqq 10^{29} \text{ cm}^2 \text{ sec}^{-1} .$$

On en déduit la durée de vie vis-à-vis des fuites

$$T_{\text{fuites}} = \frac{r^2}{2D} \approx 3 \times 10^8 \text{ ans}$$

avec $D = 10^{29}$ cm^2 sec^{-1}
et $r = 15$ kpc, rayon de la Galaxie.

La puissance injectée sous forme de rayons cosmiques peut être aussi déterminée:

modèle (I)	3,0 à 13×10^{40} ergsec^{-1},
modèle (II)	1,5 à 6×10^{40} ergsec^{-1}.

5.4 Cas des électrons

Ginzburg et Syrovatskii [1] ont calculé le spectre des électrons secondaires dans les modèles I et II en utilisant la densité des rayons cosmiques mesurée sur Terre et les données suivantes:

$r_{galaxie} = 12$ kpc	volume $V = 2 \times 10^{68}$ cm^3
$n = 0{,}01$ cm^{-3}	$W_{ph} = 1$ eVcm^{-3}
$D = 1{,}4 \times 10^{29}$ cm^{2}sec^{-1}	$H = 3 \times 10^{-6}$ oersted
($T_{diff} = 1{,}5 \times 10^8$ ans)	

Après avoir tenu compte des pertes Compton et synchrotron, ils constatent que pour le modèle (II) le nombre calculé d'électrons secondaires est 3 fois inférieur au nombre observé par EARL [4]. L'hypothèse secondaire pour les électrons exige donc d'augmenter T_{diff}.

D'autre part, le rayonnement synchrotron est 10 fois trop faible avec la valeur de T_{diff} admise par GINZBURG et SYROVATSKII [1]; pourtant une valeur $H = 10^{-5}$ gauss donnerait une valeur convenable. Etant données les erreurs expérimentales, nous ne pensons pas que les conclusions de GINZBURG et SYROVATSKII [1] puissent éliminer seules l'hypothèse des électrons secondaires, et qu'il reste nécessaire de faire de bonnes mesures du rapport e^+/e^-.

6. Conclusion

En fait, pour pouvoir connaître l'origine des électrons cosmiques, il faudrait:

1) Perfectionner le modèle galactique.

2) Mesurer le rapport (e^+/e^-) avec beaucoup plus de certitude.

3) Mesurer le spectre des électrons dans le domaine de 0,25 GeV à 10 GeV, aux fins de comparaison avec le spectre du rayonnement synchrotron de la Galaxie.

Références

[1] GINZBURG, V. L. et SYROVATSKII, S. I.: 1964, *The Origin of Cosmic Rays*, Pergamon Press, Londres.

[2] POLLACK, J. B. et FAZIO, G. G.: 1965, 'Nuclear Interactions of Cosmic Rays as the Source of the Synchrotron Radiating Particles of our Galaxy', *Astroph. J.* **141**, 730.

[3] HAYAKAWA, S. et MATSUOKA, M.: 1964, 'Origin of Cosmic X-Rays', *Suppl. Progr. Theoret. Phys.* **30.**

[4] EARL, J.: 1961, 'Cloud-Chamber Observation of Primary Cosmic-Ray Electrons', *Phys. Rev. Letters* **6**, 125.

[5] LEROUX, E.: 1961, 'Etude théorique du rayonnement synchrotron des radio-sources', *Ann. Astrophys.* **24**, 71.

[6] CLINE, T. L., LUDWIG, G. H., et MCDONALD, F. B.: 1964, 'Detection of Interplanetary 3- to 12-MeV Electrons', *Phys. Rev. Letters* **13**, 786.

Bibliographie supplémentaire

BLAND, C. J. *et al.*: 1966, 'Sign Ratio and Absolute Flux of Cosmic-Ray Electrons', *Phys. Rev. Letters* **17**, 813.

BOISCHOT, A. et LEQUEUX, J.: 1964, 'Etude radioélectrique des restes de supernovae', *Ann. Astrophys.* **27**, 514.

Bryant, D. A., Cline, T. L., Desai, U. D., et McDonald, F. B.: 1965, 'Studies of Solar Protons with Explorers XII and XIV', *Astroph. J.* **141,** 478.
Hopper, V. D.: 1964, *Cosmic Radiation and High Energy Interactions,* Academic Press, Londres.
Johnson, H. M. et McLeod, J. M.: 1963, 'The Spatial Distribution of Supernovae in Galaxies', *Publ. Astron. Soc. Pacific* **75,** 123.
Jokipii, J. R. et Davis, L. Jr.: 1964, 'Acceleration of Electrons near the Earth's Bow Shock', *Phys. Rev. Letters* **13,** 739.
Morrison, P.: 1961, 'The Origin of Cosmic Rays', *Handbuck der Physik* **XLVI/1,** 1.
Sandström, A. E.: 1964, *Cosmic Ray Physics,* North Holland, Amsterdam.
Schatzman, E.: 1963, 'On the Acceleration of Particles in Shock Fronts', *Ann. Astrophys.* **26,** 234.
Slee, O. B., Solomon, L. H., et Patston, G. E.: 1963, 'Radio Emission from Flare Star V371 Orionis', *Nature* **199,** 991.
Wild, J. P., Smerd, S. F., et Weiss, A. A.: 1963, 'Solar Bursts', *Ann. Rev. Astron. Astroph.* **1,** 291.

THEORETICAL ASPECTS OF CELESTIAL GAMMA-RAYS

R. LÜST and K. PINKAU
Max-Planck-Institut für Physik und Astrophysik, Institut für extraterrestrische Physik, 8046 Garching bei München, Germany

1. Introduction

This article is concerned with cosmic electromagnetic radiation with a quantum energy exceeding 0.2 MeV. However, no sharp distinction will be made between X-rays and gamma-rays.

Gamma-ray astronomy is still in its early stages since the observation of the very low intensities expected must be performed beyond the range of the earth's atmosphere. Thus, only few observations have been performed, and few of them have yielded positive results.

Several authors have considered the theoretical aspects of celestial gamma-rays (see, for example, [1, 2, 18, 30]). The interactions of particles and fields with matter and with fields that produce and absorb gamma rays are well known. Therefore, any measurement of gamma rays yields information about particles, matter and fields in various parts of the universe. These processes of gamma ray production may take place in extended or localized regions, the latter are also called "point sources".

2. Particles, Fields and Matter in the Galaxy and in Intergalactic Space

1. Nucleons

The cosmic ray spectrum which we measure at the earth is modulated during the solar cycle below particle rigidities of about 3 GV. One may hope to have measured the galactic cosmic ray spectrum during the minimum of solar activity.

The results of Ormes and Webber [3] show that the cosmic ray proton spectrum can be approximated well by the formula

$$J(>E) = \frac{0.7}{(1.5+E)^{1.5}} \quad \text{protons}\,\text{cm}^{-2}\text{sec}^{-1}\,\text{ster}^{-1}$$

In addition, a flux of α-particles of about 13% of that of protons has been recorded, and also particles of larger charge than two are found [4].

2. Electrons

Electrons occur in the primary radiation with an intensity which is about 2% of that of protons. Their energy spectrum has recently been a source of considerable discussion [5, 6, 7, 8, 9]. The measurements are summarized in Figure 1, taken from the thesis of L'Heureux [10].

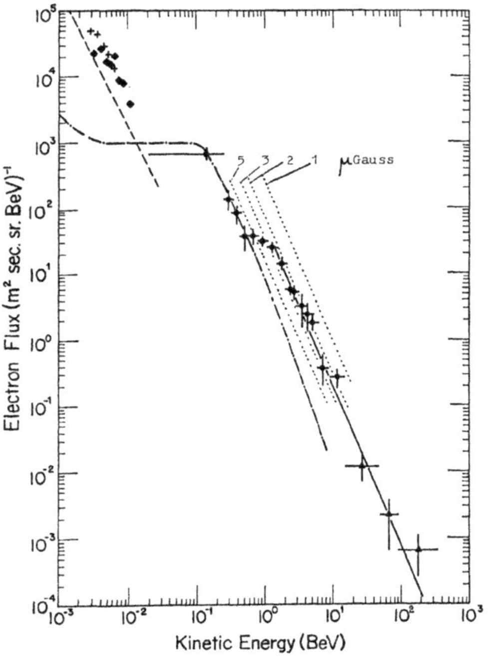

Fig. 1. Differential electron energy spectrum according to L'HEUREUX [10]. The dot-dashed curve is the spectrum predicted by the secondary collision theory according to RAMATY and LINGENFELTER [6]. This theory does not provide a sufficiently large flux of electrons. At low energies, knock-on electrons are produced. The dotted curves are electron spectra required to produce the non-thermal radio emission by the synchrotron process if the magnetic field were 1, 2, 3, and 5 μgauss respectively according to FELTEN [11].

The interesting shape of that spectrum makes it likely to be the result of several processes the relative importance of which changes with energy.

The equilibrium spectrum of electrons $\eta(E)$ is determined [9] by the source term, $q(E)$, the energy loss $-\mathrm{d}E/\mathrm{d}t$ and the leakage time τ after which they leave the region of confinement:

$$\frac{\eta(E)}{\tau} - \frac{\partial}{\partial E}\left[\frac{\mathrm{d}E}{\mathrm{d}t}\eta(E)\right] = q(E).$$

The shape of the energy spectrum of electrons is then dependent upon the leakage time, and also the shape of the energy loss function. For example, the ionization loss is, at sufficiently high energies, independent of E, while synchrotron and Compton effect losses follow an E^2 law. The electron spectrum will be steeper, if the latter energy losses dominate.

Since the relative importance of the factors dominating the equilibrium spectrum of electrons changes with the location considered (galactic disc, halo, intergalactic space), one cannot a priori expect that the electron spectrum measured near the earth is relevant throughout the galaxy or in intergalactic space.

3. Magnetic fields

Assuming that the electron spectrum at the earth is also valid in the galactic halo, the magnetic field strength in the halo can be evaluated by a measurement of the non-thermal radio emission from high galactic latitudes. This radio emission is being interpreted as due to the synchrotron effect, and the magnetic field necessary to produce the effect is about 3 μgauss.

Figure 1 shows, which electron intensities are required to produce the measured radio emission as a function of magnetic field strength [11]. It can be seen that this will not change much, even if the halo electron spectrum is somewhat different from that close to the earth.

The value of about 3 μgauss derived above is also compatible with the value derived for the magnetic field under the assumption that magnetic field energy density and cosmic ray energy density in the halo are about equal.

4. Matter

Garmire and Kraushaar [2] have compiled data about the concentration of atomic Hydrogen in the galaxy. Their results are shown in Figure 2, which gives contours of

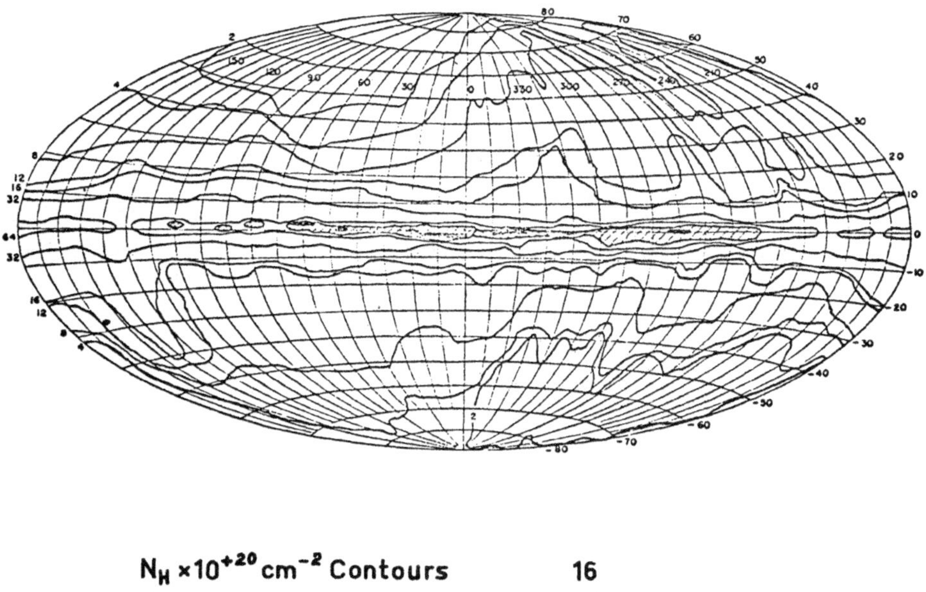

Fig. 2. Contours of constant N_H, the number of atomic hydrogen atoms cm^{-2} along the line of sight. From Garmire and Kraushaar [2].

constant N_H, the number of Hydrogen atoms cm^{-2} in various directions given in galactic coordinates.

Additional information about matter in the galaxy and in intergalactic space may also be taken from the publication of GINZBURG and SYROVATSKII [1], and their Table I is reproduced below.

Table I gives values of L, $N(L)$, $M(L)$ and ρ_{ph} in the galaxy and metagalaxy. Here, L is the effective path length, $N(L)$ the number of atoms and $M(L)$ the mass of the gas along the path L in a column of 1 cm^2 in cross-section. ρ_{ph} is the starlight photon energy density ($T \approx 5000\,°K$).

TABLE I

	L, [cm]	$N(L)$, [cm^{-2}]	$M(L)$, [gcm^{-2}]	ρ_{ph}, [$eVcm^{-3}$]
Galaxy:				
in the direction of the center	7×10^{22}	3×10^{22}	6×10^{-2}	0.2
in the direction of the anticenter	1.5×10^{22}	6×10^{21}	1.2×10^{-2}	
in the direction of the pole (including the contribution from the halo)	–	3×10^{20}	6×10^{-4}	
averaged over all directions (incl. the contribution of the halo)	–	8×10^{20}	1.6×10^{-3}	0.4*)
halo (in the direction of the pole)	3.5×10^{22}	1.5×10^{20}	3×10^{-4}	
Meta-galaxy	$R_{ph} = 5 \times 10^{27}$	5×10^{22}	0.1	2×10^{-3}

*) 0.1 $eVcm^{-3}$ according to other sources [8].

The figures given for intergalactic space are given for a line of sight of 5×10^{27} cm. This corresponds to a radius where the Doppler shift has reached a factor of two.

The energy density of the microwave field ($T \approx 3\,°K$) is of the order of 0.4 $eVcm^{-3}$. [23]

3. Gamma-Ray Production

In this chapter, various gamma ray production processes which may occur are discussed and the intensities and spectral shape due to these processes are estimated.

1. Interactions of Particles with Matter

The interactions of nucleons and antinucleons, of electrons and positrons with matter may yield gamma rays via the following processes:

a) The production of π^0-mesons in strong interactions of nucleons and antinucleons and the subsequent decay of these π^0-mesons into two gamma rays.

b) The radiation of gamma rays in bremsstrahlung processes of electrons in the coulomb field of nuclei.

c) The annihilation of positrons and electrons into two gamma rays.

d) The excitation of nuclei with subsequent decay via emission of gamma rays.

a) *π^0-meson production and decay*

π^0-mesons decay into two gamma rays with a mean lifetime of about 10^{-16} sec. In the rest frame of the π^0-meson, each quantum receives an energy of half the neutral pion's rest mass, i.e. an energy of 68 MeV. In the laboratory system, the probability for a π^0-meson of total energy U_{π^0} to produce a gamma ray with energy W in $\mathrm{d}W$ is constant between the two limits [12]

$$W_{\substack{\max\\ \min}} = \tfrac{1}{2}\left(U_{\pi^0} \pm \sqrt{U_{\pi^0}^2 - m_{\pi^0}^2 c^4}\right).$$

π^0-mesons are produced abundantly in strong interactions. The properties of the gamma ray spectrum resulting from an arbitrary spectrum of π^0-mesons have been discussed, for example, by OEHME [12] and SVENSSON [13]. The results of these considerations may be outlined as follows.

If the π^0-mesons are produced isotropically, and if the production spectrum extends down to zero kinetic energy of the π^0-mesons, the resulting gamma ray spectrum received will have an absolute maximum at the energy

$$\frac{m_{\pi^0} c^2}{2}.$$

If the π^0-mesons all have total energies above a value $U > m_{\pi^0} c^2$, the resulting spectrum will have a flat top extending from the energies $W_{\max}(U)$ to $W_{\min}(U)$.

Let W_1 and W_2 be the two photon energies belonging to the same differential intensity on either side of the maximum, then

$$W_1 W_2 = \frac{m_{\pi^0}^2 c^4}{4}.$$

Finally, the production spectrum of π^0-mesons may be derived from that of the gamma ray quanta from the following relation

$$\left|\frac{\mathrm{d}N_\gamma(W)}{\mathrm{d}W}\right| = \frac{2}{W} N_{\pi^0}\left(W + \frac{m_{\pi^0}^2 c^4}{4W}\right).$$

In order to derive gamma ray spectra, one thus has to obtain information about various modes of production of neutral pions.

i) *Production of neutral pions in thin layers of matter by cosmic rays.* Neutral pions may be produced by cosmic rays upon collision with matter. The number of quanta produced at a distance r with energy W in $\mathrm{d}W$ per unit time, volume and solid angle is given by

$$S(W)\,\mathrm{d}W = n(r)\,\mathrm{d}W \int_{E_{\min}}^{E_{\max}} \frac{2\mathrm{d}E}{\sqrt{E^2 + 2E m_{\pi^0} c^2}} \int_E^\infty \mathrm{d}E' \sigma_{\pi^0}(E', E)\, I_{\mathrm{CR}}(E')$$

where

$$\frac{2\mathrm{d}E}{\sqrt{E^2 + 2E m_{\pi^0} c^2}}$$

is the decay probability of a neutral pion of energy E into one gamma ray of energy W, and the kinetic energies $E_{\substack{max\\min}}$ are defined as the inverse of the definition for $W_{\substack{max\\min}}$ given above.

$\sigma_{\pi^0}(E', E)$ is the energy-dependent cross-section for neutral pion production by cosmic ray particles of energy E', and $I_{cr}(E')$ is the differential cosmic ray spectrum. $n(r)$ is the density of matter at distance r.

Data about the cross-section have been collected by POLLACK and FAZIO [14] and these authors have computed the integral value of the source strength for gamma rays, i.e. the expression

$$\int_0^\infty S(W)\,\mathrm{d}W\,.$$

Using the cosmic ray spectrum measured near the earth, and the data of cross-sections collected by themselves, POLLACK and FAZIO [14] arrived at an integral source strength of

$$6.3 \times 10^{-27} \times n(r) \quad \mathrm{cm}^{-3}\,\mathrm{sec}^{-1}\,\mathrm{ster}^{-1}.$$

for proton-proton-collisions, and

$$3.4 \times 10^{-27} \times n(r) \quad \mathrm{cm}^{-3}\,\mathrm{sec}^{-1}\,\mathrm{ster}^{-1}$$

for proton-alpha and alpha-proton-collisions. The total source strength is thus given by

$$9.7 \times 10^{-27} \times n(r) \quad \mathrm{cm}^{-3}\,\mathrm{sec}^{-1}\,\mathrm{ster}^{-1}.$$

The quantity $n(r)$ is the number of Hydrogen nuclei per cm^3. The total intensity received from any direction of the sky can then be obtained by multiplying the expression above with the value

$$\int_0^L n(r)\,\mathrm{d}r$$

along the line of sight.

No account has been taken yet of the shape of the spectrum which is the result of the variation of the cross-section $\sigma_{\pi^0}(E', E)$ with energy, and the decay of the neutral pions into gamma rays. However, cosmic rays also impinge on the atmosphere of the earth, and one thus may take the gamma ray spectrum in the upper atmosphere as a good guide for the shape of the spectrum expected to originate from collisions of cosmic ray particles with matter in space. Figure 3 shows the differential production spectrum of gamma rays due to neutral pion production and decay per gram of matter traversed. This has been arrived at by using the gamma ray spectrum in the upper atmosphere due to neutral pion production by cosmic rays in air, and normalizing it to the integral value calculated by Pollack and Fazio above. The spectrum of Figure 3 may now be multiplied by the amount of matter traversed by the cosmic

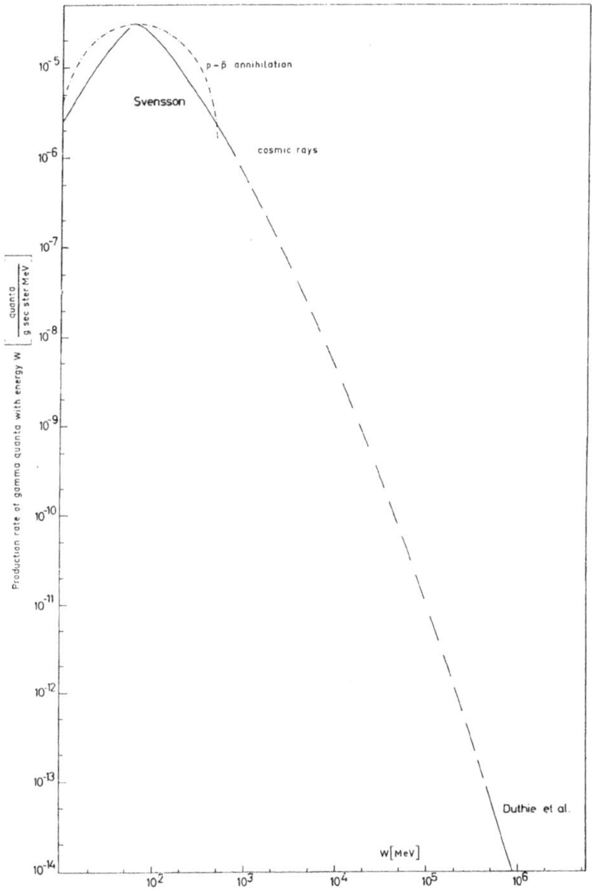

Fig. 3. Rate of gamma-ray production per g cm^{-2} due to π^0-meson decay according to PINKAU [36]. At low energies, data for production by cosmic rays are compiled from SVENSSON [13] and POLLACK and FAZIO [14]. The curve is then extrapolated to the value given by DUTHIE *et al.* [31]. π^0-meson production by proton-antiproton-annihilation has been considered by FRYE and SMITH [15], and their results have also been indicated, normalized to the same maximum value as that deduced for production by cosmic rays.

radiation, measured in gram per cm^2 to yield the intensity of gamma rays per sec ster cm^2. It should be mentioned here that absorption and self-absorption of cosmic rays and produced gamma rays in the source has been neglected and that the results thus apply only to thin layers of matter. Figure 4 shows the gamma ray flux that could be excepted from the direction of the galactic centre, together with other data of the X-ray and γ-ray flux from extended sources.

ii) *Production of neutral pions by annihilation of antiprotons.* Pions are also the result of annihilation of matter with antimatter. The cross-section for this process to occur

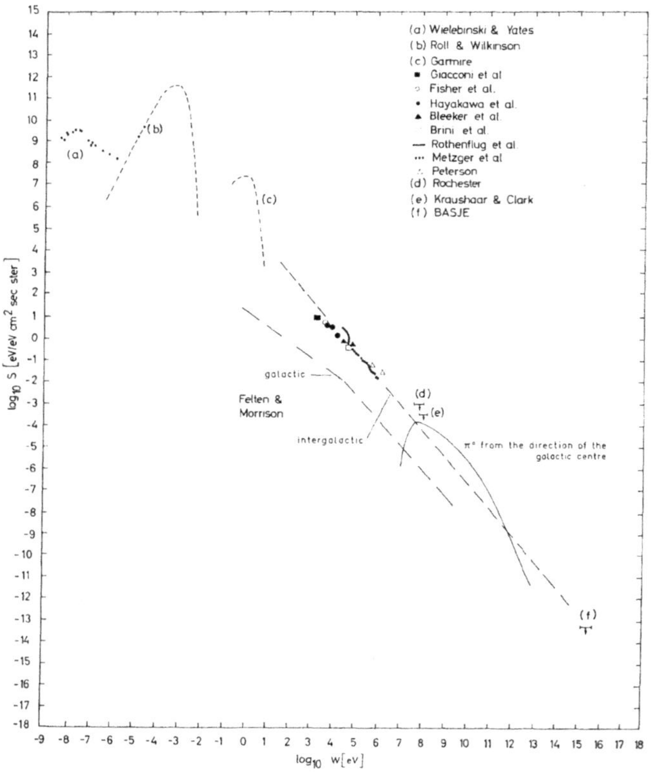

Fig. 4. Electromagnetic flux from extended sources. The data have been compiled from WIELEBINSKI and YATES [32], ROLL and WILKINSON [33], GARMIRE [34], FELTEN and MORRISON [8], and SUGA *et al.* [35]. From PINKAU [36].

shows a variation with

$$\sigma = \frac{\sigma_0}{\beta} = \frac{5 \times 10^{-26}}{\beta} \quad [\mathrm{cm}^2]$$

where β is the relative velocity of the annihilation partners. This process therefore predominantly occurs at small velocities. The average lifetime τ of an antiproton in a proton gas with density n is

$$\tau = \frac{1}{n\sigma_0 c} \quad [\mathrm{sec}].$$

The total source strength (integral) is given by

$$4\,\frac{n\bar{n}\sigma_0 c}{4\pi}$$

where it has been assumed that about 4 pions are produced per annihilation process.

The spectrum of gamma rays from this process has been calculated by FRYE and SMITH [15] from accelerator data, and it is shown in fig. 3 for comparison with the spectrum expected to arise from nuclear interactions of the cosmic radiation. The shape is distinctly different, since the pions produced have very low velocities. The spectrum from p-$\bar{p}$-annihilation shown in Figure 3 is normalized to the same maximum as that from the cosmic radiation.

b) *Bremsstrahlung processes*

Bremsstrahlung processes are of importance only if electrons are considered as primary particles. The cross-section depends on the square of the particle mass and is thus negligible for protons and heavier particles.

The theoretical cross-sections for bremsstrahlung processes depend on the energy of the radiating electron, and on the fact, whether or not the screening of the nucleus by the atom's electrons has to be taken into account.

In general, one may write for the probability for an electron of kinetic energy E traversing a thickness of $\mathrm{d}x$ gcm^{-2} to radiate a photon of energy W

$$\sigma(W)\,\mathrm{d}W = 4\alpha\frac{N}{A}Z^2 r_e^2\frac{\mathrm{d}W}{W}F(U, v)$$

where $v = W/U$ and U is the total energy of the electron. The form of the function $F(U, v)$ depends upon the electron's energy and the screening conditions, it is given by ROSSI [16]. For "complete screening", $F(U, v)$ is essentially given by

$$F(U, v) = \ln(183Z^{-1/3}).$$

The quantity X_0, the "radiation length", may be defined by the equation

$$X_0 = \left[4\alpha\frac{N}{A}Z^2 r_e^2 \ln(183Z^{-1/3})\right]^{-1}$$

in the notation of ROSSI [16]. Then, the energy loss of an electron under the assumption of complete screening is

$$-\frac{\mathrm{d}E}{\mathrm{d}x} = X_0 U.$$

The value of X_0 for hydrogen is 66 gcm^{-2} [1].

If electrons pass through ionized gas, no screening occurs. Rather, the shielding of a point charge is determined by the Debye-length

$$\lambda_D = 6.9\sqrt{\frac{T}{n_e}} \qquad T\text{: temperature, } n_e\text{: electron density.}$$

Since, in the computation of the radiation length, the screening due to the Thomas-Fermi model with a radius $a_0 Z^{-1/3}$ was used, the radiation length in a plasma will be given by

$$X_{pc} = X_0 \frac{\ln \frac{a_0}{\lambda_c}}{\ln \frac{\lambda_D}{\lambda_c}}.$$

where λ_c is the Compton-wavelength of the electron. X_{pc} is thus of the order of 10 g cm^{-2} in the H II regions of the galaxy.

Bremsstrahlung quanta may be produced, for example, by a Maxwellian distribution of electrons in a hot plasma. This case has been considered by many authors [17, 18]. In the first approximation, the bremsstrahlung quanta follow an exponential energy spectrum

$$\exp\left[-\frac{W}{kT}\right]; \quad k\text{: Boltzmann's constant, } T\text{: Temperature of plasma.}$$

For example, the X-ray source Sco X-1 shows such spectral behaviour. It should be mentioned, however, that the spectrum is not purely exponential, and that deviations from such a law must be expected.

Bremsstrahlung from collisions of electrons with interstellar matter yields a spectrum like

$$W^{-m}\,\mathrm{d}W$$

if the electron spectrum follows a power law with exponent m.

c) *Positron annihilation*

Positrons are created in collisions of cosmic ray particles with matter through the π–μ–e–decay chain. The collision partners, nuclei of the cosmic radiation and the target nuclei, are positively charged, and thus an overabundance of positrons results from this production process, at least in certain energy intervals [6]. The cross-section for annihilation of positrons with energy E_+ with electrons at rest is [19]

$$\sigma = \pi r_e^2 \frac{1}{\gamma + 1}\left[\frac{\gamma^2 + 4\gamma + 1}{\gamma^2 - 1} \ln\{\gamma + \sqrt{\gamma^2 - 1}\} - \frac{\gamma + 3}{\sqrt{\gamma^2 - 1}}\right]$$

where

$$\gamma = \frac{E_+}{m_e c^2}.$$

At very high energy, this formula becomes

$$\sigma = \pi r_e^2 \frac{1}{\gamma}[\ln 2\gamma - 1]\,\delta(W - E_+).$$

The delta function is an approximate expression for the fact that one gamma ray takes away practically all the positron's energy. The cross-section for annihilation diverges for $\gamma \to 1$. However, the rate of annihilations per second in a substance with N atoms of charge Z per unit volume is then given by

$$R = ZN\sigma v_+ = ZN\pi r_e^2 c \quad [\mathrm{sec}^{-1}].$$

About 80% of the positrons created with an energy exceeding 100 MeV are slowed down by ionization loss before they annihilate. The slowing down time by ionization loss is about 10^9 years [1]. The escape time from the galaxy, on the other hand, has recently been estimated to be about 10^8 years (Ramaty and Lingenfelter [9], see there for further discussion). Thus, most of the positrons will have left the galaxy before they can be brought to rest. The whole situation has been discussed by Ginzburg and Syrovatskii [1].

d) *Nuclear gamma rays*

The production of nuclear gamma rays in space by the collisions of cosmic ray particles with interstellar matter has been considered by Ginzburg and Syrovatskii [1], for example. One may assume that the target nucleus emits one gamma ray after one collision with a high energy particle of the cosmic radiation.

Under this assumption, the gamma ray production rate is given by

$$I \approx 3 \times 10^{-5} M(L) \frac{\text{photons}}{\text{cm}^2\,\text{sec}\,\text{ster}}.$$

A completely different source of nuclear gamma rays may lie in the supernovae in which a number of highly excited nuclei are produced which subsequently decay, emitting gamma rays amongst other particles. This problem has been dealt with in detail by Haymes *et al.* [20]. Figure 5 shows a gamma ray spectrum which one may be able to measure in the future, and which would yield information about the elements present in these sources.

2. Interactions of particles with fields

Interactions of particles with fields that may lead to the creation of gamma rays are

a) Magnetic bremsstrahlung (synchrotron radiation),
b) Inverse Compton effect, and
c) Photoproduction of π^0-mesons in collisions of particles with radiation fields.

a) *Magnetic bremsstrahlung*

Electrons gyrating in a magnetic field emit electromagnetic radiation contained in a narrow cone around the electron's direction of flight. Thus, the electron can be seen only in a state moving towards the observer and the radiation frequency is shifted towards the blue by the Doppler effect.

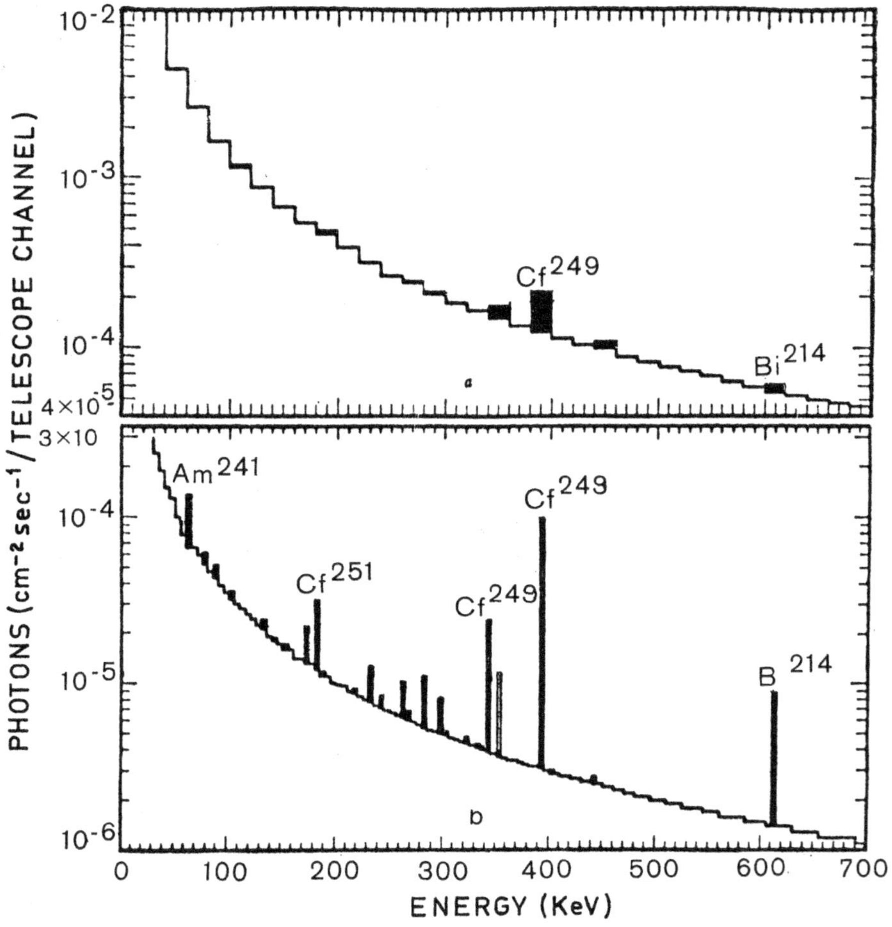

Fig. 5. Nuclear gamma rays from supernovae due to Haymes *et al.* [20]. (a) Upper curve shows the experimental result of viewing the predicted gamma spectrum from the Crab Nebula superimposed on the continuous sky background with a gamma-ray telescope having an opening of 0.1 steradians and an energy resolution of 20 keV. (b) Lower curve shows the result of viewing the same photon distribution as in (a) above, but with a telescope having an opening angle of 0.01 steradians and an energy resolution of 5 keV. Note how many more lines become visible.

The magnetic bremsstrahlung spectrum of a single electron has an asymmetric frequency distribution with the centre of gravity close to the characteristic frequency [21]

$$\nu_s = \frac{3e}{4\pi m_e c}\gamma^2 H_\perp = 4.2 \times 10^{-6}\gamma^2 H_\perp$$

with $H_\perp$ in [μgauss] and ν_s in [MHz].

The maximum of the frequency distribution occurs at the value

$$\nu_m \simeq 0.29\nu_s .$$

The radiation power emitted by a single electron due to the magnetic bremsstrahlung process is given by

$$P_s = \tfrac{2}{3} r_e^2 c^2 \gamma^2 H_\perp^2 = 9.89 \times 10^{-16} \gamma^2 H_\perp^2 \quad \left[\frac{\text{eV}}{\text{sec}}\right]$$

with $H_\perp$ in [μgauss].

This is the energy loss suffered by the electron. It is proportional to γ^2 and this dependency has an influence on the equilibrium spectrum of electrons, as has been discussed in section 1.2. If the electron flux density is of the form

$$n(\gamma)\,\mathrm{d}\gamma = n_0 \gamma^{-m}\,\mathrm{d}\gamma$$

one obtains a synchrotron power spectrum of the form [8]

$$(I_\nu)_s = 4.8 \times 10^{-20} (4.9 \times 10^2)^{3-m} n_0 R H^{(1+m)/2} \nu^{(1-m)/2}$$

with R [light years], H [μgauss], ν [MHz], n_0 [cm^{-3}], and $(I_\nu)_s$ [watts m^{-2} ster^{-1} Hz^{-1}]. This equation is in hybrid units useful for radio astronomy.

It is apparent from this equation that magnetic bremsstrahlung is probably of little importance for gamma ray production, since very large electron energies and magnetic fields are required to produce quanta with energy exceeding 0.2 MeV. However, magnetic bremsstrahlung is important to correlate radio measurements with the electron spectra in the halo (Section 1.2., Figure 1 and Figure 4).

b) *Inverse Compton effect*

The inverse Compton effect is likely to lead to the production of high energy photons. In this process, the Compton collisions of relativistic electrons with low energy photons are considered, and these photons show a large energy gain after the collision.

The correlation between scattered photon energy W and electron Lorentz factor γ after collision with a photon of a radiation field with temperature T is given by

$$W = 3.1 \times 10^{-4} \gamma^2 T \quad [\text{eV}].$$

Thus, electrons with $\gamma \approx 500$ suffice to produce gamma rays from starlight photons.

A single electron radiates a power of [8]

$$P_c = \tfrac{4}{3} \sigma_T c \gamma^2 \rho = 2.66 \times 10^{-14} \gamma^2 \rho \quad \left[\frac{\text{eV}}{\text{sec}}\right]$$

with $\sigma_T = 6.6 \times 10^{-25}$ [cm^2] and ρ the energy density [eV/cm^3] of the low energy photons. Again, the energy loss of the electron depends on γ^2 as in the synchrotron process.

The power spectrum of gamma rays produced by an electron spectrum of the form

$$n_0 \gamma^{-m}\, \mathrm{d}\gamma$$

is given by

$$(I_W)_c = 10^3 (56.9)^{3-m} n_0 R \rho T^{(m-3)/2} W^{(1-m)/2}$$

with n_0 [cm^{-3}], R [light years], T [°K], W [eV], ρ [eV/cm^3] and

$$(I_W)_c \text{ in eV eV}^{-1}\,\text{cm}^{-2}\,\text{sec}^{-1}\,\text{ster}^{-1}.$$

In contrast to the process of magnetic bremsstrahlung, where the magnetic fields are confined to the galaxy, inverse Compton processes may also take place in intergalactic space and thus contribute considerably to the gamma ray flux measured.

c) *Photopion production*

Particles of the cosmic radiation may produce π^0-mesons due to the photoproduction process with low energy photons. Protons above an energy of 10^{17} eV may produce π^0-mesons [22] with photons of 2 eV energy. If photons of the microwave field are considered (7×10^{-4} eV), the threshold is 10^{20} eV [23]. GREISEN [23] has pointed out, that the cosmic ray spectrum may end at about 10^{20} eV due to energy loss of cosmic ray particles in the photopion process. The microwave field possesses the high energy density of about 0.4 eV cm^{-3} while the starlight energy density is about 0.1 eV cm^{-3} in the halo, and perhaps 10^{-2} eV cm^{-3} in intergalactic space.

Due to this process, the intensity of gamma rays above an energy of 3×10^{16} eV will be about 4×10^{-3} of that of protons of the same energy. About 10^{-17} gamma rays cm^{-2} sec^{-1} are expected to be received at the earth above an energy of about 10^{17} eV [22].

4. Gamma-Ray Absorption

Gamma rays may be absorbed by

a. Compton effect and
b. pair production.

Both processes will be discussed only very briefly here.

a) *Compton effect*

An upper limit to the Compton absorption is given by the Thomson scattering cross-section

$$\sigma_T = \frac{8\pi}{3} r_e^2 = 6.6 \times 10^{-25}\ \text{cm}^2 .$$

The mean free path for Thomson scattering in interstellar space will then be about

$$\frac{1}{10\sigma_T} = 1.5 \times 10^{23}\ \text{cm}$$

since the electron density is less than 10 cm^{-3}. This corresponds to five galactic radii, i.e. the Compton absorption process can be neglected.

b) *Pair production*

The mean free path for pair production on matter with complete screening is $\frac{9}{7}\,X_0$. Thus, a photon traverses on average about 85 gcm^{-2} before producing an electron pair. By comparison with Table I one sees that the galaxy offers about 7×10^{-4} mean free paths for pair production in the direction of the galactic centre.

On the other hand, pair production on thermal photons may become quite important. This process has been discussed recently by JELLEY [24] in the light of the discovery of the 3°K microwave field. The results are summarized in Figure 6 which shows the data as compiled by JELLEY [24].

These results show that gamma rays between 10^{14} and 10^{18} eV will be strongly absorbed by the microwave field, in fact, the galactic centre could probably not be seen in the light of gamma rays of this energy. Jelley has assumed a microwave energy

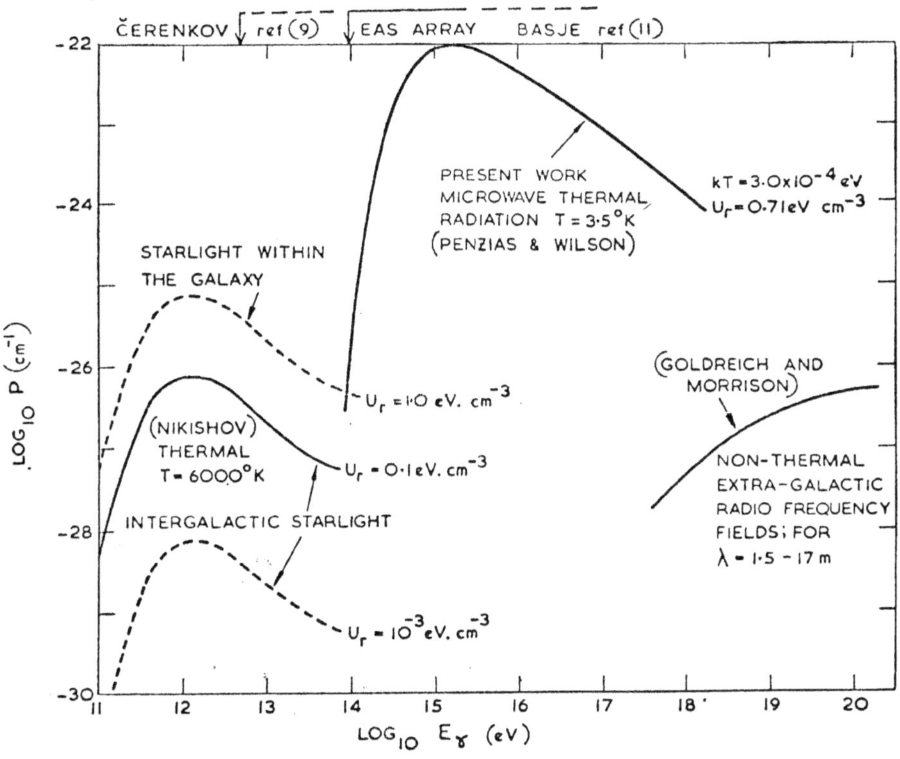

Fig. 6. JELLEY'S [24] results for the probability P of absorption of gamma rays of energy E_γ by photon pair production on starlight, the microwave field, and non-thermal radio frequency fields. The energy density of the microwave field may be closer to 0.4 eV cm^{-3} and the microwave curve should then be lowered correspondingly [8, 23].

density of 0.71 eV cm^{-3}, while it is probably closer to 0.4 eVcm^{-3}. Since the mean free path for pair production is inversely proportional to the energy density, Jelley's values should probably be lowered somewhat.

5. Discussion

1. Gamma rays from extended sources

Gamma-ray production in interstellar and intergalactic space must be considered as an extension of the same problem in the X-ray interval.

The dominating process in the X-ray region appears to be the inverse Compton effect. It has been pointed out by Felten and Morrison [8], that the measurements of the isotropic X-ray flux can be accounted for by assuming that electrons exist in intergalactic space with an intensity of 10^{-3} of that inside the galaxy. These electrons will have a steeper spectrum than the electrons in the halo, because the inverse Compton energy loss will begin to dominate at lower energies. Therefore, the X-rays, produced by these intergalactic electrons via the inverse Compton process will also follow a steeper spectrum than one which would have been created from galactic electrons. On the other hand, only galactic electrons produce the nonthermal radio emission via the synchrotron process, and the radio spectrum is therefore flatter than the X-ray spectrum, although both are produced by electrons of similar energy (but, as pointed out above, in different regions of space).

The isotropic electromagnetic radiation spectrum is shown in Figure 4 and the X-ray results have been extrapolated linearly towards higher energies.

Beyond several 10^7 eV, the π^0-meson decay gamma rays begin to be important. It can be seen from Figure 4, where the π^0-meson contribution from the direction of the galactic centre has been inserted, that this process should show up above the background of inverse Compton gamma rays. This contribution has been obtained by multiplying the production spectrum of Figure 3 with the amount of matter in the direction of the galactic centre as derived from Table I. The gamma ray intensities expected from other directions may be obtained in the same way.

At 1 GeV, 3×10^{-5} quanta $cm^{-2} sec^{-1} ster^{-1} GeV^{-1}$ are expected from the galactic centre, and 8×10^{-7} quanta $cm^{-2} sec^{-1} ster^{-1} GeV^{-1}$ averaged over all directions. The intensity of protons at 1 GeV is about 9×10^{-2} protons $cm^{-2} sec^{-1} ster^{-1} GeV^{-1}$ Therefore, the ratio of gamma rays to protons at this energy is 3.3×10^{-2} from the galactic centre, and 9×10^{-4} on average.

At 3×10^7 GeV, as pointed out above, this ratio would have decreased to a value of 4×10^{-3}, if gamma ray absorption could be neglected at these energies.

2. Gamma rays from point sources

The gamma-ray flux from some point sources has been estimated by Garmire and Kraushaar [2] using the following argument.

All the sources listed are also radio sources. If it is assumed that the radio emission is due to magnetic bremsstrahlung of electrons, and if these electrons are entirely the result of the π–μ–e-decay chain where the pions are produced in heavy particle collisions in the source, π^0-mesons will also be produced in the same collisions. Then, the energy leaving the source in form of π^0-meson decay gamma rays should be about twice as much as that leaving it in form of synchrotron radio noise. The estimates of the following table are based on this assumption.

TABLE II

Object	Radio flux [erg cm^{-2} sec^{-1}]	Magnetic energy [eV cm^{-3}]	Photon density [eV cm^{-3}]	Gamma ray flux $W > 100$ MeV [cm^{-2} sec^{-1}]
Virgo A	2×10^{-11}	20	1	6×10^{-9}
Centaurus A	1×10^{-10}	1	1	3×10^{-8}
Cygnus A	1×10^{-10}	1000	0.025	4×10^{-8}
Taurus A	5×10^{-8}	250	1.7	1.5×10^{-5}

Recently, the importance of measurements of the flux of gamma rays from quasi-stellar radio sources has been emphasized [25–29], Hoyle, Burbidge and Sargent [25] have argued that, if the synchrotron process is responsible for the continuum radiation of these objects, and if the observed rapid variations are taken into account, a very high radiation energy density will result if cosmological distances are assumed. This will result in a very large flux of inverse Compton effect gamma rays produced by the radiating electrons on the quanta of the synchrotron radiation. In fact, the inverse Compton effect sets severe boundary conditions on the models possible for the quasi-stellar radio sources. Rees and Sciama [26] could show that inclusion of self-absorption relaxes the requirements on the model somewhat, while Hillier [28] pointed out that the Compton quanta will cease to gain energy from the electrons if their energies become comparable, or they will begin to feed back energy into the electron component by the pair production process. A measurement of the flux from quasi-stellar radio sources in the gamma-ray region could possibly be decisive in explaining their nature.

References

[1] Ginzburg, V. L. and Syrovatskii, S. I.: 1965, *Soviet Phys. Usp.* **7**, 696.
[2] Garmire, G. and Kraushaar, W. L.: 1965, *Space Sci. Rev.* **4**, 123.
[3] Ormes, J. F. and Webber, W. R.: 1965, *Proc. Lond. Cosm. Ray Conf.* **I**, 349.
[4] Waddington, C. J.: 1965, *Proc. Lond. Cosm. Ray Conf.* **I**, 462.
[5] Daniel, R. R. and Stephens, S. A.: 1966, *Phys. Rev. Letters* **17**, 935.
[6] Ramaty, R. and Lingenfelter, R. E.: 1966, *J. Geophys. Res.* **71**, 3687.
[7] Cowsik, R., Yash Pal, Tandon, S. N., and Verma, R. P.: 1966, *Phys. Rev. Letters* **17**, 1298.
[8] Felten, J. E. and Morrison, P.: 1966, *Astrophys. J.* **146**, 686.
[9] Ramaty, R. and Lingenfelter, R. E.: 1966, *Phys. Rev. Letters* **17**, 1230.
[10] L'Heureux, J.: 1966, EFINS Preprint 66–109.
[11] Felten, J. E.: 1966, *Astrophys. J.* **145**, 589.
[12] Oehme, R.: 1953, in: Heisenberg, *Kosmische Strahlung*, Springer.
[13] Svensson, G.: 1958, *Arkiv Fysik* **13**, 347.

[14] POLLACK, J. B. and FAZIO, G. G.: 1963, *Phys. Rev.* **131,** 2684.
[15] FRYE, G. M. and SMITH, L. H.: 1966, *Phys. Rev. Letters* **17,** 733.
[16] ROSSI, B.: 1965, *High Energy Particles*, Prentice-Hall.
[17] ELWERT, G.: 1961, *J. Geophys. Res.* **66,** 391.
[18] GOULD, R. J. and BURBIDGE, G. R.: 1965, *Ann. Astroph.* **28,** 171.
[19] HEITLER, W.: 1954, *The Quantum Theory of Radiation*, Oxford.
[20] HAYMES, R. C., CRADDOCK, W. L., and CLAYTON, D. D.: 1966, *Space Res.* **6,** 80.
[21] GINZBURG, V. L. and SYROVATSKII, S. I.: 1964, *The Origin of Cosmic Rays*, Oxford.
[22] HAYAKAWA, S. and YAMAMOTO, Y.: 1963, *Prog. Theor. Phys.* **30,** 71.
[23] GREISEN, K. I.: 1966, *Phys. Rev. Letters* **16,** 748.
[24] JELLEY, J. V.: 1966, *Phys. Rev. Letters* **16,** 479.
[25] HOYLE, F., BURBIDGE, G. R., and SARGENT, W. L.: 1966, *Nature* **209,** 751.
[26] REES, M. J. and SCIAMA, D. W.: 1966, *Nature* **211,** 805.
[27] HOYLE, F. and BURBIDGE, G. R.: 1966, *Nature* **212,** 1223.
[28] HILLIER, R. R.: 1966, *Nature* **212,** 1334.
[29] WOLTJER, L.: 1966, *Astrophys. J.* **146,** 597.
[30] GINZBURG, V. L. and SYROVATSKII, S. I.: 1965, *Space Sci. Rev.* **4,** 267.
[31] DUTHIE, J. G., FOWLER, P. H., KADDOURA, A., PERKINS, D. H., and PINKAU, K.: 1962, *Nuovo cimento* **24,** 122.
[32] WIELEBINSKI, R. and YATES, K. W.: 1965, *Nature* **205,** 581.
[33] ROLL, P. G. and WILKINSON, D. T.: 1966, *Phys. Rev. Letters* **16,** 405.
[34] GARMIRE, G.: 1965, *Proc. Lond. Cosm. Ray Conf.* **I,** 315.
[35] SUGA, K., ESCOBAR, I., MURAKAMI, K., DOMINGO, V., TOYODA, Y., CLARK, G., LAPOINTE, M.: 1963, *Proc. Jaipur Cosm. Ray Conf.* **4,** 9.
[36] PINKAU, K.: 1966, *Mitteilungen der Max-Planck-Gesellschaft* **6,** 328.

COSMIC RADIATION*

P. L. MARSDEN
University of Leeds, England

1. General Characteristics of the Cosmic Radiation

It is important to realize at the outset that the intensity of cosmic radiation recorded at the earth is controlled, particularly for low energy particles (i.e. $E < 10^{10}$ eV) by the degree of solar activity, the so-called solar modulation process (to be discussed later). In fact it is not absolutely clear whether at any time during the solar cycle the full galactic intensity of the cosmic radiation is recorded.

1.1 Composition of cosmic radiation

1.1.1 *Nuclear component*

The composition of cosmic rays, as measured on earth, depends on the composition at the source, acceleration processes at the source and what happens to the particles as they propagate across the galaxy. The figures given in Table I refer to sunspot minimum conditions, when the intensity recorded at the earth may be expected to approximate to the true galactic value.

Approximately 95% of the particles are protons, about 4.5% are α-particles, and all the heavier nuclei constitute in total less than 0.5% of the flux. Recent results are given in Table I.

TABLE I

Nuclear composition of cosmic ray flux[a]

Group of atomic nuclei	Z	I Intensity[b] $(\mathrm{m^2\,sec\,sr})^{-1}$	% by number	% by mass	I/I_H	I/I_H in Universe
protons	1	1510 ± 150	94	74	680	2260–6830
α-particles	2	89 ± 3	5.5	17.4	46	258–1040
L (Li, Be, B)	3–5	2 ± 0.2			1.0	10^{-5}
M (C, N, O, F)	6–9	5.6 ± 0.2			3.0	2.6–10
H ($\bar{A} = 31$)	10–19	1.9 ± 0.3	0.5	8.6	1.0	1
VH ($\bar{A} = 51$)	20–28	0.7 ± 0.16			0.3	0.06

[a] after Waddington (these data refer to particles with an energy/nucleon $> 2.5 \times 10^9$ eV).
[b] the units of intensity etc. are defined in the appendix. $\bar{A}$ average atomic weight of nuclei in a group.

The following comments are warranted concerning this table:

1) Proton measurements are difficult to make, due to corrections necessary for albedo, etc. There are various ways in which albedo particles arise and these occur

* This article is based upon the lecture notes of J. Baruch and F. Engström and presented material and views generally accepted in 1965.

more easily in the case of protons than in the case of helium. Also the detection methods used make it easier to identify particles other than protons.

2) Intensities are fairly sensitive to solar activity.

3) The heavier nuclei are grouped to improve the statistical reliability of the observations.

4) Data are becoming available on the isotopic abundances of hydrogen and helium.

5) It is difficult to correct for fragmentation in space, particularly in relation to the L group; but the presence of the L group is definitely established. The measured abundances contrast markedly with the universal values. This is an extremely important feature leading directly to estimates of the amount of material traversed by cosmic rays on their path through the galaxy and also of their lifetime.

6) The light elements H and He are less abundant in the cosmic radiation than in the universe.

1.1.2 *Electron component*

Evidence from cosmic radio-emission suggests that there should be a flux of high energy electrons.

Present data due largely to EARL [3] give as the electron flux

$$I_e(E \geqq 0.5 \text{ GeV}) = 32 \pm 10 \text{ electrons}/(\text{m}^2 \text{ sec sr})$$

Observations of MEYER and VOGT [6] suggest greater intensities, particularly at lower energies. It cannot be said at the present time to what extent these latter observations given information on the true primary galactic flux rather than on solar electrons or secondaries.

Data from AGRINIER *et al.* [1] give I_e $(E \geqq 4.6 \text{ GeV}) = 6.5 \pm 1.5/(\text{m}^2 \text{sec sr})$ consistent with a reasonable extrapolation of Earl's data. In general the electron flux is approximately 1% of the proton flux for $E \geqq 1$ GeV.

The electron-positron ratio is an important but extremely difficult quantity to measure. It gives information as to whether the electron component arises from the nuclear interactions of the cosmic rays during their passage through the galactic medium, or whether the electrons are coming predominantly from the source regions. An equal abundance of electrons and positrons would favour the former; an excess of electrons over positrons would be expected in the latter case.

One experiment, that due to DE SHONG *et al.* [2] gives

$$\frac{N^+}{N^- + N^+} = 0.16$$

in the energy range $0.3 \leqq E \leqq 1$ GeV, or

$$\frac{N^-}{N^+} \simeq 5.$$

1.1.3 *Gamma rays*

A possible source is from nuclear interactions of cosmic rays with interstellar and intergalactic material which create π^0 mesons. The flux will depend upon the thickness of matter traversed. Not many experimental data are available, but according to Kraushaar and Clark

$$I_\gamma(E \geqq 50 \text{ MeV}) = (3.7\text{–}11) \times 10^{-4} \text{ photons cm}^{-2}\text{sec}^{-1}\text{sr}^{-1}.$$

No anisotropy was found.

1.1.4 *Neutrinos*

These arise from reactions such as

$$\pi^\pm \to \mu^\pm + \nu$$
$$\mu^\pm \to e^\pm + \bar{\nu} + \nu.$$

1.2 ENERGY SPECTRA

It is convenient to divide the cosmic radiation into three energy ranges. These are:

(i) low energies, say $E \leqq 10^{10}$ eV

(ii) medium energies, $10^{10} \leqq E \leqq 10^{15}$ eV

(iii) very high energies, $E > 10^{15}$ eV

The reason for this will become apparent. In the energy range 10^{10} to 10^{14} eV, the intensity is distributed over the particle energies according to a power law (cf. Appendix):

$$j(E)\,\mathrm{d}E = KE^{-\gamma}\,\mathrm{d}E$$

where

$$\gamma \approx 2.5.$$

At greater energies the value of γ increases and at extreme energies may decrease again.

In the low energy range the primary spectrum is complicated principally owing to solar modulation effects which seriously reduce the intensity of particles of $E \simeq 10^9$ eV at solar maximum relative to solar minimum. Some of these effects will be discussed later. Recent results on this part of the spectrum were obtained by ORMES and WEBBER [7] using scintillation Čerenkov counter techniques, who find that there is a maximum in the proton spectrum at about 0.8 GeV, and in the α-particle spectrum at 1.2 GeV. This splitting is important. The extent to which these maxima occur in the true galactic spectrum is not clear at present.

The shapes of the spectra for protons and α-particles for $E \geqq 3 \times 10^9$ eV are the same.

At still lower energies, MCDONALD *et al.* [5] have determined the spectrum of protons in the energy range 15–75 MeV. The spectrum is a reasonable extension of that found by Ormes and Webber.

For the energy interval 10^{10}–10^{14} eV the main characteristic is a spectrum of slope $\gamma = 2.5$, this value having been derived from data obtained from photographic emulsions; at higher energies, data come from observations of extensive air showers. It is thought that the spectra of the various components, α-particles, L etc., have the same form as the proton spectrum up to approximately 10^{14} eV.

It has now become clear that there is a considerable steepening in the energy spectrum at energies higher than 10^{15} eV. Thus γ approaches 3 at $E \simeq 10^{17}$ eV.

The change in slope at 10^{15} eV has caused much speculation; possible explanations which have been discussed are:

(i) change in the nature of the elementary nuclear interactions at ultra high energies, possibly over a limited energy range;

(ii) a genuine change in the galactic spectrum, possibly due to loss of particles from the galaxy. McCuster [4] has deduced from a study of multiple-cored showers that at high energies the primaries are predominantly heavy particles. This may be consistent with a change in the source spectrum at these energies.

It is true to say that at the present time no satisfactory solution has been proposed to explain the change in slope.

Finally, it is worthwhile pointing out that the cosmic ray data obtained from extensive air showers, show that particles with energies up to 10^{20} eV exist. Particles with this energy cannot be retained within the galaxy for a long time. They may be expected to enter and leave our galaxy.

1.3 Anisotropy of cosmic radiation

The degree of anisotropy is defined by the quantity

$$\delta = \frac{I_{max} - I_{min}}{I_{max} + I_{min}}$$

where I_{max} is the largest and I_{min} the smallest observed intensity as function of direction of incidence.

Table II gives the currently accepted values of δ (as function of energy) which are upper limits.

TABLE II

Anisotropies in cosmic ray flux

E (eV) (approximately)	δ
10^{14}	$< 0.1\%$
10^{16}	$< 1.0\%$
10^{17}	$< 3.0\%$

However, there are some slight indications that at lower energies (approximately 10^{11} eV), a slight anisotropy may exist in the direction of the centre of the galaxy.

Whether a significant anisotropy exists at any energy is at the present time a completely open question. The increase in the upper limit for the anisotropy is not significant, but reflects the increasing uncertainty of the data.

2. Cosmic Rays in the Solar System

2.1 Methods of monitoring the intensity of cosmic rays

There is much more direct information about cosmic rays in the solar system than in the galaxy. On the other hand, this does not necessarily mean that we know more about the movement of particles in the solar system than in the galaxy.

At present we have obtained most of our observational data from groundbased recorders, the most useful type being the neutron monitor, followed by the μ-meson telescope. Both types of instrument respond to secondary particles produced in the earth's atmosphere by the primary particles.

If one considers the cascade produced by an incident primary particle one can distinguish three kinds of secondaries at sea-level: the soft component (electrons, positrons and gamma rays), the nucleon component (nucleons and a few pions) and the hard component (μ-mesons).

The neutron monitor is a device which is mainly sensitive to the nucleon component: it consists of a lattice of BF_3 neutron counters surrounded by layers of lead and a thick layer of paraffin. The nucleon component of the secondary rays produces "stars" in the lead and the neutrons coming from these events are captured in the counters.

Typical counting rates of IGY-type neutron monitors are 50000/h for a high latitude station; the IQSY-type monitors (which are of larger area) give higher counting rates, about 750000/h, thereby increasing the statistical accuracy.

The μ-meson telescope consists of two layers of detectors, operated in coincidence and separated by at least 10 cm of lead. The first detectors used were Geiger counters, but it has now become common practice to use slabs of plastic scintillator material viewed by photomultipliers. Data from μ-meson telescopes are not quite as useful as data from neutron monitors, mainly because the corrections for atmospheric influence are more complicated.

Figure 1 shows the spectrum of the rigidity of the primary particles responsible for the observed counts in various detectors. Neutron monitors respond mainly to primaries of a few GeV/c, whereas μ-meson telescopes respond to somewhat higher primary momenta (approximately 15 GeV/c). Now one also makes direct measurements of the primary particles from flights at the top of the atmosphere or beyond.

2.2 Variations of cosmic ray intensity

The modulation processes cause intensity changes which can be recorded at sea-level. The major variations are as follows (this list is not exhaustive):

2.2.1 *11 year variation*

The main features are:

(i) a reduction in the intensity of the cosmic radiation as solar activity increases; for example, with the Leeds neutron monitor a reduction of approximately 20% occurred during the past solar cycle. The inverse correlation with solar activity suggests that the process is one of exclusion of galactic particles;

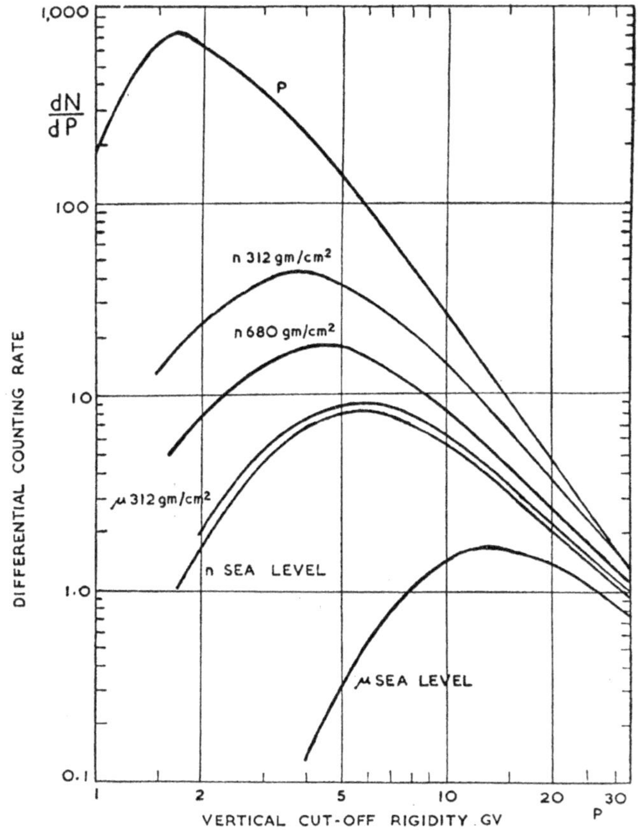

Fig. 1. Differential response curves. Differential counting rates as a function of vertical cut-off rigidity for various detectors. All curves normalised such that $N(15, x, t_0) = 100$. n, neutron monitor, μ, ionising component detector. P, primary radiation. (From Quenby and Webber: 1959, *Phil. Mag.* **4**, 654.)

(ii) equatorial stations with high threshold rigidities recorded only a decrease of about 6%, i.e. the modulation process is more effective for the lower energy particles;

(iii) there is a time lag between the maxima of cosmic ray intensity and solar activity;

(iv) emulsion and balloon experiments show that both protons and α-particles are affected.

The conclusion from these observations is that the solar modulation process *excludes* lower energy particles from the earth. Also, balloon measurements over the same period show that the primary protons and α-particles are modulated in a different way at low rigidities.

2.2.2 *Diurnal variation*

Superimposed on the long-term variation there is a daily variation of the cosmic radiation intensity.

(i) This is a small (0.2–0.3%) variation related to solar time. The solar diurnal variation exists on most days but it is somewhat irregular and may be absent, whilst on other days it is enhanced.

(ii) The maximum intensity occurs at about 1300 h LST and after allowing for the deviation of the particles in the earth's magnetic field, the direction of the maximum intensity lies at about 90° East of the earth-sun line, and remains roughly fixed (i.e. when there is a maximum) in this direction as the earth revolves around the sun.

(iii) The minimum intensity occurs predominantly in the morning. Again allowing for the magnetic deviation the direction of this sink is roughly 45° West of the earth-sun line. Other sinks are also present under certain disturbed conditions.

There has been some discussion lately on the question of whether in fact we have passed through the cosmic ray maximum in the current solar cycle. The minimum solar activity occurred around August–September 1964, and most cosmic ray recording stations were anxiously watching their records to see whether the cosmic ray intensity had passed through a maximum. It appears though that we have passed through the cosmic ray maximum in May of the year 1965.

2.2.3 *Forbush effect*

(i) This is a relatively large intensity decrease (up to 10–15%) which occurs rapidly, i.e. within a few hours to about 1 day, followed by a slower recovery lasting perhaps 10–20 days.

(ii) Satellite observations show that this effect occurs in space and therefore is not a terrestrial phenomenon.

(iii) The decrease frequently occurs in association with magnetic storms, i.e. is associated with solar outbursts.

(iv) The Forbush effect has a tendency to recur at 27-day intervals corresponding to the solar rotational period.

2.2.4 *Cosmic ray flares*

The most striking cosmic ray solar phenomenon is the cosmic ray flare. Some events are very large, particularly the one recorded in 1956. Cosmic ray flares may be rapid, and in most cases, large increases in cosmic ray intensity are associated with optical flares.

For cosmic ray flares one is in the happy situation of knowing the source of the particles. This important information leads to data on the trajectory by which particles move from the sun to the earth. The most important point is that there is a preponderance of westerly flares.

To account for this phenomenon, one must find some sort of guiding mechanism for the particles to be bent to the earth instead of travelling radially away from the sun. The flares which occur on the eastern side of the sun seldom produce a cosmic ray effect detectable on the earth. It is also important that the rigidity spectrum of flare particles is very steep with a spectral index typically of about 7.

2.3 Theory of cosmic ray modulation

In all these variations the sun exerts the controlling influence and the type of mechanism or model required to explain these effects is now considered.

In the case of the 11 year variation we are looking for some process which operates to exclude galactic particles at times of high solar activity. On a shorter time scale the same is true for Forbush decrease effects. Thus some sort of deflecting or decelerating mechanism is required.

Probably the most significant observation is that the modulation process is less effective at higher energies, suggesting that magnetic fields are responsible. These magnetic effects could be either stationary and/or timevarying.

(a) *Stationary fields* are more efficient in screening if compared with non-stationary fields of the same energy content. However, assuming reasonable field strengths, it is realized that the cosmic ray particles could pass readily through a stationary magnetic field.

(b) *Fields which vary in space and time*: $\mathrm{d}B/\mathrm{d}t$ produces an e.m.f. which could operate to reduce the energy and hence the intensity of the cosmic radiation. It is doubtful whether this process alone could cause sufficient modulation. However, moving magnetic irregularities can act as scattering centres for cosmic ray particles. This idea is basic to Parker's modulation theory which at present is widely accepted.

Parker [9] sets out by considering the equilibrium of the solar corona and comes to the conclusion that, since the thermal conductivity of the corona is large, there must be a large extension of high temperature outwards, so much so that the solar plasma will be able to escape from the solar gravitational field. In consequence, there will be a continuous outflux of solar plasma. Parker estimates that at the earth's orbit, typical plasma velocities will be several 10^2 km/sec, and densities about 10 particles/cm^3 with temperatures of about 10^5 °K.

Parker also shows that the plasma, in view of its high electrical conductivity drags out the solar magnetic field. If the sun were stationary this would give rise to a radial field system, but in effect the solar rotation causes the magnetic field lines to be drawn out into Archimedean spirals. The direction of the field lines at the earth's orbit is typically about 45° with respect to the solar radius vector for reasonable values of plasma velocity.

As the plasma moves outwards from the sun, its momentum per unit volume decreases and eventually it is slowed down by the interstellar fields. The distance at which this occurs is far from certain, and values between 2 and 100 AU have been suggested. In this region the magnetic fields would become tangled and act as scattering centres, i.e. form a "magnetic barrier". The possibility of field irregularities occurring between the barrier and the sun are not excluded, and Axford has extended Parker's theory to take this into account. If the scattering centres have a scale size l and field B associated with them in which the radius of curvature of the cosmic ray particle is R, then the angular deflection θ is given approximately by $\theta = l/R$ so that the angular deviation Δ after n collisions is $\Delta = ln^{1/2}/R$ or for $\Delta = 1$ radian, i.e. after $(R/l)^2$ collisions, the original direction of motion is lost and we can treat the problem as a random walk phenomenon and apply simple diffusion theory with a characteristic mean free path λ, which will depend upon the energy of the cosmic ray particle.

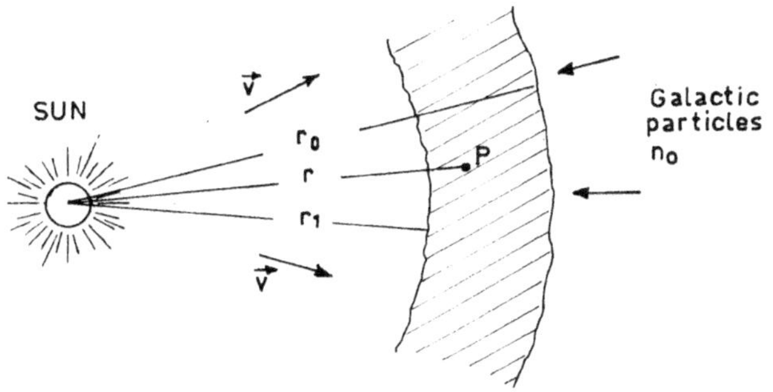

Fig. 2. Schematic drawing of magnetic barrier.

Parker considers the diffusion of galactic particles through a barrier composed of magnetic irregularities under spherically symmetrical conditions and for a barrier thickness $(r_0 - r_1)$. Throughout the barrier there will be a gradient in the density of cosmic radiation so that at a point P, at distance r from the sun, the flux of particles due to the gradient is given by

$$\phi = -\tfrac{1}{3}\lambda\nabla(wn), \tag{1}$$

where w is the velocity of cosmic ray particles, and n their density. This equation holds if the scattering centres are stationary; however, if they are moving with velocity $\mathbf{v}$ there is a contribution to the flux by sweeping cosmic ray particles away from the sun. Under these conditions the flux equation is

$$\phi = \mathbf{v}n - \tfrac{1}{3}\lambda\nabla(wn). \tag{2}$$

If there are no sinks or sources within the barrier the equation of continuity requires that

$$\frac{\partial n}{\partial t} = -\operatorname{div}\phi$$

or

$$\frac{\partial n}{\partial t} = -\nabla(\mathbf{v}n) + \tfrac{1}{3}\lambda\nabla^2(wn). \tag{3}$$

For steady state conditions such as will prevail approximately during the 11-year cycle, $\partial n/\partial t \approx 0$ and for radial symmetry Equation (3) gives, after integration, the magnitude of the gradient in the barrier:

$$\frac{1}{n}\frac{\partial n}{\partial r} = -\frac{3v}{w\lambda}. \tag{4}$$

Further integration gives the density of cosmic radiation n at distance r, when the galactic intensity is n_0:

$$n = n_0 \exp\left(-\frac{3v}{w\lambda}(r_0 - r)\right) = n_0 \exp\left(-\frac{\text{const}}{w}\right). \tag{5}$$

Equation (5) holds for low values of particle energy, when each scattering centre by itself can give a large deviation or scattering of the cosmic ray particle. On the other hand, at higher energies multiple scattering is required to deflect the particles. Under these conditions the density of cosmic radiation is given by:

$$\left.\begin{aligned} &n = n_0 \exp\left(-\frac{\text{const}}{E^2}\right) \\ &\text{or (if } E \gg mc^2\text{)} \\ &n = n_0 \exp\left(-\frac{\text{const}}{P^2}\right) \end{aligned}\right\} \tag{6}$$

where E is the total energy of the cosmic ray particles and P is the rigidity.

The transition from large scattering to multiple scattering occurs at about 1 GeV.

Thus at low energies the modulation is velocity-dependent and at higher energies the process becomes rigidity- or energy-dependent.

It can also be seen qualitatively that the modulation is greatest at the low energies, although it is not known whether the modulation is effect is exactly as predicted by Parker.

It also follows from Equation (5) that for small values of w, i.e., for low energy particles, the modulation varies appreciably with particle velocity w.

Recently, satellite experiment measuring protons in the 15–75 MeV region has given results which are in fair agreement with Parker's solar wind modulation theory (IMP-1 satellite).

The actual measurements, when plotted on a $\log E$–$\log N$ scale, show a linear relationship with positive index 1.5 (E is the *kinetic* energy). On the assumption of a primary spectrum with negative index of 2.5, ionization losses by the particles in outer space corresponding to traversal of 2 to 3 g cm^{-2} of matter, and modulation according to Equation (5), i.e., $1/w$ dependence, the calculated intensity in the same energy range (shown by the dashed line) agrees reasonably well with the measured values, although the slope is slightly different (see Figure 3). However, different assumptions on the character of the primary radiation (different slope, etc.) or slight deviations from the $1/w$ dependence in the modulation mechanism would make the agreement very good. One can add that these intensity measurements were consistent with the ORMES and

WEBBER [7] data under quite reasonable assumptions as to their extrapolation (see below).

Another check on the $1/w$ modulation can be made by investigating the intensity variations of protons and α-particles. The point is that for particles of the same rigidity the velocity of a proton is about twice that of an α-particle. With the $1/w$ dependence in the modulation function this should cause significant differences in proton

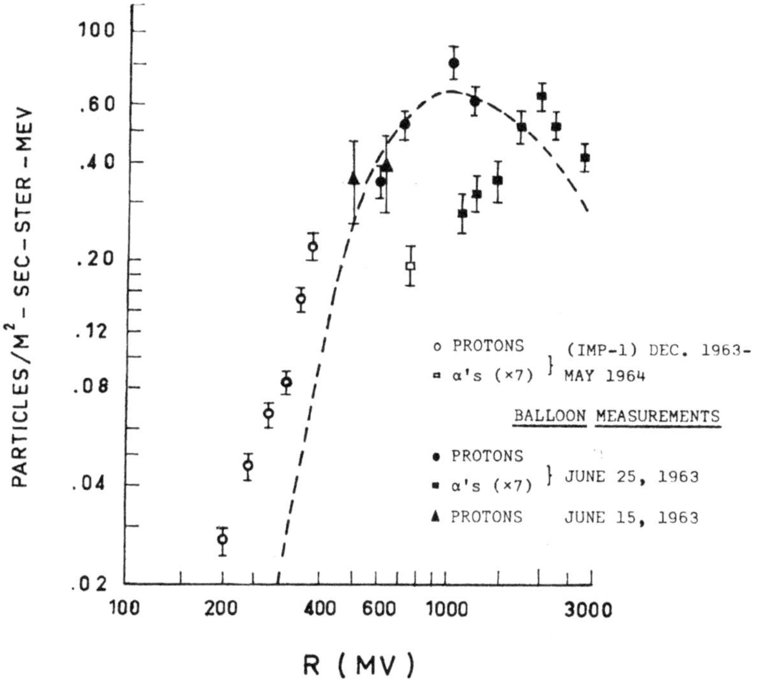

Fig. 3. IMP proton and α rigidity spectra. Shown for comparison are the balloon data of Fichtel *et al.*, and of Balasubrahmanyan *et al.* (Taken from MCDONALD and LUDWIG: 1964, *Phys. Rev. Letters* **13**, 784.)

and α-particle rigidity spectra. Ormes and Webber using balloon-borne equipment have measured proton and α-particle fluxes at rigidities of about 1 GV and reported a significant difference in the spectra. In particular, the proton spectrum exhibits a maximum at 0.8 GV while the maximum of α-particles was found at 1.2 GV. However, the sensitivity of this experiment was insufficient to determine the modulation function.

The $1/w$ modulation function would also affect the electron energy spectrum. If a satisfactory method for investigating low energy electrons can be worked out it may provide a very good check on the modulation mechanism. The point is that electrons become relativistic at much lower energies (a few MeV) than nucleons and their velocities are essentially constant. Thus if the $1/w$ modulation is operative it would only affect electrons of energy below a few MeV. Therefore the maximum in their spectrum may be expected at a few MeV.

The solar diurnal variation may also be explained by the Parker model. Axford who investigated the effect of irregularities in the spiral heliocentric magnetic lines of force came to the conclusion that scattering of cosmic rays by these irregularities would give cosmic rays a "rotational" velocity component. He obtained values of a few hundred km/sec at the earth's orbit. Since this exceeds the earth's velocity, cosmic rays overtake the earth, causing a regular diurnal variation with a maximum at 1800 h LST and a minimum at 0600 h LST as the earth rotates.

SARABHAI [8], using the IMP-1 magnetic data, has investigated the angle between the magnetic lines of force and the sun-earth radius vector. He found an angle of about 45° between these two vectors. A study of the daily variation recorded by the Deep River neutron monitor on 83 selected quiet days showed (after correction for the deflection of the particles in the earth's magnetic field) that the minimum occurred preferentially at 0300 h, i.e., in the direction of the magnetic lines of force.

We would thus have two sinks of cosmic radiation, one at 0600 h and one at 0300 h. This would be a happy situation as it would enable one to move the position of the mean sink in order to explain changes in the daily variation in relation with solar activity by altering the relative strengths of the two sinks. Sarabhai gives the following explanation for this phenomenon: cosmic rays spiralling along the magnetic lines of force may be scattered by irregularities in the magnetic field, some of them would be scattered into the sun, thus removing them from that particular solid angle.

Another prediction of the Parker model is the *density gradient* of cosmic rays, given by equation (4):

$$\frac{1}{n}\frac{\partial n}{\partial r} = -\frac{3v}{\lambda w}$$

According to the equation, the numerical value of the density gradient is about 10% per AU. Measurements by the Mars and Venus probes give an estimate of this quantity, which is however deduced over a relatively small distance, astronomically speaking (about 0.4 AU). Different people studying these data have given results as 14 ±7% per AU and 0 ±10% per AU which are not in contradiction with the model. The uncertainty in these data is obviously very large.

Some of the observations on *Cosmic Ray Flare Events* are easily explained by the Parker model. The preponderance of westerly flares observed would stem from a guiding mechanism of the magnetic lines of force, bending the fairly low energy particles into a trajectory which crosses the orbit of the earth. However, Gold's "magnetic bottle" theory may also explain cosmic ray flare phenomena. Gold postulates a pre-flare emission of plasma which is confined by lines of force in a "magnetic bottle", into which the flare then discharges its cosmic rays. However, the non-quantitative nature of Gold's model makes it difficult to compare its predictions with actual experimental observations.

Forbush decreases can also be explained on the basis of the solar wind model.

Emission of magnetic plasma clouds from the sun causes a shock wave system to be set up in the general plasma stream. This shock wave will sweep particles away, causing a sharp decrease in the intensity observed at the earth as the shock front passes over the earth. The time scale, rate of decrease of intensity, etc. can be explained by suitably chosen variations in the parameters of the model.

The preceding discussion has been made from the point of view of cosmic ray variations, but there are also strong solar-physical indications that the Parker model is approximately correct. It should be kept in mind that there is direct experimental evidence from the measurements of the direction and strength of the magnetic field from the IMP-1 satellite. These measurements give values which are quite consistent with Parker's predictions, namely, a magnetic field of the order of a few gammas which is inclined by about 45° with respect to the earth-sun line. There have also been plasma probe experiments which have measured the density and velocity of the solar wind giving velocities of 500 km/sec and densities of a few particles/cm^3, quite in agreement with values Parker has predicted. All these results strongly support the Parker model.

Appendix

In order to interpret the measurements to be discussed it is necessary to define a number of quantities.

1. The *unidirectional intensity,* I, is the number of particles incident upon unit area placed at right angles to the direction of observation per unit solid angle: thus the units of intensity are particles $(\text{cm}^2\,\text{sec}\,\text{sr})^{-1}$, e.g. the flux of cosmic radiation in space is 0.18 $\text{cm}^{-2}\,\text{sec}^{-1}\,\text{sr}^{-1}$.

2. The *omnidirectional intensity,* I_0, is the number of particles incident upon unit area from all directions, and if the surface is flat and the radiation isotropic, then the omnidirectional intensity is $I_0 = \pi I$.

3. *Energy spectra*: the particles comprising the cosmic radiation have a wide range of energies from some 10 MeV up to more than 10^{19} eV, the flux falling off with increasing energy. There is a distribution of particle numbers with respect to energy or simply an energy spectrum. The quantity $\mathrm{d}I/\mathrm{d}E$ is used to represent the number of particles/cm^2 sec sr(unit energy),

$$j(E) \equiv \frac{\mathrm{d}I}{\mathrm{d}E} = \frac{\mathrm{d}N}{\mathrm{d}A\,\mathrm{d}t\,\mathrm{d}\Omega\,\mathrm{d}E}.$$

In general, one finds:

$$j(E) = KE^{-\gamma}.$$

4. Instead of energy spectra momentum spectra or, what is almost the same, *rigidity spectra* are very often used. For a charged particle gyrating in a magnetic field the equation holds:

$$B\rho = \frac{pc}{Ze} \equiv P$$

where B is the magnetic field strength, ρ = radius of curvature of trajectory, Ze the charge, and p the momentum. The quantity $B\rho$ is known as the magnetic rigidity, P. It is expressed in gauss/cm which is not very convenient as one usually expresses energies in electron volts.

Since, dimensionally, $(p) = (E/c)$, momenta may be expressed in (eV/c), and then as $P = pc/Ze$, the magnetic rigidity P may be expressed in volts (V) or Gigavolts (GV).

Numerically, the equation holds:

$$\rho\,(\mathrm{cm}) = \frac{1}{300}\,\frac{P\,(\mathrm{volts})}{B\,(\mathrm{gauss})}$$

5. In order to interpret cosmic ray experiments performed at the earth it is necessary to specify the motion of cosmic ray particles in the earth's magnetic field. In particular, it has been shown by Störmer that particles with a rigidity less than a critical value, P_c (so-called *threshold rigidity*), are unable to reach the earth's surface. P_c is a function of geomagnetic latitude and direction of incidence of the particle.

For vertically arriving particles $P_c\,(\mathrm{GV}) = 14.9 \cos^4 \phi$, where ϕ is the geomagnetic latitude. A perfect dipole field is assumed in the derivation of the above equation.

Thus, from the magnetic poles to the equator, P_c varies from 0 to 15 GV. To treat the field as dipolar is a crude approximation, particularly for latitudes less than 45°, where, instead of a well-defined cut-off rigidity, the critical rigidities are smeared out to give rise to penumbral effects.

References

[1] Agrinier, B., Koechlin, Y., Parlio, B., Boella, G., Degli Antoni, G., Dilworth, C., Scarsi, L., and Sironi, G.: 1964, 'Investigation of Primary Electrons at 46° Geomagnetic Latitude', *Proc. 8th Intl. Conf. on Cosmic Rays (Jaipur, Dec. 2–14, 1963)*, Tata Institute of Fundamental Research, Bombay.

[2] De Shong, J. A., Hildebrand, R. M., and Meyer, P.: 1964, 'Ratio of Electrons to Positrons in the Primary Cosmic Radiation', *Phys. Rev. Letters* **12**, 3.

[3] Earl, J. A.: 1961, 'Cloud-Chamber Observations of Primary Cosmic-Ray Electrons', *Phys. Rev. Letters* **6**, 125.

[4] McCuster: 1965, Private communication.

[5] McDonald, F. B. and Ludwig, G. H.: 1964, 'Measurement of Low-Energy Primary Cosmic-Ray Protons on IMP-1 Satellite', *Phys. Rev. Letters* **13**, 783.

[6] Meyer, P. and Vogt, R.: 1961, 'Electrons in the Primary Cosmic Radiation', *Phys. Rev. Letters* **6**, 193.

[7] Ormes, J. and Webber, W. R.: 1964, 'Measurements of Low-Energy Protons and Alpha-Particles in the Cosmic Radiation', *Phys. Rev. Letters* **13**, 106.

[8] Sarabhai, V., Pai, G. L., and Wada, M.: 1965, 'Anisotropy of Galactic Cosmic Rays and the Interplanetary Magnetic Field', *Nature* **206**, 703.

[9] Parker, E. N.: 1963, *Interplanetary dynamical processes*, John Wiley, New York.

SOME UNRESOLVED PROBLEMS IN THE CORPUSCULAR COSMIC RADIATION

B. PETERS*
The Niels Bohr Institute, University of Copenhagen, Denmark

In planning this talk, I have selected a few topics in cosmic ray research which have not yet found the attention which they deserve. Satellites and space stations have provided opportunities to make measurements which were too difficult in the past. Lunar exploration may increase these opportunities further during the coming decades. The technological developments are giving added importance to the corpuscular radiation as a channel of communication through which we may explore the close and the distant surroundings of the earth.

1. The Archaeology of Cosmic Rays

One field of research which is almost entirely dependent on cosmic radiation is the evolutionary history of the earth's surroundings. Electromagnetic radiation reaching the earth's neighbourhood leaves few if any traces which can be pursued into the past, but cosmic radiation, being able to produce nuclear transmutations, leaves permanent marks.

An example is the abundance of light elements in the earth's crust. According to prevailing views on the nuclear processes which must have been at play in synthesizing the elements of which the planetary system is composed, the presence of large amounts of lithium, beryllium, boron and deuterium is difficult to understand. These elements are highly unstable under the conditions of temperature and pressures which must have prevailed at the time of synthesis. Greenstein, Hoyle and Fowler have concluded that they must have been created during an intense bombardment of the earth's surface by energetic particles at an early stage in the history of the solar system. We have here an indication that solar particle emission may have been much more powerful some billions of years ago than it is today.

The question arises, whether one can hope to trace the history of solar activity back in more detail so as to get reliable information on the sun thousands, millions or hundreds of million years ago. Almost no work has yet been done on this most important question. But considerable work is being done on a technically closely related problem, the history of galactic corpuscular radiation. I shall therefore discuss briefly research of this nature, which is being carried out by Geiss in Bern, Schaeffer in Heidelberg, Arnold in La Jolla, Anders in Chicago, and others. The studies are based on chemical transformations produced by cosmic rays in meteorites. There exists by now an impressive list of isotopes, radioactive as well as stable, which are due to the

* Present address: Danish Space Research Institute, Lundtoftevej 7–9, Lyngby, Denmark.

J. G. Emming (ed.), Electromagnetic Radiation in Space.

bombardment of meteorites by cosmic ray particles; their concentrations are closely interrelated and correlated with the chemical nature of the meteorite so that their genesis as products of nuclear transformations induced by cosmic ray particles is beyond doubt.

In principle, the abundance of any such isotope, if stable, permits a calculation of the total radiation dose to which the meteorite was exposed beginning with the time when it split off from a larger parent body. The concentration of an isotope which is radioactive permits integration of the radiation dose over a certain part of this period, roughly over the last mean life of the nucleide.

Since transmutation products with mean lives ranging from days or months to 4×10^6 years have been detected, it is in principle possible to reconstruct C.R. intensity over the entire span of the last, say, 10^7 years. It is also possible to find the time when the meteorite was released into space, presumably as the result of the break-up of an asteroid, as well as the time of its arrival on earth. When the trajectory of atmospheric entry is known one may even obtain a comparison of corpuscular radiation intensity at perihelion with that in the neighbourhood of the earth (e.g., by comparing the activity due to A^{39} (383 years) and A^{37} (49.5 days)). Furthermore, the variation of isotope concentration with depth below the meteorite surface gives information on the energy spectrum of particles and enables one to study the variation of particle flux with time for different energy groups.

The results so far obtained show that during the last 10^7 years there has been no major change in galactic cosmic ray intensity or in its energy spectrum. One must remember, however, that in this type of work short term fluctuations will not be detected as long as the total integrated radiation dose is not affected appreciably.

The observed constancy of radiation level over such a span of time is perhaps not too surprising, since it is likely that the galactic particles which we measure today and which presumably have been created at distances as far as the galactic center have a large spread in travel times, so that substantial intensity variations in less than 10^7 years can hardly be expected.

Little is known as yet about galactic cosmic ray flux in earlier periods, i.e., several hundreds or thousands of millions of years ago. It has been suggested that the sudden extinction of the huge reptiles which populated the earth some 300 million years ago may have been connected with a strong temporary increase in the radiation level which affected mutation rates. There are many reasons for studying cosmic ray intensity in the past; the possible connection with evolution by itself provides an incentive for trying to reconstruct the history of corpuscular radiation in detail and as far back into the past as possible.

At the moment there is only one isotope available for studying very early periods. K^{40} ($\tau = 1.83 \times 10^9$ years) in iron meteorites has been identified as cosmogenic. However, not many iron meteorites are old enough to represent suitable objects for investigations of the radiation level in these early periods. Non-iron meteorites are still less likely to lend themselves to this work, since most of them seem to have broken off from their parent bodies only comparatively recently, i.e. 20–30 million years ago.

In addition to transmutations produced in meteorites, isotopes produced in the earth's atmosphere can be used to trace cosmic ray intensity into the past. Here one is dealing mostly with effects produced by high energy particles, i.e. particles capable of penetrating the earth's magnetic field. Some of the cosmic ray produced isotopes, those with appropriate chemical properties and long mean lives, reach the deep ocean sediments; examples are Al^{26} ($\tau = 1.07 \times 10^6$ years) and Be^{10} ($\tau = 3.9 \times 10^6$ years). Where these have accumulated undisturbed for long periods, a complete record may have been preserved of past cosmic ray intensity, which could be used for studying not only average radiation levels, but short-lived variations as well.

The question arises whether it is possible to carry out similar historical studies on the corpuscular radiation of solar origin.

Consider the effects to be expected from solar particles whose average energy is, at least in our era, much lower than that of the galactic radiation. They play a minor role in isotope production on the earth because they are practically excluded from reaching the atmosphere except for the uppermost layers over the poles. As a result, the present day global inventory of cosmogenic radioactive isotopes (e.g. of natural C^{14} ($\tau = 8320$ years) produced by neutron capture in N^{14}) is in good accord with what is to be expected from galactic radiation alone. From the C^{14} inventory one can conclude that solar particles accounted for less than 25% of neutron production in the atmosphere during the last ~ 10000 years. Beyond the magnetosphere, however, it must be different; on exposed surfaces, transformation produced by solar particles cannot be rarer than those produced by galactic radiation. One has estimated that the largest as yet observed particle burst to originate at the sun (November 12, 1961) produced 10^{10} protons/cm^2 with energy greater than 20 MeV; these arrived at the earth over a period lasting for about 3 days. Compare this with the highest flux of galactic particles close to the earth, which occurs in the polar stratosphere during solar minimum; it is for any three day period smaller by at least a factor of 100000. Thus, even if a solar burst of similar magnitude occurred only at the rate of once every century, most of the nuclear transmutations produced in matter exposed in outer space near the earth's orbit would have to be due to solar rather than to galactic particles.

One must expect then that meteorites possess a thin skin of material whose composition has been modified considerably. If one could sample it, one could hope to study the history of flare activities for the past 10 million years or more. But actually meteorites are not well suited for this purpose for several reasons. Probably they spent most of their life at a large distance from the sun; the effect of solar bursts may therefore be much smaller than it is near the earth's orbit. Furthermore, the skin of the meteorite usually melts and is blown off during entry into the atmosphere: while some of this outer layer is believed to have been recovered in the form of nickel-iron spherules in deep sea sediments, it seems difficult to obtain adequate samples for a detailed study.

A good chance for investigating the history of the sun's activity may come when samples of lunar matter can be examined in detail, provided that there exist, on the moon, regions where surface material has been exposed for long periods without being removed by erosion or covered by accretion. The study of slightly deeper layers on the moon would make it possible to replace meteorites with lunar matter as radiation meters also for galactic particles.

The same techniques which are now being applied to meteorites could then be extended to more elements and one may eventually overcome the limitation implicit in a dependence on rare and haphazard sampling.

Perhaps it will then be possible to extend the study of corpuscular radiation intensity back into the period where important changes occurred in the flora and fauna of our planet or even back into periods more than 10^9 years ago when the structure of the milky way, and therefore probably its cosmic ray contents, differed appreciably from the present.

It does not seem unduly optimistic to predict that some day the study of corpuscular radiation from the galaxy and the sun by such methods will cover a considerable fraction of its history.

2. Galactic Corpuscular Radiation

Next let us discuss information which could be obtained by observing more closely the contemporary galactic particle flux.

Not only does the galactic cosmic radiation give rise to nuclear transmutations when intercepted by matter, but it is itself subject to nuclear transmutation as a result of collisions in the process of acceleration and in transit through space. It therefore carries information about its own history and the conditions of the interstellar regions through which it has passed.

A study of such changes, as detailed as that which is being carried out in meteorites, is still not within the range of existing technical possibilities, but it can now be developed much further than previously and may lead to important new insights. In restricted segments of the energy spectrum, the flux of certain individual light elements and of groups of heavier elements is known to within 20%, but over a large part of the energy spectrum little detailed knowledge has yet been obtained about the prevalence of the various species of nucleides, in particular about nucleides of atomic number above $Z \approx 30$. There is rough information on time variations for the radiation as a whole but not for its various constituents.

It seems possible to construct detectors which will measure with very good statistics the intensity of each of the elements present at all energies from about 10^7 to $\sim 10^{12}$ eV/nucleon. The problem is no longer whether suitable apparatus can be built, but rather what priority in research time and payloads should be given to such studies. This in turn depends on the question: how much information is to be gained on distant space from a more detailed knowledge of the galactic corpuscular radiation. At present one knows that particles in the energy range of 1–10 GeV/nucleon have traversed a fairly small amount of matter after acceleration and before arrival. Ultimately one would like to know very much more, apart from extending such information to all energy ranges. A more detailed analysis may reveal by what processes and in what physical surroundings the particles were accelerated, through what part of space they have travelled, how long they have been on their way, and also of course the particular chemical composition prevailing in the source or sources of galactic cosmic rays.

Answers to such questions will have to be based on a minute investigation of the flux of isotopes of individual elements in various energy ranges and on its variation in time. Of major importance here are isotopes of low cosmic abundance, radioactive isotopes, and isotopes which for various theoretical reasons may be expected to reflect particular properties of the processes of nucleogenesis or of acceleration, in short the isotopes whose abundance may be expected to be most sensitive to the processes which have intervened to select and separate particles of various masses and charges from the surrounding matter, accelerate them and transport them to the solar system.

From this point of view one can list some of the groups which are of particular interest:

a) Elements with very high and very low ionization potential; their relative abundances may be expected to be sensitively related to the temperature and density of the source region where the initial acceleration took place. For instance, the flux ratios Na/Ne or K/Ar would be instructive.

b) Isotopes unstable in the presence of low energy neutrons. They may be rare in the source region before acceleration and could have arisen later in energetic nuclear interactions of heavy nuclei with stellar or interstellar matter. Examples are D, He^3 and the isotopes of Li, Be, and B.

c) Abundance ratios between particles of odd and even atomic numbers.

d) Other elements which are known from earlier surveys of the composition to be rare in cosmic rays and therefore likely to be strongly affected by spallation reactions occurring in space. This applies to S, Cl, Ar, K, but also to F and perhaps to N.

e) Elements which are comparatively rare in the planetary system and in stellar envelopes, in particular the elements heavier than iron.

f) Elements which are copiously produced in collisions of low energy such as D and He^3.

g) Elements which are lying just below the most prominent abundance peaks (i.e. carbon and iron), whose production should therefore depend in a particularly sensitive manner on the amount of matter traversed.

h) Isotopes which would be expected to be abundant if γ,n reactions played an important role in the source regions, e.g. B^{11} produced from the reaction C^{12} (γ,n) $C^{11} \rightarrow B^{11}$.

Existing techniques are not yet sufficiently developed to measure directly the abundance of individual isotopes of an element present in the primary radiation, except in the case of hydrogen and helium. Nevertheless, it seems feasible to obtain information on the isotopic composition indirectly by making use of modulation effects which depend on the ratio of atomic weight to atomic number (A/Z) and may differ therefore for the different isotopes of the same element. Thus, if the abundance ratios between neighbouring elements were measured with high precision, for example during Forbush decreases and at different periods in the solar cycle, effects due to the isotopic composition of various elements may show up. One can estimate the order of magnitude of effects to be expected from the following table:

Isotope	A/Z
H	1.00
D	2.00
He^3	1.50
He^4	2.00
Li^6	2.00
Li^7	2.33
Be^7	1.75
Be^9	2.25
Be^{10}	2.50
B^{10}	2.00
B^{11}	2.20
C^{12}	2.00
C^{13}	2.17
Fe^{58}	2.23

In this connection I would like to draw attention in particular to the need of investigating in greater detail the beryllium component of the primary beam, because it may lead to the establishment of a time scale, i.e. permit an estimate of how many years the primary particles have travelled in space before they reached the solar system. Collisions of heavy primary nuclei in the interstellar gas lead to the various isotopes of lithium, beryllium and boron in definite ratios, which can be determined in the laboratory. The three beryllium isotopes Be^7, Be^9, and Be^{10} are possibly produced in comparable amounts, but only Be^9 is stable. Be^{10} decays into B^{10}, and Be^7 decays into Li^7 so that the observed flux ratio Be/B and Be/Li will in general differ from the ratio in which these spallation fragments are produced.

Since Be^{10} has a mean life τ of 3.9 million years, the flux ratios at low energy are sensitive to travel times of magnitude 1 to 20 million years, and at 10 GeV/nucleon the relevant time scale extends to about 200 million years. Together with the amount of matter traversed, this provides information on gas density. As an example, let us assume that at 10 GeV/nucleon, the flux of various spallation products such as Li, Be, and B indicated that the primaries have traversed a total of 3 g/cm^2 of hydrogen in space, and let us assume further that the ratio between these elements in the primary beam proves that almost all of the Be^{10} has decayed before arrival. One would then have to conclude that the mean gas density along the trajectory from the source to the solar system is

$$\rho \leq \frac{3}{M_r c}\frac{1}{10\tau} \simeq 0.06 \quad \text{hydrogen atoms/cm}^3.$$

Conversely, if most of Be^{10} is still present, the mean gas density would have to be larger than this.

Be^7 decays by K-capture (79 days) and is therefore stable in the completely ionized state. Yet it is conceivable that electron attachment, and therefore capture, occurs during the long transit time, especially if the fragment is nonrelativistic. This may lead to a strong enhancement of the ratio Li/Be over the ratio observed in spallation reactions. Such an effect would give information on electron density and temperature in the relevant interstellar regions.

Finally, it is useful to study the primary composition at very low energies, because here ionization losses will modify spectra of the heavy elements more severely than those of the light ones. In this energy region, the amount of interstellar matter traversed could then be deduced from effects due to atomic rather than nuclear collisions.

These examples may suffice to show how a careful investigation of the chemical composition in well-defined energy bands and at various periods of the solar cycle (and also during Forbush decreases) could enhance our knowledge of the cosmic ray phenomenon.

Some major experimental difficulties still have to be overcome before such a detailed analysis of the galactic corpuscular radiation becomes possible.

1) The measurements have to be carried out at great altitude. Since the amount of material traversed by the particles before reaching the solar system is only of the order of a few g/cm^2, the additional thickness of matter which the particles can be allowed to traverse before their charge is measured, should be kept below 1 g/cm^2. The altitude reached by balloons is not sufficient.

2) If one wants to determine isotope ratios in the various elements by studying time variation of charge ratios during a Forbush decrease for particles of a given energy, one must be able to measure the flux of particles belonging to neighbouring elements within hours and with a statistical accuracy better than 1%. The detectors must therefore have an aperture of the order of 1 m^2 steradian. (Beyond the magnetosphere a detector with an acceptance of 1 m^2 steradian will register about 300000 particles of the Li, Be, B group, and about 100000 particles of the iron group per day.)

3) Complete charge resolution and accurate energy determinations are of course necessary.

4) An interpretation of precise data on composition in terms of interactions, which the particles suffered during acceleration or in transit, requires a more precise knowledge of spallation reactions than is available at present. The relevant cross-sections for the creation of various fragmentation products in the collision between heavy nuclei and hydrogen can all be studied in accelerator laboratories up to energies of 30 GeV/nucleon. At lower energies, spallation processes due to collisions with helium

and with gamma-rays can also be studied. (Hydrogen plus helium in space presumably constitute the principal target nuclei responsible for modification of the original chemical composition of the galactic particles.)

Condition 1) can be satisfied if the measurements are carried out in satellites or space platforms, conditions 4) by carrying out rather extensive auxiliary experiments on spallation reactions at various accelerators. The second and third conditions require the design of detectors which are substantially different from those used for charge analysis so far (nuclear emulsions or scintillator plus Čerenkov counter telescopes).

Emulsion detectors have the required charge and energy resolution but are not suitable for work where very large statistics are required. The scintillation plus Čerenkov counter telescopes have been used for fairly light elements $Z \lesssim 12$ and for small acceptance angles. Energy resolution is fairly good only in the non-relativistic range. The detectors which combine high charge resolution, high energy resolution and large aperture, remain to be designed. In principle this seems possible within the size and weight limitations of existing space craft.

The specific ionization can be measured accurately and fluctuations reduced by letting the particles pass through a series of scintillators. If the detector has large area and acceptance angle, high charge resolution requires an accurate knowledge of the trajectory of each particle, especially angle of incidence. It seems possible to measure these track coordinates accurately, e.g. by means of spark chambers. Finally, it is necessary to incorporate in the detector velocity selectors which accept particles whose velocity exceeds a fixed value, independent of charge or mass. Threshold Čerenkov counters may be useful in this connection.

It is clear that the required detector will be far bigger, more complicated and more expensive than any instrument which has been developed until now for the study of the composition of galactic primary cosmic radiation. Yet, such an increase in effort seems unavoidable if one wants to exploit more fully the information brought to the solar system by corpuscular radiation from remote regions of the universe.

3. The Solar Neutron Component

The third problem which I should like to mention is also difficult though less ambitious than the preceding two.

It has been known for many years that the sun can accelerate protons to energies high enough to induce nuclear reactions. Not all the accelerated particles escape, the majority is retained by the magnetic field of the sun. If accelerated to $\gtrsim 500$ MeV, they are likely to produce one or more neutrons before being slowed down again. These neutrons can escape and reach the earth.

Although the neutron flux is likely to be highest at the time of strong flare activity, it is possible and even probable that a weak neutron beam of varying intensity is emitted by the sun at all times, even during quiet periods. Neither neutron bursts nor a steady neutron flux have been discovered as yet. If the sky were patrolled by a satellite carrying a sufficiently sensitive neutron detector, the sun could be monitored through a new window; one could view the nuclear reaction rate occurring near the surface and correlate it with other solar activities.

It is difficult to predict the intensity of the expected neutron flux. Evidence for a fairly steady low energy solar proton flux below ~ 200 MeV has been reported during the last years. Simpson has suggested that these nucleons escape from the sun as neutrons but decay into protons in transit. Flamm and Lingenfelter have given a corresponding neutron flux of 0.05 particles/cm^2 sec with energies $\gtrsim 10$ MeV. The correlation with the position of the sun should make it possible to distinguish solar neutrons from the general albedo background. But since neutrons of $\gtrsim 10$ MeV are difficult to collimate, nobody has succeeded as yet in designing a detector which makes use of the unique property of completely parallel particle trajectories.

Bibliography

METEORITES AND HISTORY OF COSMIC RADIATION

[1] Fowler, W. A.: 1962, 'Nuclear Clues to the Early History of the Solar System', *Science* **135,** 1037.

[2] Fowler, W. A., Greenstein, J. L., and Hoyle, F.: 1962, 'Nucleosynthesis During the Early History of the Solar System', *Geophys. J.* **6,** 148.

[3] Anders, F.: 1963, 'Meteorites and the Early History of the Solar System', *Origin of the Solar System*, Academic Press, pp. 95–142.

[4] Geiss, J., Oeschger, H., and Schwarz, U.: 1962, 'The History of Cosmic Radiation as Revealed by Isotopic Changes in the Meteorites and on the Earth', *Space Sci. Rev.* **1,** 197.

[5] Honda, M. and Arnold, J. R.: 1964, 'Effects of Cosmic Rays on Meteorites', *Science* **143,** 203.

COMPOSITION OF COSMIC RADIATION

[6] Ginzburg, V. L. and Syrovatskii, S. I.: 1964, *The Origin of Cosmic Rays*, Pergamon Press.

[7] Ginzburg, V. L. and Syrovatskii, S. I.: 1963, *Proc. Jaipur Conf. on Cosmic Rays* **3.**

[8] Ginzburg, V. L. and Syrovatskii, S. I.: 1965, *Proc. London Conf. on Cosmic Rays.*

SOLAR NEUTRONS IN COSMIC RADIATION

[9] Lingenfelter, R. E. and Flamm, E. J.: 1964, *Science* **144,** 292.

[10] Simpson, J. A.: 1963, in: 'Semaine d'étude sur les problèmes du rayonnement cosmique dans l'espace interplanétaire', *Pontificia Academia Scientarum Scripta Varia* **25,** 323.

CELESTIAL GAMMA-RAYS – OBSERVATIONAL ASPECTS

G. W. HUTCHINSON
University of Southampton, England

Abstract. The most important characteristic parameters of celestial γ-rays, such as spectrum, angular distribution and variations in time are briefly reviewed. The various ways of observations of γ-rays are listed. The experimental results obtained so far are summarised.

We shall define gamma-ray astronomy as the study of photons with energies greater than 0.2 MeV, arriving from any extra-terrestrial source. As with all other electromagnetic radiation, these photons travel in straight lines, and conserve their energies, except when they are occasionally scattered. As yet few experimental results have been obtained, but several groups are developing experiments in this field of research.

1. Important Parameters of Celestial Gamma-Radiation

1.1 Energy spectrum

The mechanisms which are probably important for producing γ-rays result in the formation of a continuum whose intensity usually decreases with increasing energy. Only a few processes produce spectral lines, of which the most important are the decay of radio-active nuclei, and electron-positron annihilation. Since the particles in question are usually created at high energy, their lines will be Doppler-broadened unless they are brought to rest before the γ-rays are emitted. The decay of π mesons at rest would also produce a γ-ray spectral line (at 70 MeV) but these particles are nearly always created at high energies, resulting in extensive Doppler broadening of the line. A possible production mechanism of low energy π mesons (and thus a γ-ray line) is the annihilation of protons and anti-protons at low kinetic energies. If proton-antiproton annihilation occurs in space, this would produce low energy π mesons and hence a γ-ray line.

1.2 Direction of incidence

From a study of the arrival direction of γ-rays it would be possible to detect discrete sources. Possible γ-ray sources, of small angular diameter, are the discrete non-thermal radio sources, both inside and outside the galaxy, and also stars, inter-stellar clouds, and near-by normal galaxies. Since at both optical and γ-ray energies the universe is optically thin, the application of Olber's paradox indicates that discrete γ-ray sources would not produce an isotropic flux. However, if an isotropic flux was detected, this would not necessarily prove the existence of inter-galactic sources, since a large source, spread over the local galactic cluster, would also produce isotropy. The galactic disc and centre are possible anisotropic sources, but it is not easy to estimate the degree of anisotropy which may be quite small, and may thus be difficult to detect above a background of flux from the metagalaxy.

J. G. Emming (ed.), Electromagnetic Radiation in Space.

The number of photons of the isotropic flux received by a detector is proportional to the product of its opening angle and its effective area. To detect discrete sources, the detector should have a small acceptance angle, since the counting rate from most types of background radiation will be proportional to this opening angle, while the counting rate from the source will be constant if it subtends less than the acceptance angle.

1.3 Polarisation

Polarisation is not expected in the γ-ray flux. The only known mechanism which could produce polarised γ-rays is synchrotron radiation from very high energy electrons in a magnetic field whose direction is coherent over a large area of the investigated source.

1.4 Time variation

As techniques improve, it may be possible to detect time variation in the γ-ray flux. Short term effects may be produced on the sun during flares, while longer term variations may arise from such objects as the Crab nebula.

2. Vehicles

Since γ-rays are both absorbed and produced in the atmosphere, vehicles are desirable to carry experimental apparatus above the atmosphere. These include:

2.1 Balloons

Balloons can be flown for several hours duration at above 5 g/cm^2 atmospheric depth. They are best suited for experiments related to possible discrete sources, as the locally produced background of γ-rays makes the interpretation of results on the isotropic flux difficult.

2.2 Rockets

Rocket flights are short (about 5 minutes), and are generally of little use in γ-ray astronomy because the fluxes are known to be low, and hence, statistically significant results cannot be obtained with signal detectors of reasonable size.

2.3 Satellites

Satellites are generally accepted to be the most suitable vehicles for γ-ray astronomy. The long duration of the experiment (about 1 year) is the overriding advantage.

3. Ground Based Experiments

Gamma rays of very high energy ($E > 10^{12}$ eV) can be detected by their secondary effects with ground based equipment. The experiments are designed to detect the Čerenkov light produced by the shower electrons in the atmosphere. This light is directed forward in a narrow cone along the original direction of the γ-ray. Since cosmic ray protons also initiate such showers, and these will be isotropically distributed, the method is only applicable to discrete sources.

The largest showers ($E > 10^{14}$ eV) can be detected directly when they reach the ground. Such an extensive shower should be virtually free of muons when produced by a γ-ray. However, not enough is known of proton-initiated showers to rule out the possibility that these may also occasionally have an abnormally low muon content, and a shower initiated by a nuclear event of unusual type might also exhibit the same features.

4. Measurements of Isotropic Gamma-Ray Flux

TABLE I

Observer	Energy Range	Flux
Arnold, Metzger	400–800 keV	0.014 $cm^{-2} sec^{-1} str^{-1}$
Fazio, Duthrie	60–150 MeV	5×10^{-3} $cm^{-2} sec^{-1} str^{-1}$
Kraushaar	> 100 MeV	$\leqslant 3 \times 10^{-4}$ $cm^{-2} sec^{-1} str^{-1}$
Peterson	0.05–0.15 MeV	9 $cm^{-2} sec^{-1} MeV^{-1}$
Peterson	0.5–1.5 MeV	0.36 $cm^{-2} sec^{-1} MeV^{-1}$
Peterson	1.5–4.5 MeV	0.06 $cm^{-2} sec^{-1} MeV^{-1}$

Figure 1 shows measurements of the X-ray and γ-ray flux plotted on the same graph. It appears that the results lie very approximately on a straight line with a spectral index of about -2. If this is true, it is surprising since, in this range of energy, the radiation is the result of various independent production mechanisms (LÜST [1]), each of which has its characteristic spectral index. Extensive air shower, and nuclear emulsion results are not shown because of their lower reliability, but they do not contradict the general trend.

There are discrepancies in some experimental results, e.g. DUTHIE *et al.* [3] and KRAUSHAAR *et al.* [4]. The first suggests that the flux contributed by extra-galactic sources may be appreciable, while the second tends to suggest that it is small or non-existent. However, the theoretical predictions are uncertain and no definite conclusions can yet be drawn.

The detailed analysis by KRAUSHAAR *et al.* [4] of the results of the Explorer XI satellite experiment has indicated the great difficulty of interpreting the results of balloon experiments. The analysis has shown that there is a large increase in flux near the horizon due to cosmic ray albedo from the atmosphere. Also, the detecting telescope has a response at large angles to its axis due to scattering of the converted electrons in the upper counter, and this explains the variation of counting rate with geomagnetic latitude, even when the telescope is pointing away from the earth, and also the east-west asymmetry under the same conditions.

The work of METZGER *et al.* [2] and PETERSON [5], who investigated the energy range 0.5–2.3 MeV, showed a spectrum unexpectedly flat with no evidence of the

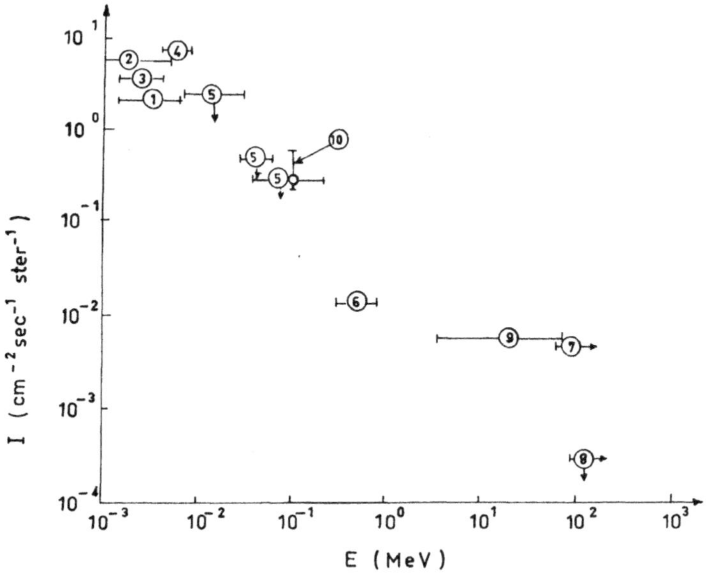

Fig. 1. Measured omnidirectional flux of celestial gamma-rays. (From BRINI *et al.*: 1965, *Nuov. cim.* **28,** 130.)

1. GIACCONI, R., *et al.*: 1963, *Phys. Rev. Lett.* **11,** 530.
2. BOWYER, S., *et al.*: 1964, *Nature* **28,** 1307.
3. BOWYER, S., *et al.*: priv. comm.
4. CLARK, D. B., *et al.*: 1964, *I.A.U. Symposium, Liège.*
5. GIACCONI, R., *et al.*: 1964, *Cospar Symposium.*
6. ARNOLD, J. R., *et al.*: 1962, *J. Geoph. Res.* **67,** 4878.
7. HAFNER, E. M., priv. comm.
8. KRAUSHAAR, W. L., priv. comm.
9. PERLOW, G. J., *et al.*: 1951, *Phys. Rev.* **81,** 552.
10. BRINI, D., *et al.*: 1965, *Nuov. cim.* **28,** 130.

electron annihilation line at 0.5 MeV, or of the proton-neutron capture line at 2.23 MeV. Doppler broadening may have been sufficient to spread out these lines until they were indistinguishable from the continuum.

5. Observation of Discrete Gamma-Ray Sources

In addition to the results listed in Table II, other results have been given by CLINE [6] DUTHIE *et al.* [3] and BRACCESI *et al.* [7]. All the experiments were flown on balloons and thus suffer from the difficulties of locally produced background. The fluxes quoted are the most significant, and are upper limits. The slight discrepancies between experiment and theory are not yet significant, owing to uncertainties in the theoretical predictions.

TABLE II

Source	Theoretical prediction		Experimental Upper Limit	
	Intensity $cm^{-2} sec^{-1}$	Energy MeV	Intensity $cm^{-2} sec^{-1}$	Energy MeV
Crab	2×10^{-5}	> 100	7×10^{-4}	> 50
	4.5×10^{-10}	> 1000	5×10^{-11}	$> 5 \times 10^{6}$
Cygnus A	5×10^{-8}	> 100	5×10^{-4}	> 50
	10^{-9}	> 1000	5×10^{-11}	$> 5 \times 10^{6}$
Virgo A	8×10^{-9}	> 100	3×10^{-4}	> 50
	10^{-9}	> 100	5×10^{-11}	$> 5 \times 10^{6}$
3C-273B	5×10^{-6} to 5×10^{-9}	$3 < E < 100$	3×10^{-10}	$> 5 \times 10^{6}$

6. Experimental Techniques

Eventually the whole γ-ray spectrum should be studied, but so far only three energy regions have been investigated. This is due both to the fact that there exist different characteristic γ-ray production processes, and to the fact that different kinds of γ-ray detectors are available. The three regions are:

1) "Nuclear" Region 0.2–10.0 MeV
2) "π^0" Region 50 MeV–1.0 GeV
3) Very high energy region $> 10^6$ MeV

For the third region, it is mainly a matter of time for the experiments now in operation to build up sufficient statistics to give definite results, which will prove valid if they show anisotropy. The following discussion will concentrate only on the nuclear and π^0 regions.

6.1 Nuclear region

Energy Measurement

The main detector in this region is the scintillation crystal, caesium iodide being preferred to sodium iodide on account of its superior mechanical properties. An anti-coincidence counter is required to eliminate the flux of charged particles. A simple device is the so-called phoswich counter, where the crystal is entirely surrounded by a plastic scintillator and the assembly is viewed by one photomultiplier. Since the two scintillators have different decay times, it is possible by electronic techniques to separate the light pulses, and to reject events due to charged particles. This also improves the energy resolution to some extent since events in which the electron leaves the crystal are also rejected. Such a phoswich (Figure 2) has been flown on the Ranger series moon probes [2]. It is usually desirable to calibrate the crystal with γ-rays of known energy, since the detector produces a spectrum of pulse heights from a mono-energetic γ-ray. The energy resolution obtainable with a crystal is typically about 0.05 MeV, so nuclear transition lines will only be observed if relatively few are important, and if they are not Doppler-broadened. A three-inch cubic crystal flown in a satellite for a year should be able to detect a line whose intensity is about 0.1% of the background. Balloons could be used to detect lines in the primary spectrum if these are different from lines produced in the atmosphere. As a balloon flight is of much shorter duration than a satellite measurement, the line would have to be stronger before it could be recognized above the background.

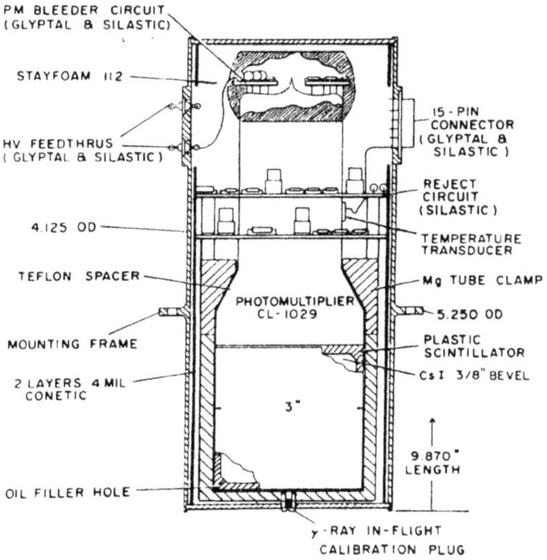

Fig. 2. Phoswich detector for observation of celestial gamma rays. (From VAN DILLA *et al.*: 1962, *Trans. I.E.E.E. Nucl. Sci.* **NS9–3,** 405.)

Direction Measurement

The scintillation crystal has no inherent directional properties. It is not possible to use a coincidence telescope since the γ-ray is essentially absorbed in the detection process. Compton scattering which does, in fact, preserve useful direction information is difficult to utilise, since both the energy and direction of the γ-ray and electron are required. Compton scattering in which the γ-ray undergoes only very small deviation has a very low cross-section. Thus simple collimation is the only practical method of making a telescope, but since a small acceptance angle is usually desirable, the weight of the collimator for a 3-inch crystal can amount to more than 10 kg, and such a mass will produce a high flux of γ-rays from interactions with charged cosmic rays. "Active" collimation, i.e. using a thick anti-coincidence shield as a collimator, is another possible solution.

It may be possible to use the moon as an occulting disc for discrete sources, with an un-shielded crystal in a lunar orbit. Such an experiment would not suffer from effects due to the earth's atmosphere. If the moon subtended an angle of $\pm 5°$, assuming such an orbit was possible, then any discrete source would be occulted for 2–3% of the time on average. If the experiment were working for 1 year, a total of about 10^8 counts would be received, about 10^6 being received while any one point was behind the moon. The statistical error of any excess in the counting rate for this region of the sky could then be as low as about 0.1%.

6.2 π^0 REGION

The flux here is known to be low, e.g. smaller than $10^{-4}\text{cm}^{-2}\text{sec}^{-1}\text{sterad}^{-1}$. For a 0.3% accuracy, at least 10^5 counts are required. If these are to be collected in a year, a sensitive area of 100 cm^2 is needed if an acceptance angle of about 1 sterad is used.

Directional Measurement

The measurement of γ-ray direction is achieved by using the direction either of the shower, or of the electron pair first produced by the γ-ray. The angle between the direction of the electron and the direction of the γ-ray is small (about 1° for a 100 MeV γ-ray) and decreases with increasing γ-ray energy. Three methods seem to be available for obtaining the directional properties.

The first method is typified by the Explorer XI experiment (KRAUSHAAR *et al.* [4]), illustrated in Figure 3. The instrument consists of two counters, some distance apart with their outputs phased in coincidence. The acceptance cone is fixed mainly by the geometric properties of the two counters, but also to some extent by the scattering of the converted electrons in the first counter.

The second method is that used by DUTHIE *et al.* [3]. As illustrated in Figure 4, Čerenkov light is produced by the electrons, and is directed out in a cone of half angle θ around the direction of motion of the electron, where θ is given by

$$\theta = \cos^{-1}\frac{V}{cn}$$

where V is the velocity of the electron, and n the refractive index of the material. There are various optical problems encountered when practical equipment is constructed. Fazio and Hafner have pointed out that care must be taken concerning the response of the instrument at large angles.

The third method is to measure the γ-ray direction using a spark chamber. This reduces the problem of electron scattering which limits the angular resolution in the two previous methods. A good detector requires a high efficiency and low electron scattering so that angular resolution can be high. However, it is impossible to satisfy these requirements simultaneously. In a spark chamber, the efficiency per gap used to calculate the direction of the γ-ray can be quite low, and the scattering therefore small, but since many gaps can be stacked together, the total efficiency of the whole chamber can be large. An additional advantage with a spark chamber is that whilst its angular resolution is high, its acceptance cone can be large, about ± 30°, so that many possible objects can be studied simultaneously.

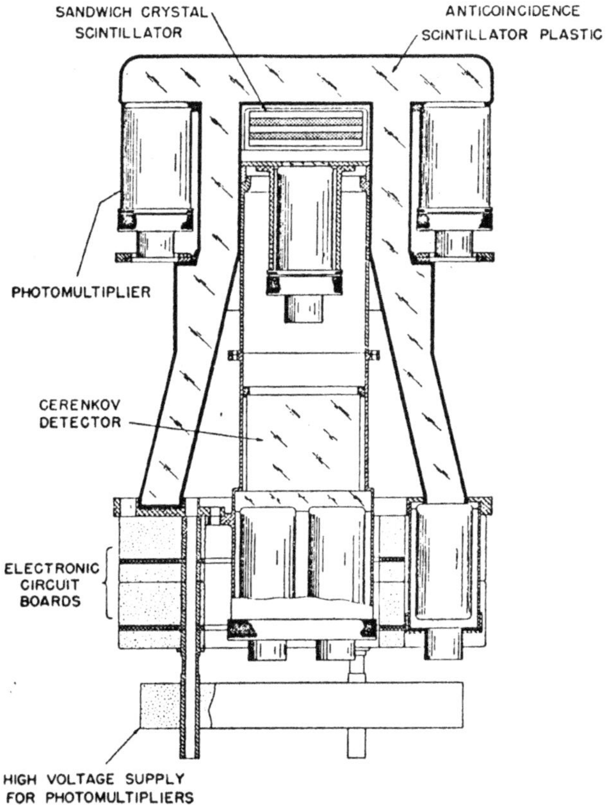

Fig. 3. Schematic drawing of gamma-ray detector. (From KRAUSHAAR *et al.*: 1965, *Astroph. J.* **141,** 845.)

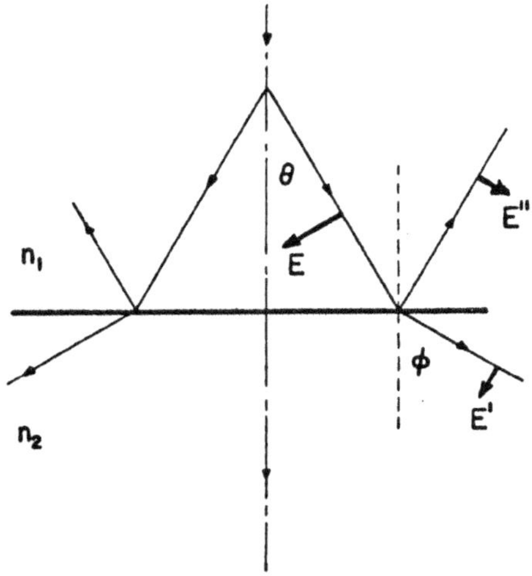

Fig. 4. Electric vectors in refraction and reflection of Čerenkov light from a relativistic charged particle incident normally on a dielectric interface. (From FAZIO and HAFNER: 1961, *Rev. Sci. Inst.* **32,** 697).

The coordinates of the sparks may be obtained by various means, as follows:

1) *Television Camera*. This means scanning the gaps optically.

2) *Wire Chamber*. Here, some or all of the plates are replaced by sets of parallel wires. Each spark causes a current to flow in the appropriate wire. This can be recorded, and later read out from a magnetic store.

3) *Magnetostrictive Read-out*. This is a version of the wire chamber above. The current in the wire causes a sonic pulse to pass along a magnetostrictive material placed at right angles to the wires. The time for the sonic pulse to reach a pick-up probe is determined by the position of the wire which sparked.

4) *Audio Read-out*. The speed of sound is quite slow by electronic standards, so the position of a spark can be established by the arrival times of its sound at various microphones. There are certain problems associated with multiple sparks, the initial

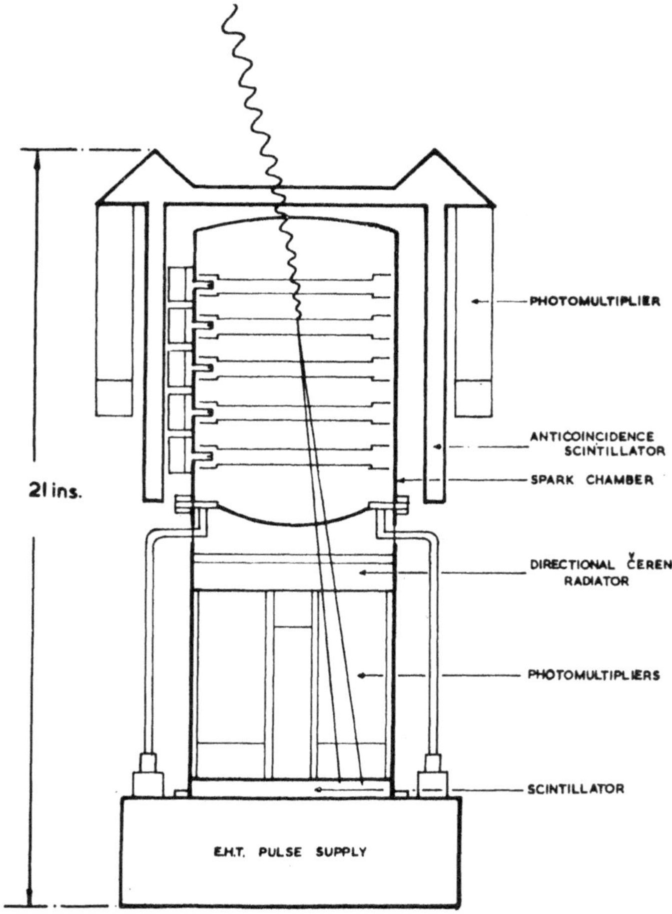

Fig. 5. Schematic diagram of gamma-ray detector (Southampton Experiment).

shock wave around the spark, and the variation of the speed of sound with temperature. However, it appears that these problems can be overcome to some extent by the use of enough microphones, although in some cases it may be difficult to obtain the coordinates of multiple sparks. An acoustic spark chamber has been built at the University of Southampton for a flight in the Satellite OGO-E (Figure 5).

References

[1] Lüst, R.: 1967, Lecture held at this summer school; see this Volume p. 231–248. Also published as ESRO SM-60.
[2] Metzger, A. E., Anderson, E. C., Van Dilla, A., and Arnold, J. R.: 1964, 'Detection of an interstellar flux of gamma rays', *Nature* **204,** 766.
[3] Duthie, J. G., Fazio, G. C., Hafner, E. M., and Kaplon, M. F.: 1963, 'Gamma rays at high altitudes', *Phys. Rev. Letters* **10,** 364.
[4] Kraushaar, W., Clark, G. W., Garmire, G., Helmken, H., Higbie, P., and Agogino, M.: 1965, 'Explorer XI experiment on cosmic gamma rays', *Astrophys. J.* **141,** 845.
[5] Peterson, L. E.: 1965, *Proc. COSPAR Symposium, Mar del Plata.*
[6] Cline, T. J.: 1961, *Phys. Rev. Letters* **7,** 109.
[7] Braccesi, A., Ceccarelli, M., and Salandin, G., 1960: *Nuov. cim.* **17,** 69.

Further Recommended Literature

A) OBSERVATIONS OF COSMIC GAMMA RAYS

Perlow, G. J. and Kissinger, C. W.: 1951, *Phys. Rev.*, **81**, 552.
Zawadski, A. *et al.* and Chudakov, A. E. *et al., Proc. Int. Conf. Cosmic Rays Jaipur* **4**.

B) ELECTRON IONISATION AND SCATTERING

Barkas, W. H. and Rosenfield, A. H.: 1961, UCRL 8030 (Rev.) **11,** 1961.
Yuan, L. C. L. and Wu, C. S.: 1961, *Methods of Experimental Physics* **5a,** Academic Press, 1961.

C) SCINTILLATION AND CERENKOV COUNTERS

Chase, R. L.: 1961, *Nuclear Pulse Spectrometry*, McGraw Hill.
Fazio, G. G. and Hafner, E. M.: 1961, *Rev. Sci. Inst.* **32,** 697.
Hutchinson, G. W.: 1960, *Prog. Nucl. Phys.* **8,** 197.
Leontic, B.: 1959, CERN report 59–14.
Ritson, D. M.: 1961, *Techniques of High Energy Physics*, Interscience.
Yuan, L. C. L. and Wu, C. S.: 1961, *Methods of Experimental Physics* **5a,** Academic Press.

D) SPARK CHAMBERS

Charpak, G.: 1962, *Nucl. Inst. and Methods* **15,** 318.
Cronin, J. W.: 1963, *Nucl. Inst. and Methods* **20,** 143.
Lillethun, E. *et al.*: 1964, *Suppl. Nuov. cim.* (*Series 1*) **2,** 310.
Rutherglen, J. G.: 1964, *Prog. Nucl. Phys.* **9,** 1.

EXPERIMENTAL OBSERVATIONS OF PRIMARY COSMIC RAY ELECTRONS*

L. SCARSI

Istituto di Scienze Fisiche, University of Milan, Italy

Abstract. In the introduction, some of the astrophysical implications of studies on the galactic cosmic ray electrons are mentioned. The various difficulties encountered in the measurement of these electrons are discussed. The experimental results which were obtained so far by balloon-borne and also by satellite-borne instruments are critically reviewed. A summary of the results is presented.

1. Introduction

Electrons, together with γ-rays, constitute for the experimenters a difficult and elusive component of the primary cosmic radiation. The difficulties arise mainly from the low value of the electron flux relative to the flux of nucleons of comparable energy or magnetic rigidity, and from the presence of a secondary electromagnetic component rapidly building up in the atmosphere as a consequence of the nuclear interactions of the incident "nucleonic" cosmic rays. On the other hand, these electrons represent an important source of information on a number of fundamental problems related to the origin of the cosmic radiation, its propagation in space and the structure of our galaxy. Their existence in the primary flux arriving at the earth is expected. In fact, apart from electrons emitted by the same sources as the nucleonic component, there must exist electrons arising from collisions of the nucleonic flux with interstellar matter (for example knock-on electrons and $\pi^{\pm}\rightarrow\mu^{\pm}\rightarrow e^{\pm}$ decay).

Some evidence for the existence of primary electrons comes from radioastronomical observations: the presence of a flux of nonthermal radiation in the radiofrequency region from MHz to some GHz and its attribution to synchrotron radiation bring us to admit the existence of relativistic electrons together with magnetic fields in interstellar space.

The direct information on galactic electrons available at present is rather fragmentary and is derived from relatively recent experiments. Some of these results are still preliminary. The presence of electrons in the primary cosmic radiation has been established by balloon-borne detectors in the GeV region (and possibly down to a few hundred MeV) and is found to be of the order of one percent of the total charged flux of same energy. The IMP observations outside the magnetosphere have shown the existence, in the 3–12 MeV range, of a strong flux at least one order of magnitude larger than that of protons of similar energy. An estimate of the charge sign ratio has also been obtained, both by direct measurement and through the east-west effect in the geomagnetic field.

We would like to give in what follows a critical analysis of the experimental data which exist at present. Some indications will be given as to the confidence with which conclusions could be drawn on the origin of the electrons and their relation with the radio-astronomical observations.

J. G. Emming (ed.), Electromagnetic Radiation in Space.

2. Measurements of Primary Cosmic-Ray Electrons

2.1 Causes of uncertainty in the measured values

Measurements on primary electrons are subject to the following uncertainties:

a) Measurements within the earth's atmosphere are affected by the secondary electrons produced in the overlying atmosphere.

b) Measurements within the magnetosphere are affected by re-entrant albedo particles.

c) Measurements of flux and energy spectrum within the solar system may not, in general, be representative of the average galactic intensity, because of modulation by interplanetary magnetic and electric fields. Further, electrons of solar origin may be present. These effects are expected to be more serious for low energies than for the region of high energy (larger than several GeV).

The majority of the data up to now have been obtained from balloon observations and are therefore subject to the uncertainties listed in (a) and (b). The satellite experiment on IMP-1 was carried out outside the magnetosphere, thus being only subject to uncertainty (c).

2.2 Electrons produced in the atmosphere

For a balloon-borne detector, (a) is the main problem to be concerned with. A secondary electromagnetic component is rapidly building up in the atmosphere as a consequence of interactions of the primary nucleons with nuclei of the air. The decay:

$$\pi^{\pm} \rightarrow \mu^{\pm} \rightarrow e^{\pm}$$

and

$$\pi^{0} \rightarrow 2\gamma \rightarrow e^{\pm}$$

are the main processes leading to electrons.

Since the primary electron flux is only a few percent of the nucleonic flux at the same rigidity, at least in the energy range extending from some tens of MeV to several GeV, the secondaries obscure the primary component already at a few g/cm^2 residual atmosphere. Furthermore, a degradation of the primary electron spectrum due to their own interaction occurs. Evidently, the first requirement is to have the detectors at an altitude as high as possible. On the other hand about 2 g/cm^2 of residual atmosphere seems to be the minimum permissible by the present balloon technique. The contamination by secondaries is energy-dependent: it decreases with increasing energies, as we will see in the following. However, about 10 g/cm^2 should be considered as maximum depth at which results can be corrected for secondaries in a meaningful way.

The two early experiments (Rossi and Hulsizer [1] for $E \gtrsim 4.5$ GeV and Critchfield, Ney and Oleksa [2] for $E \gtrsim 1$ GeV) carried out at an atmospheric depth of about 20 g cm^{-2} gave inconclusive results just for these reasons.

Important features of the secondary electron component are the marked excess of positively charged particles and a differential energy spectrum peaked around

20–50 MeV. The intensity of the particle flux and the energy spectrum depend on altitude and, at a given period of solar activity, on the local geomagnetic cut-off.

Corrections can be made in two ways:

a) By measuring the growth curve of the secondary plus primary electron flux in the upper layers during balloon ascent and extrapolating to zero residual atmosphere or by estimating the secondary contamination from the flux of γ-rays measured in the same experiment.

b) With a "calculation" approach, computing the number of particles arising from nuclear collisions occurring in the atmosphere above the payload. The first approach has the advantage of being experimental. On the other hand, the "growth curve" method suffers from limited statistics because of the relatively short time of observation during ascent of the balloon in the successive layers of the atmosphere. Another drawback is the necessity to assume a shape of the extrapolation curve.

The second approach is limited by the accuracy with which flux, energy spectrum and composition of the nucleonic primary cosmic rays are known, as well as the interactions with the atoms in the atmosphere.

While some experimenters have followed both lines (as a check), others have only employed the first or the second method.

A third method of discriminating against secondaries is to observe a net discontinuity in the energy distribution at the value corresponding to the geomagnetic cut-off. This method is only possible at geomagnetic latitudes for which the cut-off is sharply defined.

As a general remark, we can say that the difficulties arising in the interpretation of the available experimental results, especially for establishing the energy spectrum of primary electrons below 1 GeV, arise almost all from this type of correction. For these reasons, it is of interest to consider in some detail the characteristics of secondary electrons in the upper atmosphere.

Estimates of the energy spectrum of the electrons produced in the atmosphere have been published by L'Heureux and Meyer [3] and by Daniel and Stephens [4]. A general calculation has recently been made by G. Perola and L. Scarsi (to be published) based on the results of accelerator measurements for proton energies up to 28 GeV and on cosmic ray data at higher energies. The secondary electron energy spectra have been obtained for solar minimum conditions and for various depths in the atmosphere and various values of the geomagnetic cut-off. Some data will be of interest when discussing the experiments on primary electrons shown in Figure 1, where curve (b) represents the computed electron differential energy spectrum (for $E > 200$ MeV) for a depth of 4 $\mathrm{g\,cm^{-2}}$ at a geomagnetic location equivalent to Fort Churchill.

2.3 Re-entrant albedo

Experiments on balloons or on satellites orbiting inside the magnetosphere are affected by re-entrant albedo electrons (cause (b) of paragraph 2.1). Re-entrant albedo electrons arise from secondary electrons produced in the atmosphere which are ejected

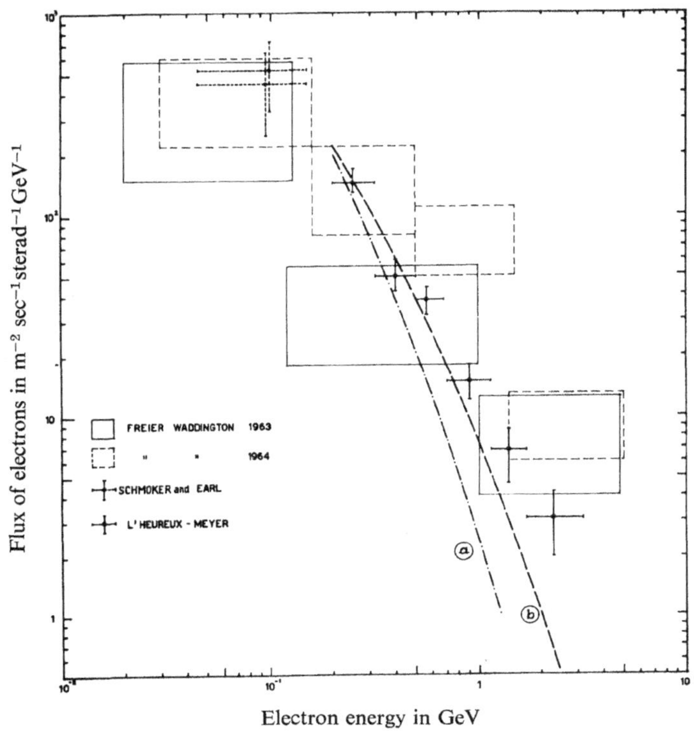

Fig. 1. Measured flux of primary cosmic ray electrons (crosses and squares) and calculated flux of secondary electrons (curve (a) after L'Heureux and Meyer, curve (b) after Perola and Scarsi).

upwards in the magnetosphere; those with magnetic rigidity higher than the local cut-off will be lost in space, whereas those with rigidity less than the local geomagnetic cut-off will be trapped in semi-stable orbits and eventually re-enter the atmosphere at a geomagnetic latitude similar to that at which they were emitted.

In principle, in a stationary state the flux of re-entrant particles should be the mirroring part of the flux of ejected particles but in reality the situation is more complicated. Only approximate estimates can be given because of the intrinsic difficulty of a problem which goes through two steps, i.e., first the computation of the outgoing flux and, second, the trapping and re-entry conditions.

Few data are available on this subject. A calculation has been done by Bland [5] for 4.5 GV cut-off and solar minimum condition. Measurements of the total flux of atmospheric secondaries and re-entrant albedo electrons at 5 $\mathrm{g\,cm^{-2}}$ have been published by Schmoker and Earl [6] in the rigidity interval from Fort Churchill to Texas and by Daniel and Stephens [4] for the equatorial region.

The energy spectrum of the re-entrant albedo electrons is limited to rigidities below the local geomagnetic cut-off. Thus at rigidities above this value, the flux of primary electrons is not contaminated by re-entrant albedo electrons. However, some uncertainty remains in the energy region near the cut-off in view of the limited resolution in the measurement of the electron energy and, for high geomagnetic latitudes, owing to the temporal fluctuations of the cut-off rigidity.

2.4 Modulation of the primary electron flux by interplanetary magnetic and electric fields

Any direct experimental observation yields flux and spectrum of the electrons as they move inside the solar system and in the vicinity of the earth. Extrapolation to the true galactic electrons requires taking into account the perturbations caused by local magnetic and electric fields. The earth is immersed in an interplanetary magnetic field of solar origin whose interaction with the general galactic field is poorly known (an electric field could also be present but we will not consider it for the moment). If the interplanetary field is stronger than the interstellar field, it would be possible that electrons as well as the other charged particles of the galactic cosmic radiation are partly excluded from interplanetary space. This "screening" will be more effective for particles of low magnetic rigidity; an exception could exist for particles of such low rigidity that the gyro-radius is of the same order of magnitude as the field inhomogeneities. Analogous considerations apply to the magnetic fields in the disk and in the halo (if the halo exists at all), bearing on the relationship between the flux of electrons responsible for the isotropic radio emission in the disk and the electron flux in the halo. It should be noted that even small differences in the magnetic fields of two adjacent regions can upset the isotropy of the electron flux. In fact, flux isotropy in a region of high magnetic field intensity does not necessarily imply that there exists isotropy in regions of lower field strength.

The information available at present on the effects of local fields in the solar system applies only to the time-dependent part of the modulation of the nucleonic component. The extrapolation of this information, if only to derive the time-dependent part of the modulation of electrons, is not possible with any degree of confidence at the present time. Several qualitative models of the 11-year modulation cycle exist but the experimental evidence does not permit a definite choice between them.

We will come back to this point later. Generally, it can be said that the perturbation effect will be smaller for high energy electrons than for the low energy region of the spectrum.

2.5 Experimental results

The measurements of flux and energy spectrum of primary electrons which have been performed up to Dec. 1965 are listed in Table I. There are 12 of them: 11 refer to balloon-borne detectors and 1 to a satellite flight. Apart from two (Rossi and Hulsizer [1] and Critchfield, Ney and Oleksa [2], they have all been performed in the last 6 years and some of the results are still preliminary.

Several techniques of detection have been employed:

a) *Cloud chamber*, randomly expanded, with or without magnetic field. Electrons

TABLE I

Experiments on the primary electron flux

Author	Year	Residual atmosphere $g\,cm^{-2}$	Vertical cutoff rigidity (GV)	Instrumentation used
Rossi and Hulsizer [1]	1948	> 20	4.5	Balloon flight – Ionization chamber with lead absorbers
Critchfield *et al.* [2]	1949–50	> 20	$\simeq 1$	Balloon flight – Multiplate cloud chamber
Earl [7]	1960	4–6.5	$\simeq 1$	Balloon flight – Multiplate cloud chamber
Meyer and Vogt [11]	1960	3–5	< 0.2	Balloon flight – Telescope of scintillators interleaved with lead absorbers
Schmoker and Earl [6]	1962–63	5	< 0.2	Balloon flight – Cloud chamber with magnetic field
Agrinier *et al.* [8]	1963	4–5	4.5	Balloon flight – Multiplate spark chamber
Agrinier *et al.* [9]	1965	2–3	4.5	Balloon flight – Multiplate spark chamber
Waddington and Freier [10]	1963–64	2–3	< 0.2	Balloon flight – Nuclear emulsions
Rubstov [13]	1964	7.5	2.5	Balloon flight – Telescope of scintillators and lead absorbers
Daniel and Stephens [4]	1963	10.2	16	Balloon flight – Nuclear emulsions
Bleeker *et al.* [12]	1965	10	2.2–2.6	Balloon flight – Telescope of scintillator and Čerenkov detectors. Al and Pb absorbers
Cline *et al.* [14]	1963–64	–	–	dE/dx *vs.* E scintillator telescope IMP-1 satellite outside magnetosphere

and gamma-rays are recognized by their characteristic shower development in the multiplate chamber [2–7] or by curvature and energy loss, if the chamber has only one plate and a magnetic field [6].

This technique is very good with regard to the identification of events but it suffers from the severe limitation of the effective time of exposure and, consequently, from very limited statistics.

b) *Multiplate spark chamber*, triggered by a telescope counter, with photographic recording of the events [8, 9].

The pictures of the events show less details than the cloud chamber pictures but the instrument has a definite advantage as regards the effective time of exposure.

c) *Photographic emulsions* [4–10]. Although in principle a visual technique allows the most detailed investigation of single events, the method has the drawback that the plates have to be scanned, which results in a poor and somewhat distorted statistics.

d) *Counter techniques* [1, 11, 12, 13, 14]. The arrangements are different according to the range of energy investigated. Electron showers are discriminated against nuclear interactions by their average development as a function of the amount of matter traversed. In some cases additional discrimination is achieved by using Čerenkov gas counters.

We will discuss first the balloon experiments, for which the problem of correction for atmospheric secondaries is of paramount importance in he interpretation of the results.

2.5.1 *Balloon Flights at Fort Churchill*

Four of the balloon flights have been carried out at Fort Churchill, Manitoba [3, 6, 10, 11] where the geomagnetic cut-off rigidity [15] is smaller than 0.2 GV, which should allow the investigation of the primary energy spectrum down to the 100 MeV region. In Figure 1 are shown the total incident fluxes as recorded by the following experimenters: L'Heureux and Meyer [3] at 4 $g cm^{-2}$ residual atmosphere in 1964; Schmoker and Earl [6] with detectors at almost the same atmospheric depth (4–5 $g cm^{-2}$) in August 1962 and July 1963; Waddington and Freier [10] with detectors at 2–3 $g cm^{-2}$ in 1963 and 1964. We are not considering here the first experiment carried out by Meyer and Vogt in 1960 both because of the large uncertainty by which the results are affected and because they were obtained in a period very near to solar maximum. In Figure 1 is shown the measured and calculated flux of atmospheric electrons in the energy interval 50 MeV$<E<$150 MeV at 5 $g cm^{-2}$ as given by Schmoker and Earl.

Curve (a) indicates the secondary electron spectrum above 200 MeV as proposed by L'Heureux and Meyer for 4 $g cm^{-2}$. Curve (b) represents the secondary spectrum as computed by Perola and Scarsi for the same atmospheric depth.

As it appears from the diagram, the energy spectra of curve (a) and (b), although converging below 200 MeV, show a significant discrepancy at higher energies; in the GeV region the values of curve (b) are about three times higher than those of curve (a). At any rate it is evident that, at Fort Churchill, at an atmospheric depth of approximately 4 $\mathrm{g\,cm^{-2}}$ and at solar minimum, the total recorded flux of electrons below about 300 MeV can be entirely accounted for by atmospheric secondaries. If curve (b) is adopted, it is noted that the contamination by secondary electrons is very large up to 1 GeV.

Obviously, the energy spectrum of the primary electrons depends on the corrective factors assumed. In Figure 2 the differential primary electron spectrum is given. The value in the range 50–150 MeV is from Schmoker and Earl who give only an upper limit, the average value being zero; the data of L'Heureux and Meyer were corrected according to curve (a) of Figure 1. The authors propose for the energy interval 500 MeV–3.5 GeV the exponential law (curve (c) in Figure 2):

$$\frac{\mathrm{d}J}{\mathrm{d}E} = 11 \times E^{-1.6} \quad (\mathrm{m^2\,sec\,sterad\,GeV})^{-1}$$

Alternative values of the primary flux as deduced by application of curve (b) of Figure 1 are also shown (the asymmetry on the error limits follows from the energy dependence of the correction); in this case, the differential primary energy spectrum is consistent with a turnover in the 500 MeV–1 GeV region.

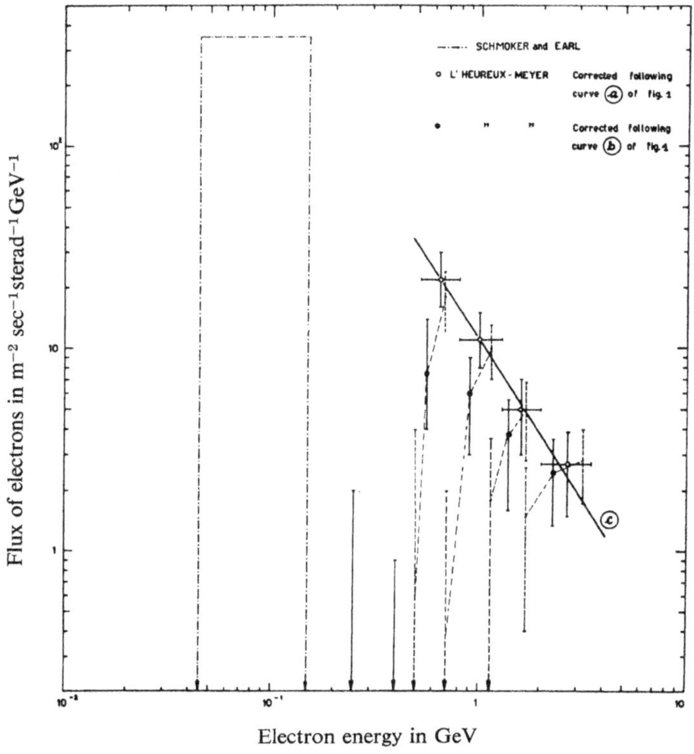

Fig. 2. Flux of primary cosmic ray electrons.

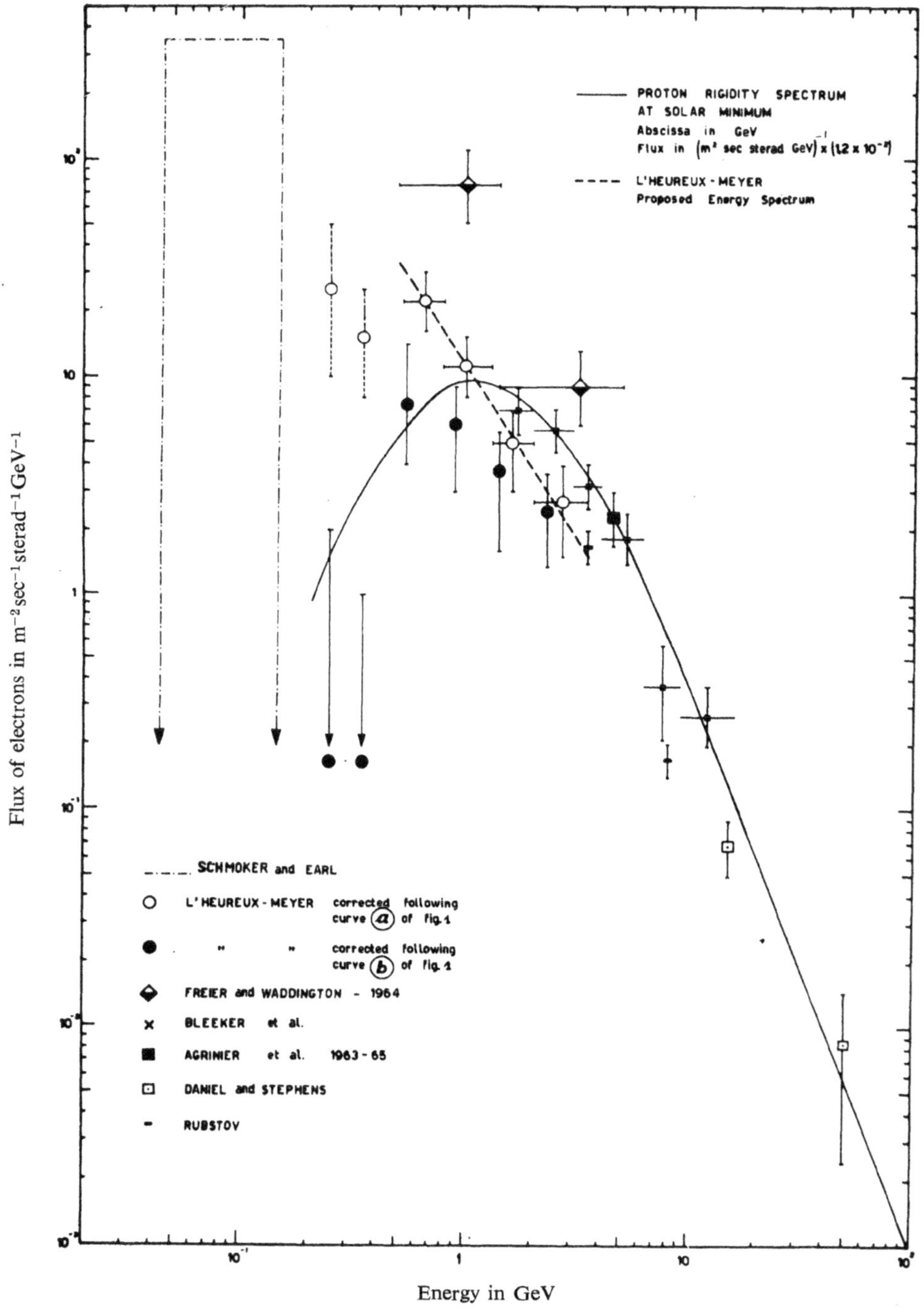

Fig. 3. Flux of primary cosmic ray electrons and protons.

The data of Freier and Waddington are not shown in Figure 2 but appear together with the other results in Figure 3. It should be noted that the error limits here are not to be taken literally since they refer to samples of only two or three events.

2.5.2 *Flights at Minneapolis*

Two balloon flights have been carried out near Minneapolis (vertical cut-off, 1.39 GV [15]). In 1950, Critchfield, Ney and Oleksa [2] obtained results which were inconclusive because of the low altitude at which the observations were carried out.

In 1960, Earl [7] had a balloon-borne multiplate cloud chamber floating at an average atmospheric depth of 4.5 $\mathrm{g\,cm^{-2}}$. In the flight during which the cloud chamber was randomly expanded, he registered 11 electrons with $E > 500$ MeV corresponding to an integral flux of (32 ± 10) electrons $\mathrm{m^{-2}\,sec^{-1}\,sterad^{-1}}$, $(3 \pm 1)\%$ of the proton flux value. Also 6 gamma photons were detected, 5 of which with $E > 500$ MeV.

Quenby and Wenk [16] and Shea *et al.* [15] give for the effective vertical cut-off rigidity at Minneapolis 1.34 GV and 1.39 GV, respectively. If there is no systematic underestimate of the energy in Earl's experiment, a fraction of the recorded electron flux could possibly be due to re-entrant albedo, and only events with $E > 1$ GeV should be considered as primaries. In this case the primary electron flux at Minneapolis for the 1960 period, i.e. near solar maximum, would be about 1.5% of the proton flux, corresponding roughly to 15 (electrons $\mathrm{m^{-2}\,sec^{-1}\,sterad^{-1}}$) for $E > 1$ GeV.

2.5.3 *Flights at 2.2–20.5 GV vertical cut-off rigidity*

Two experiments have been performed, both using counter telescopes: one by Rubstov [13] in 1964 and the other by Bleeker *et al.* [12] in 1965.

Rubstov had a flight with a ceiling at an atmospheric depth of 7.5 $\mathrm{g\,cm^{-2}}$. He gives the following values for the integrated flux of primary electrons:

$$N(>3.5\ \mathrm{GeV}) = (4 \pm 0.5) \quad (\mathrm{m^2\,sec\,sterad})^{-1}$$
$$N(>8.0\ \mathrm{GeV}) = (0.9 \pm 0.2) \quad (\mathrm{m^2\,sec\,sterad})^{-1}$$

The quoted error represents only the statistical uncertainty. An additional error could arise from the discrimination procedure and from the extrapolation of the data to the top of the atmosphere. The energy determination was made by estimating the size of the shower development.

The second experiment [12] was carried out at about 10 $\mathrm{g\,cm^{-2}}$ of residual atmosphere. The primary electron differential energy spectrum has been obtained in the 2–15 GeV range. The contribution of the secondary component has been deduced by analysis of the growth curves of both, gamma rays and electrons with atmospheric depth and by the relative intensity of the gamma-ray flux at ceiling, checked against the computed background. The contribution of spurious events was also determined by analyzing the discontinuity in the energy distribution at the value corresponding to the local rigidity. The cut-off detector was calibrated with electron and proton beams from accelerators. The geometrical acceptance factor, on the other hand, was only estimated. The differential flux values shown in Figure 3 were obtained assuming the geometrical factor to be 20 cm^2 sterad as proposed by the authors.

2.5.4 *Flights at 4.5 GV vertical cut-off rigidity*

The experiments of ROSSI and HULSIZER [1] in 1948, and of CRITCHFIELD, NEY OLEKSA [2] were carried out at too low an altitude to give acceptable results.

Balloon flights in Southern France have been made by the Saclay-Milan group at 4–5 $g cm^{-2}$ in 1963 [8] and between 2 and 3 $g cm^{-2}$ in 1965 [9]. The detector employed consists of a multiplate spark chamber triggered by a scintillation counter telescope and photographed by a stereo-camera. The discrimination properties and an empirical relation between the size of the electron showers observed and the electron energy have been obtained with a series of calibration exposures to beams from accelerators.

The primary electron integral flux values obtained are:

$$N(>4.5\ \text{GeV}) = 6^{+2.8}_{-1.7} \quad (\text{m}^2\ \text{sec sterad})^{-1}$$

and

$$N(>4.5\ \text{GeV}) = (8 \pm 2) \quad (\text{m}^2\ \text{sec sterad})^{-1}$$

for the 1963 and the 1965 flight, respectively; the latter value is based on more than 200 events; its relatively large error takes into account the uncertainty associated with the preliminary nature of the analysis of the pictures obtained.

In the energy range observed the contamination by atmospheric electrons is negligible, as is proved by the energy distribution of the observed events.

2.5.5 *Flights in the equatorial region*

DANIEL and STEPHENS [4] had a stack of emulsions flown under 10 $g cm^{-2}$ of residual atmosphere over Hyderabad, India, (vertical geomagnetic cut-off $\simeq 16$ GV) in April 1963.

The number of events identified as due to primary electrons with energy greater than the cut-off value was 12, 5 of which with $E > 50$ GeV. The contribution of atmospheric electrons was estimated to be negligible.

Taking into account the "scanning" efficiency at the various energies the authors deduce the following values for the integral primary electron flux at the top of the atmosphere:

$$N(>16\ \text{GeV}) = (0.68 \pm 0.2) \quad (\text{m}^2\ \text{sec sterad})^{-1}$$
$$N(>50\ \text{GeV}) = (0.3 \pm 0.2) \quad (\text{m}^2\ \text{sec sterad})^{-1}$$

2.5.6 *Experiments on satellites*

Only one satellite experiment has been made up to now. The measurement has been carried out by CLINE, LUDWIG and MCDONALD [14] with detectors on board of the highly eccentric-orbit satellite IMP-1 (Explorer XVIII) which had an apogee of 193000 km, well outside the magnetosphere. The detector was a simple scintillator telescope allowing the discrimination of light nuclei, protons and electrons from the measurement of dE/dx and E of particles coming to rest.

So far, the differential energy spectrum of electrons in the range 2.7–7.5 MeV as measured during the first orbit has been published; the total flux in this energy range was 210 ± 10 electrons m^{-2} sec^{-1} $sterad^{-1}$. During subsequent orbits the total flux was found to be larger, with increases of typically 100 electrons m^{-2} sec^{-1} $sterad^{-1}$ in the energy range 3–12 MeV. Evidence that the electrons observed are not of local origin (but result from interactions in the satellite) was derived from the temporal variation of the flux which shows statistically significant increases in correspondence with the recurrent minima of the Kp index.

Definite conclusions on the origin of these electrons (solar or galactic) are very difficult to draw on the basis of this experiment only. Since the measured flux fluctuated in time, additional observations are needed in order to establish the average flux and energy spectrum.

2.6 Summary of the experimental results on the primary electron flux

We now can try to summarise the experimental evidence which is available on the primary electron energy spectrum. In doing that, we will limit ourselves to the experiments carried out in the period 1962–1965, which might be considered as corresponding approximately to solar minimum.

Table II taken from the "Rapporteur paper" of Waddington [17] at the 1965 Cosmic Ray Conference in London, summarises the values proposed for the exponent γ by the experimenters who assume that the differential electron energy spectrum is of the form:

$$\frac{dJ}{dE} = KE^{-\gamma}$$

For $E > 3.5$ GeV, the data given by the experimenters refer to integral fluxes and they are all consistent with a power law, up to about 50 GeV, with γ between 2 and 3.

Representing these data in differential form we assume for γ, as a first approximation, the value 2.6, which is equal to the proton differential rigidity spectrum in this range. With this assumption the data are displayed in Figure 3 together with the results obtained at lower energies ($E > 50$ MeV).

In cases where the correction for the atmospheric background leads to different values, depending on whether curve (a) or (b) of Figure 1 is adopted, both values are

TABLE II

Proposed values for γ

Author	Differential (γ)	Integral ($\gamma - 1$)	Energy range (GeV)
L'Heureux and Meyer [3]	1.6 ± 0.6	–	0.5–3.5
Waddington and Freier [10]	1.6	–	0.5–5.0
Bleeker *et al.* [12]	1.9 ± 0.4	–	2–15
Rubstov [13]	(2.8)	1.8	3.5–8
Agrinier *et al.* [8, 9]	(3.0 ± 0.5)	2.0 ± 0.5	4.5
Daniel and Stephens [4]	(2.2)	$\simeq 1.2$	16–50

given. In Figure 3 are also shown the power law proposed by L'HEUREUX and MEYER [3] to represent the primary electron spectrum in the energy interval 500 MeV–3.5 GeV and the proton differential rigidity spectrum (for solar minimum) scaled down by a factor (1.2×10^{-2}). It appears from the figure that between 4 and 50 GeV the experimental points fit reasonably well into this proton differential rigidity spectrum. They are also consistent with an electron flux of 1.2% of the proton flux of equal rigidity. For $E \lesssim 4$ GeV the electron spectrum becomes less steep. L'Heureux and Meyer who assume curve (a) of Figure 1 to represent the atmospheric contamination propose a power law with $\gamma = (1.6 \pm 0.6)$ from 3.5 GeV down to at least 500 MeV and consequently an increasing proportion of electrons at low energies. On the other hand, if all the available experimental data are used* together and the L'Heureux-Meyer results are corrected using curve (b) of Figure 1, the primary spectrum could be described by a form similar to the one representing the proton rigidity spectrum, with a constant proportion of electrons of about 1%, down to at least 100 MeV electron energy or 100 MV proton rigidity.

Using the IMP-1 points in the 3–10 MeV range for extrapolating the energy spectrum from the GeV region is a very risky procedure owing to the complete lack of information on what happens in the wide energy region in between.

Below 1 GeV, balloon-borne experiments will be of little help to clarify the situation unless the ceiling atmospheric depth is reduced to a value smaller than about 2 g cm^{-2}. Satellite observations are scheduled for the next few years (see Table III), covering the energy region down to 50 MeV.

3. Measurements of the Relative Fraction of Electrons and Positrons in Primary Cosmic Rays

The electron-positron ratio is an important factor in determining the origin of the primary electron component. A strong negative excess would indicate a predominant role of primary sources in common with the nucleonic cosmic rays, while a comparable fraction of electrons and positrons would suggest that nuclear collision processes are the main production mechanism.

TABLE III

Scheduled satellite experiments on primary electrons of energy larger than 50 MeV

Group	Satellite	Year	Energy range
Leeds (Marsden)	ESRO II	1968	500 MeV–6 GeV
Chicago (Meyer)	OGO E	1968	50–100 MeV
Dutch group (Wapstra)	OGO E	1968	500 MeV–> 6 GeV
Milan-Saclay (Dilworth-Labeyrie)	HEOS I	1968–69	50–600 MeV

* It should be noted that the error shown for the Waddington and Freier data are only indicative, the number of events being only 2 or 3.

Experiments aimed at the direct determination of the sign of the charge by observing the deflection of the particle trajectories in a magnetic field have been carried out by the Chicago group (DE SHONG *et al.* [18] in 1963 and HARTMAN *et al.* [19] in 1964) with a telescope based on spark chambers and by SCHMOKER and EARL [6] with a cloud chamber (in 1962–63). The apparatus were flown on balloons from Fort Churchill.

The experiment of Schmoker and Earl covers the energy range $50 < E < 150$ MeV; the results are inconclusive. Since the atmospheric depth was about 5 $\mathrm{g\,cm^{-2}}$, most, if not all, of the registered particles were secondaries. Furthermore, the statistics were extremely poor (only 4 events, 2 of which identified as positrons).

The data obtained by De Shong *et al.* and by Hartman *et al.* are summarised in Table IV.

TABLE IV

Fraction of positrons for various energy intervals

Time	Average atm. depth ($\mathrm{g\,cm^{-2}}$)	Energy interval (MeV)	$\frac{N^+}{N^+ + N^-}$
July–August 1963	4	50–100	0.31 ± 0.12
		100–300	0.38 ± 0.07
		300–1000	0.16 ± 0.04
August 1964	6–4	40–100	0.79 ± 0.40
		100–300	0.43 ± 0.21
		300–1000	0.30 ± 0.14
		1000–3000	0.33 ± 0.16

Also taking into account the warning of Hartman *et al.* that the energy intervals defined in the first experiment are not very accurate (energy losses in the liquid Čerenkov counter lead to an under-estimate of the average electron energy), the interpretation of the 1963 results is rather puzzling, unless we admit that the energy was underestimated by a factor of more than 3 (at least 6 for the first interval). Because of the atmospheric contamination we would only expect the $N^+/(N^+ + N^-)$ ratio to be smaller than 0.5 above, say, 300 MeV even if *all* primary electrons are negative. For the experiment of 1964, for which the atmospheric depth was somewhat greater (4–6 $\mathrm{g\,cm^{-2}}$), the measured value of $N^+/(N^+ + N^-)$ might be considered significant for the 1000–3000 MeV interval and indicative for the 300–1000 MeV interval. In both cases an excess of negative electrons in the primary flux is suggested.

An alternative way of determining the positron-electron ratio is related to the East-West effect. This method is suitable at intermediate and low geomagnetic latitudes and it can furnish data in a region of energy where the magnetic field intensity necessary to produce an observable deflection on the trajectories is too high to make the direct analysis competitive. The dependence of the primary flux on the direction of arrival has been examined in 1965 by the Milan-Saclay Group [9] and by DANIEL and STEPHENS [4] in 1963.

The results of Agrinier *et al.* refer to a vertical cut-off rigidity of about 4.5 GV. In their experiment the axis of the spark chamber was inclined at 38° with respect to the vertical, and rotated in azimuth at about 1 r.p.m. For each event triggering the chamber the orientation of the axis of the apparatus was measured by a flux-gate magnetometer. Analyzing the dependence of the flux of particles on the orientation of the detector in the geomagnetic field the following results were obtained (R denotes the ratio of fluxes from the West to the East quadrant):

$$R_p = \frac{\text{Proton flux from West}}{\text{Proton flux from East}} = 1.30 \pm 0.03$$

$$R_{Z \geqq 2} = \frac{\text{Flux of } Z \geqq 2 \text{ particles from West}}{\text{Flux of } Z \geqq 2 \text{ particles from East}} = 1.38 \pm 0.17$$

$$l/R_e = \frac{\text{Flux of electrons from East}}{\text{Flux of electrons from West}} = 1.52 \pm 0.28$$

The results imply the existence of a negative excess in the primary electron flux; the analysis is based on 210 electron events. Using the proton ratio to determine the ratio of cut-off rigidities at West and East, the following values are obtained for $N^+/(N^+ + N^-)$ as a function of the exponent assumed for the energy spectrum $N(>E) = kE^{-(\gamma-1)}$:

$(\gamma - 1)$	$N^+/(N^+ + N^-)$
< 2.5	< 0.2
$\geqslant 2.5$	< 0.4

Taking for $(\gamma - 1)$ the value 1.6 as suggested by Figure 3 we can deduce that the primary electron flux between about 4 GeV and at least 15 GeV (the fraction of events with higher energies gives a minor contribution) is predominantly negatively charged.

Daniel and Stephens [4] arrive at the opposite conclusion, at least for energies larger than about 15 GeV. These authors flew a balloon which carried a stack of nuclear plates oriented in the East-West direction. Analyzing the distribution of the events as function of the direction of arrival and of particle energy, they found 4 events with energy greater than the corresponding cut-off rigidity for electrons but less than the one for positrons and only one event in the opposite case. This would indicate, if taken literally, that $N^+/(N^+ + N^-) = \frac{4}{5}$ and consequently an overwhelming proportion of positrons. Before reaching the conclusion that there is a change in the composition of the electron component between the energies around 1 GeV and that above 20 GeV, one should possibly keep in mind the words of Daniel and Stephens themselves: "In view of the small numbers with which we are dealing at present these results should be treated with caution...".

References

[1] Rossi, B. and Hulsizer, R. I.: 1948, 'Search for Electrons in the Primary Cosmic Radiation', *Phys. Rev.* **73,** 1402.

[2] Critchfield, C. L., Ney, E. P., and Oleksa, S.: 1952, 'Soft Radiation at a Balloon Altitudes', *Phys. Rev.* **85,** 461.

[3] L'Heureux, J. and Meyer, P.: 1965, 'Flux and Energy Spectrum of Primary Cosmic-Ray Electrons', *Phys. Rev. Letters* **15,** 93.

[4] Daniel, R. P. and Stephens, S. A.: 1965, 'Electron Component of the Primary Cosmic Radiation at Energies > 15 GeV', *Phys. Rev. Letters* **15,** 769.

[5] Bland, C. J.: 1965, 'An Estimate of the Contribution from Re-entrant Albedo to the Measurement of Primary Electrons', *Space Res.* **5,** North-Holland, Amsterdam.

[6] Schmoker, J. W. and Earl, J. A.: 1965, 'Magnetic Cloud-Chamber Observations of Low-Energy Cosmic-Ray Electrons', *Phys. Rev.* **138,** 300.

[7] Earl, J. A.: 1961, 'Cloud-Chamber Observation of Primary Cosmic-Ray Electrons', *Phys. Rev. Letters* **6,** 125.

[8] Agrinier, B., Koechlin, Y., Parlier, B., Boella, G., Degli Antoni, G., Dilworth, C., Scarsi, L., and Sironi, G.: 1964, 'Flux of Primary Cosmic-Ray Electrons of Rigidity above 4.5 BV', *Phys. Rev. Letters* **13,** 377.

[9] Agrinier, B., Koechlin, Y., Parlier, B., Vasseur, J., Bland, C. J., Boella, G., Degli Antoni, G., Dilworth, C., Scarsi, L., and Sironi, G.: 1967, *Proc. of Intern. Conf. on Cosmic Rays, London 1965*, to be published.

[10] Waddington, C. J. and Freier, P. S.: 1967, 'Contribution of the University of Minnesota (Techn. Report CR 82, p. 35), *Proc. of Intern. Conf. on Cosmic Rays, London 1965*, to be published.

[11] Meyer, P. and Vogt, R.: 1961, 'Electrons in the Primary Cosmic Radiation', *Phys. Rev. Letters* **6,** 193.

[12] Bleeker, J. A., Burger, J. J., Scheepmaker, A., Swanenberg, B. N., and Tanaka, Y.: 1967, *Proc. of Intern. Conf. on Cosmic Rays, London 1965*, to be published.

[13] Rubstov, V. I.: 1967, *Proc. of Intern. Conf. on Cosmic Rays, London 1965*, to be published.

[14] Cline, T. L., Ludwig, G. H., and McDonald, F. B.: 1964, 'Detection of Interplanetary 3- to 12-MeV Electrons', *Phys. Rev. Letters* **13,** 786.

[15] Shea, M. A., Smart, D. F., and McCracken, K. G.: 1965, 'A Study of Vertical Cutoff Rigidities Using Sixth Degree Simulations of the Geomagnetic Field', *J. Geophys. Res.* **70,** 4117.

[16] Quenby, J. J. and Wenk, G. J.: 1962, 'Cosmic-Ray Threshold Rigidities and the Earth's Magnetic Field', *Phil. Mag.* **7,** 1457.

[17] Waddington, C. J.: 1965, Contribution of the University of Minnesota (Techn. Report CR 82, p. 124), *Proc. of Intern. Conf. on Cosmic Rays, London 1965*, to be published.

[18] De Shong, J. A., Hildebrand, R. H., and Meyer, P.: 1964, 'Ratio of Electrons to Positrons in the Primary Cosmic Radiation', *Phys. Rev. Letters* **12,** 3.

[19] Hartman, R. C., Hildebrand, R. H., and Meyer, P.: 1965, 'Observation of the Cosmic-Ray Electron-Positron Ratio from 100 MeV to 3 BeV in 1964', *J. Geophys. Res.* **70,** 2713.

INDEX OF NAMES

INDEX OF SUBJECTS

GPSR Compliance

The European Union's (EU) General Product Safety Regulation (GPSR) is a set of rules that requires consumer products to be safe and our obligations to ensure this.

If you have any concerns about our products, you can contact us on ProductSafety@springernature.com

In case Publisher is established outside the EU, the EU authorized representative is:

Springer Nature Customer Service Center GmbH
Europaplatz 3
69115 Heidelberg, Germany

Batch number: 10761697

Printed by Printforce, the Netherlands